Secrets of RF
Circuit Design

Secrets of RF Circuit Design

Third Edition

Joseph J. Carr

McGraw-Hill

New York San Francisco Washington, D.C. Auckland Bogotá
Caracas Lisbon London Madrid Mexico City Milan
Montreal New Delhi San Juan Singapore
Sydney Tokyo Toronto

Library of Congress Cataloging-in-Publication Data

Carr, Joseph J.
  Secrets of RF circuit design / Joseph J. Carr.—3rd ed.
    p. cm.
  Includes bibliographical references and index.
  ISBN 0-07-137067-6
  1. Radio circuits—Design and construction. 2. Electronic circuit design. I. Title.

TK6560.C37 2000
621.384′12—dc21   00-050017

McGraw-Hill

A Division of The McGraw·Hill Companies

3 4 5 6 7 8 9 0   DOC/DOC   0 9 8 7 6 5 4 3 2

ISBN 0-07-137067-6

The sponsoring editor for this book was Scott Grillo, the editing supervisor was Sally Glover, and the production supervisor was Sherri Souffrance. It was set at York Production Services.

Printed and bound by R.R. Donnelley & Sons Company.

 This book is printed on recycled, acid-free paper containing a minimum of 50% recycled, de-inked fiber.

Contents

v

Introduction

The first two editions of *Secrets of RF Circuit Design* were met with a tremendous and heartwarming response; the sales of those books were more than satisfying—especially because the mail was so positive. Because of the positive responses to the first two editions, the publisher and I decided to make this third edition available.

In this third edition, I've incorporated a lot of the feedback from readers, received both in person and by mail. For example, the coverage of direct-conversion receivers (always a popular construction project) has been greatly expanded. In addition, chapters on Very Low Frequency (VLF) have been added for the benefit of "Lowfers" and their kin. I also added a chapter on designing and building inductance-capacitance (LC) RF filters (high-pass, low-pass, bandpass, and notch). These filters not only make projects work better, but they also keep electromagnetic interference at bay. Another feature is coverage of double-balanced mixers. These devices are now more accessible, in both integrated circuit and diode versions, and are a step ahead of the other forms of mixer circuits.

During the 20 or more years that I have written articles for electronics and amateur radio magazines, such as *Popular Electronics, Radio Electronics, Popular Communications, Nuts 'n' Volts,* and *Ham Radio,* it has become apparent that many professional and amateur electronics people consider radio-frequency (RF) electronics as something of a "hard-to-do," deep mystery.

Perhaps the major contributor to this attitude is that formulas and circuits don't seem to work on the bench as they do on paper. In other words, theory seems to depart markedly from practice as frequency increases. The reason for this seeming anomaly is that RF electronics use much higher frequencies than other electronics circuits, so the stray (ordinarily unaccounted for) capacitances and inductances must be factored into the equation. Otherwise, things do not work well. In addition, the resistance in RF circuits is not the same as it is in dc and low-frequency ac circuits because of the skin effect (see Chapter 1).

Finally, there is the problem of measurements. Instruments for dc and low-frequency ac circuits are easily obtained or built, but RF instruments are both more complicated and more expensive. This book shows some alternatives for obtaining lower-cost radio-frequency circuits and instruments or building your own. You also will find basic information on using these instruments. In the pages to follow, you will hopefully find some of those deep mysteries unraveled.

Contact me at P.O. Box 1587, Annandale, VA 22003. Comments and criticisms can be addressed either to that address or by e-mail to: carrjj@aol.com.

Secrets of RF
Circuit Design

1
CHAPTER

Introduction to RF electronics

Radio-frequency (RF) electronics differ from other electronics because the higher frequencies make some circuit operation a little hard to understand. Stray capacitance and stray inductance afflict these circuits. Stray capacitance is the capacitance that exists between conductors of the circuit, between conductors or components and ground, or between components. Stray inductance is the normal inductance of the conductors that connect components, as well as internal component inductances. These stray parameters are not usually important at dc and low ac frequencies, but as the frequency increases, they become a much larger proportion of the total. In some older very high frequency (VHF) TV tuners and VHF communications receiver front ends, the stray capacitances were sufficiently large to tune the circuits, so no actual discrete tuning capacitors were needed.

Also, skin effect exists at RF. The term *skin effect* refers to the fact that ac flows only on the outside portion of the conductor, while dc flows through the entire conductor. As frequency increases, skin effect produces a smaller zone of conduction and a correspondingly higher value of ac resistance compared with dc resistance.

Another problem with RF circuits is that the signals find it easier to radiate both from the circuit and within the circuit. Thus, coupling effects between elements of the circuit, between the circuit and its environment, and from the environment to the circuit become a lot more critical at RF. Interference and other strange effects are found at RF that are missing in dc circuits and are negligible in most low-frequency ac circuits.

The electromagnetic spectrum

When an RF electrical signal radiates, it becomes an electromagnetic wave that includes not only radio signals, but also infrared, visible light, ultraviolet light, X-rays, gamma rays, and others. Before proceeding with RF electronic circuits, therefore, take a look at the electromagnetic spectrum.

1

The electromagnetic spectrum (Fig. 1-1) is broken into bands for the sake of convenience and identification. The spectrum extends from the very lowest ac frequencies and continues well past visible light frequencies into the X-ray and gamma-ray region. The extremely low frequency (ELF) range includes ac power-line frequencies as well as other low frequencies in the 25- to 100-hertz (Hz) region. The U.S. Navy uses these frequencies for submarine communications.

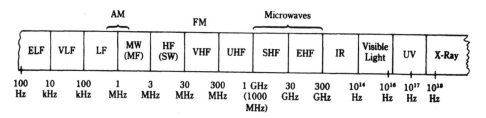

1-1 The electromagnetic spectrum from VLF to X-ray. The RF region covers from less than 100 kHz to 300 GHz.

The very low frequency (VLF) region extends from just above the ELF region, although most authorities peg it to frequencies of 10 to 100 kilohertz (kHz). The low-frequency (LF) region runs from 100 to 1000 kHz—or 1 megahertz (MHz). The medium-wave (MW) or medium-frequency (MF) region runs from 1 to 3 MHz. The amplitude-modulated (AM) broadcast band (540 to 1630 kHz) spans portions of the LF and MF bands.

The high-frequency (HF) region, also called the *shortwave bands* (SW), runs from 3 to 30 MHz. The VHF band starts at 30 MHz and runs to 300 MHz. This region includes the frequency-modulated (FM) broadcast band, public utilities, some television stations, aviation, and amateur radio bands. The ultrahigh frequencies (UHF) run from 300 to 900 MHz and include many of the same services as VHF. The microwave region begins above the UHF region, at 900 or 1000 MHz, depending on source authority.

You might well ask how microwaves differ from other electromagnetic waves. Microwaves almost become a separate topic in the study of RF circuits because at these frequencies the wavelength approximates the physical size of ordinary electronic components. Thus, components behave differently at microwave frequencies than they do at lower frequencies. At microwave frequencies, a 0.5-W metal film resistor, for example, looks like a complex RLC network with distributed L and C values—and a surprisingly different R value. These tiniest of distributed components have immense significance at microwave frequencies, even though they can be ignored as negligible at lower RFs.

Before examining RF theory, first review some background and fundamentals.

Units and physical constants

In accordance with standard engineering and scientific practice, all units in this book will be in either the CGS (centimeter-gram-second) or MKS (meter-kilogram-second) system unless otherwise specified. Because the metric system depends on using multiplying prefixes on the basic units, a table of common metric prefixes (Table 1-1) is provided. Table 1-2 gives the standard physical units. Table 1-3 gives physical constants of interest in this and other chapters. Table 1-4 gives some common conversion factors.

Table 1-1. Metric prefixes

Metric prefix	Multiplying factor	Symbol
tera	10^{12}	T
giga	10^{9}	G
mega	10^{6}	M
kilo	10^{3}	K
hecto	10^{2}	h
deka	10	da
deci	10^{-1}	d
centi	10^{-2}	c
milli	10^{-3}	m
micro	10^{-6}	u
nano	10^{-9}	n
pico	10^{-12}	p
femto	10^{-15}	f
atto	10^{-18}	a

Table 1-2. Units of measure

Quantity	Unit	Symbol
Capacitance	farad	F
Electric charge	coulomb	Q
Conductance	mhos	
Conductivity	mhos/meter	Ω/m
Current	ampere	A
Energy	joule (watt-second)	j
Field	volts/meter	E
Flux linkage	weber (volt/second)	
Frequency	hertz	Hz
Inductance	henry	H
Length	meter	m
Mass	gram	g
Power	watt	W
Resistance	ohm	Ω
Time	second	s
Velocity	meter/second	m/s
Electric potential	volt	V

Table 1-3. Physical constants

Constant	Value	Symbol
Boltzmann's constant	1.38×10^{-23} J/K	K
Electric chart (e^-)	1.6×10^{-19} C	q
Electron (volt)	1.6×10^{-19} J	eV
Electron (mass)	9.12×10^{-31} kg	m
Permeability of free space	$4\pi \times 10^{-7}$ H/m	U_0
Permitivity of free space	8.85×10^{-12} F/m	ϵ_0
Planck's constant	6.626×10^{-34} J-s	h
Velocity of electromagnetic waves	3×10^8 m/s	c
Pi (π)	3.1416	π

Table 1-4. Conversion factors

1 inch	= 2.54 cm
1 inch	= 25.4 mm
1 foot	= 0.305 m
1 statute mile	= 1.61 km
1 nautical mile	= 6,080 feet (6,000 feet)[a]
1 statute mile	= 5,280 feet
1 mile	= 0.001 in = 2.54×10^{-5} m
1 kg	= 2.2 lb
1 neper	= 8.686 dB
1 gauss	= 10,000 teslas

a Some navigators use 6,000 feet for ease of calculation. The nautical mile is 1/360 of the Earth's circumference at the equator, more or less.

Wavelength and frequency

For all wave forms, the velocity, wavelength, and frequency are related so that the product of frequency and wavelength is equal to the velocity. For radiowaves, this relationship can be expressed in the following form:

$$\lambda F \sqrt{\epsilon} = c, \qquad (1\text{-}1)$$

where

λ = wavelength in meters (m)

F = frequency in hertz (Hz)

ϵ = dielectric constant of the propagation medium

c = velocity of light (300,000,000 m/s).

The dielectric constant (ϵ) is a property of the medium in which the wave propagates. The value of ϵ is defined as 1.000 for a perfect vacuum and very nearly 1.0 for dry air (typically 1.006). In most practical applications, the value of ϵ in dry air is taken to be 1.000. For media other than air or vacuum, however, the velocity of propagation is slower and the value of ϵ relative to a vacuum is higher. Teflon, for example, can be made with ϵ values from about 2 to 11.

Equation (1-1) is more commonly expressed in the forms of Eqs. (1-2) and (1-3):

$$\lambda = \frac{c}{F\sqrt{\epsilon}} \qquad (1\text{-}2)$$

and

$$F = \frac{c}{\lambda\sqrt{\epsilon}}. \qquad (1\text{-}3)$$

[All terms are as defined for Eq. (1-1).]

Microwave letter bands

During World War II, the U.S. military began using microwaves in radar and other applications. For security reasons, alphabetic letter designations were adopted for each band in the microwave region. Because the letter designations became ingrained, they are still used throughout industry and the defense establishment. Unfortunately, some confusion exists because there are at least three systems currently in use: pre-1970 military (Table 1-5), post-1970 military (Table 1-6), and the IEEE and industry standard (Table 1-7). Additional confusion is created because the military and defense industry use both pre- and post-1970 designations simultaneously and industry often uses military rather than IEEE designations. The old military designations (Table 1-5) persist as a matter of habit.

Skin effect

There are three reasons why ordinary lumped constant electronic components do not work well at microwave frequencies. The first, mentioned earlier in this chapter, is that component size and lead lengths approximate microwave wavelengths.

**Table 1-5. Old U.S. military
microwave frequency bands
(WWII–1970)**

Band designation	Frequency range
P	225–390 MHz
L	390–1550 MHz
S	1550–3900 MHz
C	3900–6200 MHz
X	6.2–10.9 GHz
K	10.9–36 GHz
Q	36–46 GHz
V	46–56 GHz
Q	56–100 GHz

**Table 1-6. New U.S. military microwave
frequency bands (Post-1970)**

Band designation	Frequency range
A	100–250 MHz
B	250–500 MHz
C	500–1000 MHz
D	1000–2000 MHz
E	2000–3000 MHz
F	3000–4000 MHz
G	4000–6000 MHz
H	6000–8000 MHz
I	8000–10000 MHz
J	10–20 GHz
K	20–40 GHz
L	40–60 GHz
M	60–100 GHz

**Table 1-7. IEEE/Industry standard
frequency bands**

Band designation	Frequency range
HF	3–30 MHz
VHF	0–300 MHz
UHF	300–1000 MHz
L	1000–2000 MHz
S	2000–4000 MHz
C	4000–8000 MHz
X	8000–12000 MHz
Ku	12–18 GHz
K	18–27 GHz
Ka	27–40 GHz
Millimeter	40–300 GHz
Submillimeter	>300 GHz

The second is that distributed values of inductance and capacitance become significant at these frequencies. The third is the phenomenon of skin effect. While dc current flows in the entire cross section of the conductor, ac flows in a narrow band near the surface. Current density falls off exponentially from the surface of the conductor toward the center (Fig. 1-2). At the critical depth (δ, also called the depth of penetration), the current density is $1/e = 1/2.718 = 0.368$ of the surface current density.

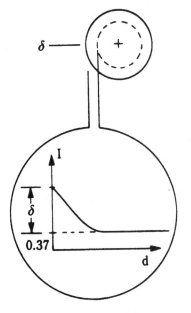

1-2
In ac circuits, the current flows only in the outer region of the conductor. This effect is frequency-sensitive and it becomes a serious consideration at higher RF frequencies.

The value of δ is a function of operating frequency, the permeability (μ) of the conductor, and the conductivity (σ). Equation (1-4) gives the relationship.

$$\delta = \sqrt{\frac{1}{2\pi F \sigma \mu}} \qquad (1\text{-}4)$$

where
δ = critical depth
F = frequency in hertz
μ = permeability in henrys per meter
σ = conductivity in mhos per meter.

RF components, layout, and construction

Radio-frequency components and circuits differ from those of other frequencies principally because the unaccounted for "stray" inductance and capacitance forms a significant portion of the entire inductance and capacitance in the circuit. Consider a tuning circuit consisting of a 100-pF capacitor and a 1-μH inductor. According to an equation that you will learn in a subsequent chapter, this combination should resonate at an RF frequency of about 15.92 MHz. But suppose the circuit is poorly laid out and there is 25 pF of stray capacitance in the circuit. This capacitance could come from the interaction of the capacitor and inductor leads with the chassis or with other components in the circuit. Alternatively, the input capacitance of a transistor or integrated circuit (IC) amplifier can contribute to the total value of the "strays" in the circuit (one popular RF IC lists 7 pF of input capacitance). So, what does this extra 25 pF do to our circuit? It is in parallel with the 100-pF discrete capacitor so it produces a total of 125 pF. Reworking the resonance equation with 125 pF instead of 100 pF reduces the resonant frequency to 14.24 MHz.

A similar situation is seen with stray inductance. All current-carrying conductors exhibit a small inductance. In low-frequency circuits, this inductance is not sufficiently large to cause anyone concern (even in some lower HF band circuits), but as frequencies pass from upper HF to the VHF region, strays become terribly important. At those frequencies, the stray inductance becomes a significant portion of total circuit inductance.

Layout is important in RF circuits because it can reduce the effects of stray capacitance and inductance. A good strategy is to use broad printed circuit tracks at RF, rather than wires, for interconnection. I've seen circuits that worked poorly when wired with #28 Kovar-covered "wire-wrap" wire become quite acceptable when redone on a printed circuit board using broad (which means low-inductance) tracks.

Figure 1-3 shows a sample printed circuit board layout for a simple RF amplifier circuit. The key feature in this circuit is the wide printed circuit tracks and short distances. These tactics reduce stray inductance and will make the circuit more predictable.

Although not shown in Fig. 1-3, the top (components) side of the printed circuit board will be all copper, except for space to allow the components to interface with the bottom-side printed tracks. This layer is called the "ground plane" side of the board.

1-3
Typical RF printed circuit layout.

Impedance matching in RF circuits

In low-frequency circuits, most of the amplifiers are voltage amplifiers. The requirement for these circuits is that the source impedance must be very low compared with the load impedance. A sensor or signal source might have an output impedance of, for example, 25 Ω. As long as the input impedance of the amplifier receiving that signal is very large relative to 25 Ω, the circuit will function. "Very large" typically means greater than 10 times, although in some cases greater than 100 times is preferred. For the 25-Ω signal source, therefore, even the most stringent case is met by an input impedance of 2500 Ω, which is very far below the typical input impedance of real amplifiers.

RF circuits are a little different. The amplifiers are usually specified in terms of power parameters, even when the power level is very tiny. In most cases, the RF circuit will have some fixed system impedance (50, 75, 300, and 600 Ω being common, with 50 Ω being nearly universal), and all elements of the circuit are expected to match the system impedance. Although a low-frequency amplifier typically has a very high input impedance and very low output impedance, most RF amplifiers will have the same impedance (usually 50 Ω) for both input and output.

Mismatching the system impedance causes problems, including loss of signal—especially where power transfer is the issue (remember, for maximum power transfer, the source and load impedances must be equal). Radio-frequency circuits very often use transformers or impedance-matching networks to affect the match between source and load impedances.

Wiring boards

Radio-frequency projects are best constructed on printed circuit boards that are specially designed for RF circuits. But that ideal is not always possible. Indeed, for many hobbyists or students, it might be impossible, except for the occasional project built from a magazine article or from this book. This section presents a couple of alternatives to the use of printed circuit boards.

Figure 1-4 shows the use of perforated circuit wiring board (commonly called *perfboard*). Electronic parts distributors, RadioShack, and other outlets sell various versions of this material. Most commonly available perfboard offers 0.042″ holes spaced on 0.100″ centers, although other hole sizes and spacings are available. Some perfboard is completely blank, and other stock material is printed with any of several different patterns. The offerings of RadioShack are interesting because several different patterns are available. Some are designed for digital IC applications and others are printed with a pattern of circles, one each around the 0.042″ holes.

1-4 Perfboard layout.

In Fig. 1-4, the components are mounted on the top side of an unprinted board. The wiring underneath is "point-to-point" style. Although not ideal for RF circuits, it will work throughout the HF region of the spectrum and possibly into the low VHF (especially if lead lengths are kept short).

Notice the shielded inductors on the board in Fig. 1-4. These inductors are slug-tuned through a small hole in the top of the inductor. The standard pin pattern for these components does not match the 0.100-hole pattern that is common to perfboard. However, if the coils are canted about half a turn from the hole matrix, the pins will fit on the diagonal. The grounding tabs for the shields can be handled in either of two ways. First, bend them 90° from the shield body and let them lay on the top side of the perfboard. Small wires can then be soldered to the tabs and passed through a nearby hole to the underside circuitry. Second, drill a pair of ⅟₁₆″ holes (between two of the premade holes) to accommodate the tabs. Place the coil on the board at the desired location to find the exact location of these holes.

Figure 1-5 shows another variant on the perfboard theme. In this circuit, pressure-sensitive (adhesive-backed) copper foil is pressed onto the surface of the perfboard to form a ground plane. This is not optimum, but it works for "one-off" homebrew projects up to HF and low-VHF frequencies.

1-5 Perfboard layout with RF ground plane.

The perfboard RF project in Fig. 1-6 is a frequency translator. It takes two frequencies (F_1 and F_2), each generated in a voltage-tuned variable-frequency oscillator (VFO) circuit, and mixes them together in a double-balanced mixer (DBM) device. A low-pass filter (the toroidal inductors seen in Fig. 1-6) selects the difference frequency (F_2-F_1). It is important to keep the three sections (osc1, osc2, and the low-pass filter) isolated from each other. To accomplish this goal, a shield partition is provided. In the center of Fig. 1-6, the metal package of the mixer is soldered to the shield partition. This shield can be made from either 0.75″ or 1.00″ brass strip stock of the sort that is available from hobby and model shops.

Figure 1-7 shows a small variable-frequency oscillator (VFO) that is tuned by an air variable capacitor. The capacitor is a 365-pF "broadcast band" variable. I built this circuit as the local oscillator for a high-performance AM broadcast band receiver project. The shielded, slug-tuned inductor, along with the capacitor, tunes the 985- to 2055-kHz range of the LO. The perfboard used for this project was a preprinted RadioShack. The printed foil pattern on the underside of the perfboard is a matrix of small circles of copper, one copper pad per 0.042″ hole. The perfboard is held off the chassis by nylon spacers and 4-40 × 0.75″ machine screws and hex nuts.

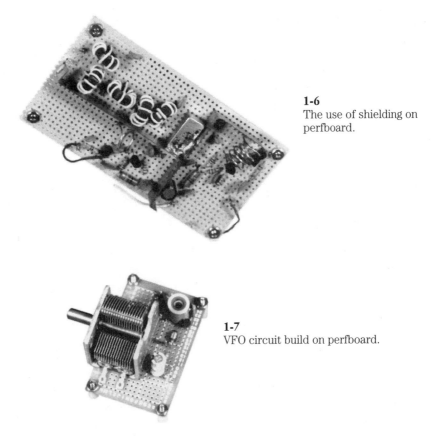

1-6
The use of shielding on
perfboard.

1-7
VFO circuit build on perfboard.

Chassis and cabinets

It is probably wise to build RF projects inside shielded metal packages wherever possible. This approach to construction will prevent external interference from harming the operation of the circuit and prevent radiation from the circuit from interfering with external devices. Figure 1-8 shows two views of an RF project built inside an aluminum chassis box; Fig. 1-8A shows the assembled box and Fig. 1-8B shows an internal view. These boxes have flanged edges on the top portion that overlap the metal side/bottom panel. This overlap is important for interference reduction. Shun those cheaper chassis boxes that use a butt fit, with only a couple of nipples and dimples to join the boxes together. Those boxes do not shield well.

The input and output terminals of the circuit in Fig. 1-8 are SO-239 "UHF" coaxial connectors. Such connectors are commonly used as the antenna terminal on shortwave radio receivers. Alternatives include "RCA phono jacks" and "BNC" coaxial connectors. Select the connector that is most appropriate to your application.

RF shielded boxes

At one time, more than 2 decades ago, I loathed small RF electronic projects above about 40-m as "too hard." As I grew in confidence, I learned a few things about

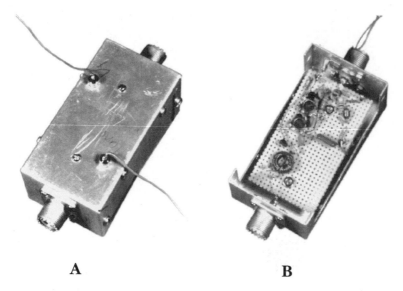

A　　　　　　　　　　　**B**

1-8 Shielded RF construction: (A) closed box showing dc connections made via coaxial capacitors; (B) box opened.

RF construction (e.g., layout, grounding, and shielding) and found that by following the rules, one can be as successful building RF stuff as at lower frequencies.

One problem that has always been something of a hassle, however, is the shielding that is required. You could learn layout and grounding, but shielding usually required a better box than I had. Most of the low-cost aluminum electronic hobbyist boxes on the market are alright for dc to the AM broadcast band, but as frequency climbs into the HF and VHF region, problems begin to surface. What you thought was shielded "t'ain't." If you've read my columns or feature articles over the years, you will recall that I caution RF constructors to use the kind of aluminum box with an overlapping flange of at least 0.25″, and a good tight fit. Many hobbyist-grade boxes on the market just simply are not good enough.

Enter SESCOM, Inc. [Dept. JJC, 2100 Ward Drive, Henderson, NV, 89015-4249; (702) 565-3400 and (for voice orders only) 1-800-634-3457 and (for FAX orders only) 1-800-551-2749]. SESCOM makes a line of cabinets, 19″ racks, rack mount boxes, and RF shielded boxes. Their catalog has a lot of interesting items for radio and electronic hobbyist constructors. I was particularly taken by their line of RF shielded boxes. Why? Because it seems that RF projects are the main things I've built for the past 10 years.

Figure 1-9 shows one of the SESCOM RF shielded steel boxes in their SB-x line. Notice that it uses the "finger" construction in order to get a good RF-tight fit between the lid and the body of the box. Also notice that the box comes with some snap-in partitions for internal shielding between sections. The box body is punched to accept the tabs on these internal partitions, which can then be soldered in place for even better stability and shielding.

At first, I was a little concerned about the material; the boxes are made of hot tin-plated steel rather than aluminum. The tin plating makes soldering easy, but steel

1-9 RF project box with superior shielding because of the finger-grip design of the top cover (one of a series made by SESCOM).

is hard on drill bits. I found, however, in experimenting with the SB-5 box supplied to me by SESCOM that a good-quality set of drill bits had no difficulty making a hole. Sure, if you use old, dull drill bits and lean on the drill like Attila the Hun, then you'll surely burn it out. But by using a good-quality, sharp bit and good workmanship practices to make the hole, there is no real problem.

The boxes come in 11 sizes from 2.1″ × 1.9″ footprint to a 6.4″ × 2.7″ footprint, with heights of 0.63,″ 1.0, or 1.1″. Prices compare quite favorably with the prices of the better-quality aluminum boxes that don't shield so well at RF frequencies.

The small project in Fig. 1-10 is a small RF preselector for the AM broadcast band. It boosts weak signals and reduces interference from nearby stations. The tuning capacitor is mounted to the front panel and is fitted with a knob to facilitate tuning. An on/off switch is also mounted on the front panel (a battery pack inside the aluminum chassis box provides dc power). The inductor that works with the capacitor is passed through the perfboard (where the rest of the circuit is located) to the rear panel (where its adjustment slug can be reached).

1-10
Battery-powered RF project.

The project in Fig. 1-11 is a test that I built for checking out direct-conversion receiver designs. The circuit boards are designed to be modularized so that different sections of the circuit can easily be replaced with new designs. This approach allows comparison (on an "apples versus apples" basis) of different circuit designs.

Coaxial cable transmission line ("coax")

Perhaps the most common form of transmission line for shortwave and VHF/UHF receivers is coaxial cable. "Coax" consists of two conductors arranged concentric to each other and is called *coaxial* because the two conductors share the same center axis (Fig. 1-12). The inner conductor will be a solid or stranded wire, and the other conductor forms a shield. For the coax types used on receivers, the shield will be a braided conductor, although some multistranded types are also sometimes seen. Coaxial cable intended for television antenna systems has a 75-Ω characteristic impedance and uses metal foil for the outer conductor. That type of outer conductor results in a low-loss cable over a wide frequency range but does not work too well for most applications outside of the TV world. The problem is that the foil is aluminum, which doesn't take solder. The coaxial connectors used for those antennas are generally Type-F crimp-on connectors and have too high of a casualty rate for other uses.

The inner insulator separating the two conductors is the dielectric, of which there are several types; polyethylene, polyfoam, and Teflon are common (although the latter is used primarily at high-UHF and microwave frequencies). The velocity factor (V) of the coax is a function of which dielectric is used and is outlined as follows:

Dielectric type	Velocity factor
Polyethylene	0.66
Polyfoam	0.80
Teflon	0.70

1-11
Receiver chassis used as a "test bench" to try various modifications to a basic design.

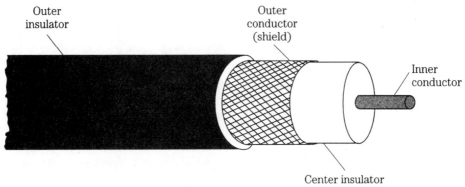

1-12 Coaxial cable (cut-away view).

Coaxial cable is available in a number of characteristic impedances from about 35 to 125 Ω, but the vast majority of types are either 52- or 75-Ω impedances. Several types that are popular with receiver antenna constructors include the following:

RG-8/U or RG-8/AU	52 Ω	Large diameter
RG-58/U or RG-58/AU	52 Ω	Small diameter
RG-174/U or RG-174/AU	52 Ω	Tiny diameter
RG-11/U or RG-11/AU	75 Ω	Large diameter
RG-59/U or RG-59/AU	75 Ω	Small diameter

Although the large-diameter types are somewhat lower-loss cables than the small diameters, the principal advantage of the larger cable is in power-handling capability. Although this is an important factor for ham radio operators, it is totally unimportant to receiver operators. Unless there is a long run (well over 100 feet), where cumulative losses become important, then it is usually more practical on receiver antennas to opt for the small-diameter (RG-58/U and RG-59/U) cables; they are a lot easier to handle. The tiny-diameter RG-174 is sometimes used on receiver antennas, but its principal use seems to be connection between devices (e.g., receiver and either preselector or ATU), in balun and coaxial phase shifters, and in instrumentation applications.

Installing coaxial connectors

One of the mysteries faced by newcomers to the radio hobbies is the small matter of installing coaxial connectors. These connectors are used to electrically and mechanically fasten the coaxial cable transmission line from the antenna to the receiver. There are two basic forms of coaxial connector, both of which are shown in Fig. 1-13 (along with an alligator clip and a banana-tip plug for size comparison). The larger connector is the PL-259 UHF connector, which is probably the most-common form used on radio receivers and transmitters (do not take the "UHF" too seriously, it is used at all frequencies). The PL-259 is a male connector, and it mates with the SO-239 female coaxial connector.

The smaller connector in Fig. 1-12 is a BNC connector. It is used mostly on electronic instrumentation, although it is used in some receivers (especially in handheld radios).

The BNC connector is a bit difficult, and very tedious, to correctly install so I recommend that most readers do as I do: Buy them already mounted on the wire. But the PL-259 connector is another matter. Besides not being readily available already mounted very often, it is relatively easy to install.

Figure 1-14A shows the PL-259 coaxial connector disassembled. Also shown in Fig. 1-14A is the diameter-reducing adapter that makes the connector suitable for use with smaller cables. Without the adapter, the PL-259 connector is used for RG-8/U and RG-11/U coaxial cable, but with the correct adapter, it will be used with smaller RG-58/U or RG-59/U cables (different adapters are needed for each type).

1-13
Various types of coaxial connectors, cable ends, and adapters.

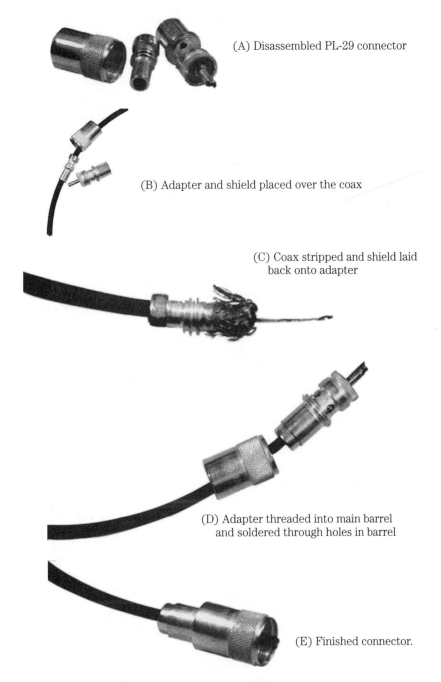

(A) Disassembled PL-29 connector

(B) Adapter and shield placed over the coax

(C) Coax stripped and shield laid back onto adapter

(D) Adapter threaded into main barrel and soldered through holes in barrel

(E) Finished connector.

1-14 Installing the PL-259 UHF connector.

The first step is to slip the adapter and thread the outer shell of the PL-259 over the end of the cable (Fig. 1-14B). You will be surprised at how many times, after the connector is installed, you find that one of these components is still sitting on the workbench . . . requiring the whole job to be redone (sigh). If the cable is short enough that these components are likely to fall off the other end, or if the cable is dangling a particularly long distance, then it might be wise to trap the adapter and outer shell in a knotted loop of wire (note: the knot should not be so tight as to kink the cable).

The second step is to prepare the coaxial cable. There are a number of tools for stripping coaxial cable, but they are expensive and not terribly cost-effective for anyone who does not do this stuff for a living. You can do just as effective a job with a scalpel or X-acto knife, either of which can be bought at hobby stores and some electronics parts stores. Follow these steps in preparing the cable:

1. Make a circumscribed cut around the body of the cable ¾″ from the end, and then make a longitudinal cut from the first cut to the end.
2. Now strip the outer insulation from the coax, exposing the shielded outer conductor.
3. Using a small, pointed tool, carefully unbraid the shield, being sure to separate the strands making up the shield. Lay it back over the outer insulation, out of the way.
4. Finally, using a wire stripper, side cutters or the scalpel, strip ⅝″ of the inner insulation away, exposing the inner conductor. You should now have ⅝″ of inner conductor and ⅜″ of inner insulation exposed, and the outer shield destranded and laid back over the outer insulation.

Next, slide the adapter up to the edge of the outer insulator and lay the unbraided outer conductor over the adapter (Fig. 1-14C). Be sure that the shield strands are neatly arranged and then, using side cutters, neatly trim it to avoid interfering with the threads. Once the shield is laid onto the adapter, slip the connector over the adapter and tighten the threads (Fig. 1-14D). Some of the threads should be visible in the solder holes that are found in the groove ahead of the threads. It might be a good idea to use an ohmmeter or continuity connector to be sure that there is no electrical connection between the shield and inner conductor (indicating a short circuit).

Warning

Soldering involves using a hot soldering iron. The connector will become dangerously hot to the touch. Handle the connector with a tool or cloth covering.

- Solder the inner conductor to the center pin of the PL-259. Use a 100-W or greater soldering gun, not a low-heat soldering pencil.
- Solder the shield to the connector through the holes in the groove.
- Thread the outer shell of the connector over the body of the connector.

After you make a final test to make sure there is no short circuit, the connector is ready for use (Fig. 1-14E).

RF components and tuned circuits

This chapter covers inductance (L) and capacitance (C), how they are affected by ac signals, and how they are combined into LC-tuned circuits. The tuned circuit allows the radio-frequency (RF) circuit to be selective about the frequency being passed. Alternatively, in the case of oscillators, LC components set the operating frequency of the circuit.

Tuned resonant circuits

Tuned resonant circuits, also called *tank circuits* or *LC circuits*, are used in the radio front end to select from the myriad of stations available at the antenna. The tuned resonant circuit is made up of two principal components: inductors and capacitors, also known in old radio books as condensers. This section examines inductors and capacitors separately, and then in combination, to determine how they function to tune a radio's RF, intermediate-frequency (IF), and local oscillator (LO) circuits. First, a brief digression is needed to discuss vectors because they are used in describing the behavior of these components and circuits.

Vectors

A vector (Fig. 2-1A) is a graphical device that is used to define the magnitude and direction (both are needed) of a quantity or physical phenomenon. The length of the arrow defines the magnitude of the quantity, and the direction in which it points defines the direction of action of the quantity being represented.

Vectors can be used in combination with each other. For example, Fig. 2-1B shows a pair of displacement vectors that define a starting position (P_1) and a final position (P_2) for a person traveling 12 miles north from point P_1 and then 8 miles east to arrive at point P_2. The displacement in this system is the hypotenuse of the

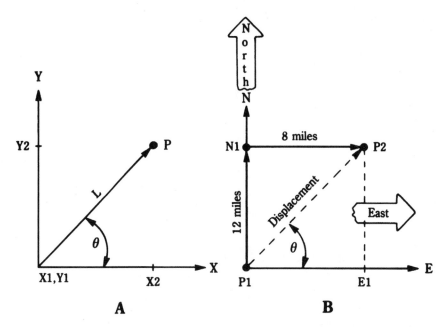

2-1 (A) Vector notation is used in RF circuit analysis. The resultant of vector X and vector Y is vector L between (X_1, Y_1) and (X_2, Y_2), or point P; (B) A way of viewing vectors is to measure the displacement of a journey that is 12 miles north and 8 miles east.

right triangle formed by the north vector and the east vector. This concept was once vividly illustrated by a university bumper sticker's directions to get to a rival school: "North 'till you smell it, east 'til you step in it."

Another vector calculation trick is used a lot in engineering, science, and especially in electronics. You can translate a vector parallel to its original direction and still treat it as valid. The east vector (E) has been translated parallel to its original position so that its tail is at the same point as the tail of the north vector (N). This allows you to use the Pythagorean theorem to define the vector. The magnitude of the displacement vector to P_2 is given by

$$P_2 = \sqrt{N^2 + E^2}. \tag{2-1}$$

But recall that the magnitude only describes part of the vector's attributes. The other part is the direction of the vector. In the case of Fig. 2-1B, the direction can be defined as the angle between the east vector and the displacement vector. This angle (θ) is given by

$$\theta = \arccos\left(\frac{E_1}{P}\right). \tag{2-2}$$

In generic vector notation, there is no natural or standard frame of reference so that the vector can be drawn in any direction so long as the user understands what it means. This system has adopted a method that is basically the same as the old-fashioned Cartesian coordinate system X-Y graph. In the example of Fig. 2-1B, the X axis is the east–west vector and the Y axis is the north–south vector.

In electronics, vectors are used to describe voltages and currents in ac circuits. They are standardized (Fig. 2-2) on a similar Cartesian system where the inductive reactance (X_L) (i.e., the opposition to ac exhibited by inductors) is graphed in the north direction, the capacitive reactance (X_C) is graphed in the south direction, and the resistance (R) is graphed in the east direction.

Negative resistance (the west direction) is sometimes seen in electronics. It is a phenomenon in which the current decreases when the voltage increases. Some RF examples of negative resistance include tunnel diodes and Gunn diodes.

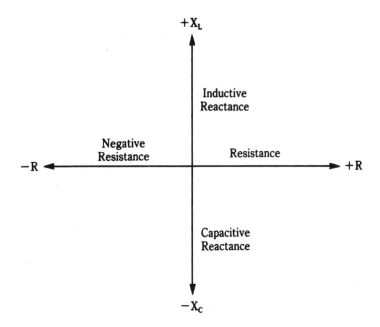

2-2 Vector system used in ac or RF circuit analysis.

Inductance and inductors

Inductance is the property of electrical circuits that opposes changes in the flow of current. Notice the word "changes"—it is important. As such, it is somewhat analogous to the concept of inertia in mechanics. An inductor stores energy in a magnetic field (a fact that you will see is quite important). In order to understand the concept of inductance, you must understand three physical facts:

1. When an electrical conductor moves relative to a magnetic field, a current is generated (or induced) in the conductor. An electromotive force (EMF or voltage) appears across the ends of the conductor.

2. When a conductor is in a magnetic field that is changing, a current is induced in the conductor. As in the first case, an EMF is generated across the conductor.

3. When an electrical current moves in a conductor, a magnetic field is set up around the conductor.

According to Lenz's law, the EMF induced in a circuit is ". . . in a direction that opposes the effect that produced it." From this fact, you can see the following effects:

1. A current induced by either the relative motion of a conductor and a magnetic field or changes in the magnetic field always flows in a direction that sets up a magnetic field that opposes the original magnetic field.
2. When a current flowing in a conductor changes, the magnetic field that it generates changes in a direction that induces a further current into the conductor that opposes the current change that caused the magnetic field to change.
3. The EMF generated by a change in current will have a polarity that is opposite to the polarity of the potential that created the original current.

Inductance is measured in henrys (H). The accepted definition of the henry is the inductance that creates an EMF of 1 V when the current in the inductor is changing at a rate of 1 A/S or

$$V = L\left(\frac{\Delta I}{\Delta t}\right) \qquad (2\text{-}3)$$

where

V = induced EMF in volts (V)
L = inductance in henrys (H)
I = current in amperes (A)
t = time in seconds (s)
Δ = "small change in . . ."

The henry (H) is the appropriate unit for inductors used as the smoothing filter chokes used in dc power supplies, but is far too large for RF and IF circuits. In those circuits, subunits of millihenrys (mH) and microhenrys (μH) are used. These are related to the henry as follows: 1 H = 1,000 mH = 1,000,000 μH. Thus, 1 mH = 10^{-3} H and 1 μH = 10^{-6} H.

The phenomena to be concerned with here is called *self-inductance:* when the current in a circuit changes, the magnetic field generated by that current change also changes. This changing magnetic field induces a countercurrent in the direction that opposes the original current change. This induced current also produces an EMF, which is called the *counter electromotive force* (CEMF). As with other forms of inductance, self-inductance is measured in henrys and its subunits.

Although inductance refers to several phenomena, when used alone it typically means self-inductance and will be so used in this chapter unless otherwise specified (e.g., mutual inductance). However, remember that the generic term can have more meanings than is commonly attributed to it.

Inductance of a single straight wire

Although it is commonly assumed that "inductors" are "coils," and therefore consist of at least one, usually more, turns of wire around a cylindrical form, it is also true that a single, straight piece of wire possesses inductance. This inductance of a wire in which the length is at least 1000 times its diameter (d) is given by the following:

$$L_{\mu H} = 0.00508b\left(Ln\left(\frac{4a}{d}\right) - 0.75\right). \tag{2-4}$$

The inductance values of representative small wires is very small in absolute numbers but at higher frequencies becomes a very appreciable portion of the whole inductance needed. Consider a 12" length of #30 wire ($d = 0.010$ in). Plugging these values into Eq. (2-4) yields the following:

$$L_{\mu H} = 0.00508\,(12\text{ in})\left(Ln\left(\frac{4 \times 1\text{ in}}{0.010\text{ in}}\right) - 0.75\right) \tag{2-5}$$

$$L_{\mu H} = 0.365\ \mu H. \tag{2-6}$$

An inductance of 0.365 μH seems terribly small; at 1 MHz, it is small compared with inductances typically used as that frequency. But at 100 MHz, 0.365 μH could easily be more than the entire required circuit inductance. RF circuits have been created in which the inductance of a straight piece of wire is the total inductance. But when the inductance is an unintended consequence of the circuit wiring, it can become a disaster at higher frequencies. Such unintended inductance is called *stray inductance* and can be reduced by using broad, flat conductors to wind the coils. An example is the printed circuit coils wound on cylindrical forms in the FM radio receiver tuner in Fig. 2-3.

2-3
Printed circuit inductors.

Self-inductance can be increased by forming the conductor into a multiturn coil (Fig. 2-4) in such a manner that the magnetic field in adjacent turns reinforces itself. This requirement means that the turns of the coil must be insulated from each other. A coil wound in this manner is called an "inductor," or simply a "coil," in RF/IF circuits. To be technically correct, the inductor pictured in Fig. 2-4 is called a *solenoid wound coil* if the length (l) is greater than the diameter (d). The inductance of the coil is actually self-inductance, but the "self-" part is usually dropped and it is simply called *inductance*.

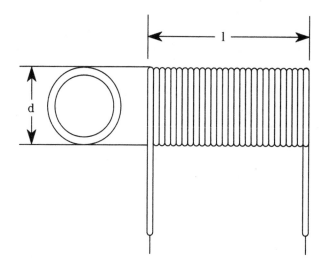

2-4 Inductance is a function of the length (l) and diameter (d) of the coil.

Several factors affect the inductance of a coil. Perhaps the most obvious factors are the length, the diameter, and the number of turns in the coil. Also affecting the inductance is the nature of the core material and its cross-sectional area. In the examples of Fig. 2-3, the core is simply air and the cross-sectional area is directly related to the diameter. In many radio circuits, the core is made of powdered iron or ferrite materials.

For an air-core solenoid coil, in which the length is greater than 0.4d, the inductance can be approximated by the following:

$$L_p = \frac{d^2 N^2}{18 d + 40 l}. \tag{2-7}$$

The core material has a certain magnetic permeability (μ), which is the ratio of the number of lines of flux produced by the coil with the core inserted to the number of lines of flux with an air core (i.e., core removed). The inductance of the coil is multiplied by the permeability of the core.

Combining inductors in series and in parallel

When inductors are connected together in a circuit, their inductances combine similar to the resistances of several resistors in parallel or in series. For inductors in which their respective magnetic fields do not interact, the following equations are used:

Series connected inductors:

$$L_{total} = L_1 + L_2 + L_3 + \cdots + L_n. \tag{2-8}$$

Parallel connected inductors:

$$L_{total} = \frac{1}{\dfrac{1}{L_1} + \dfrac{1}{L_2} + \dfrac{1}{L_3} + \cdots + \dfrac{1}{L_n}}. \tag{2-9}$$

In the special case of the two inductors in parallel:

$$L_{\text{total}} = \frac{L_1 \times L_2}{L_1 + L_2}.$$ (2-10)

If the magnetic fields of the inductors in the circuit interact, the total inductance becomes somewhat more complicated to express. For the simple case of two inductors in series, the expression would be as follows.

Series inductors:

$$L_{\text{total}} = L_1 + L_2 \pm 2M,$$ (2-11)

where

M = mutual inductance caused by the interaction of the two magnetic fields. (Note: $+M$ is used when the fields aid each other, and $-M$ is used when the fields are opposing.)

Parallel inductors:

$$L_{\text{total}} = \frac{1}{\left(\dfrac{1}{L_1 \pm M}\right) + \left(\dfrac{1}{L_2 \pm M}\right)}.$$ (2-12)

Some LC tank circuits use air-core coils in their tuning circuits (Fig. 2-5). Notice that two of the coils in Fig. 2-5 are aligned at right angles to the other one. The reason for this arrangement is not mere convenience but rather is a method used by the radio designer to prevent interaction of the magnetic fields of the respective coils. In general, for coils in close proximity to each other the following two principles apply:

1. Maximum interaction between the coils occurs when the coils' axes are parallel to each other.
2. Minimum interaction between the coils occurs when the coils' axes are at right angles to each other.

For the case where the coils' axes are along the same line, the interaction depends on the distance between the coils.

2-5 Inductors in an antique radio set.

Inductor circuit symbols

Figure 2-6 shows various circuit symbols used in schematic diagrams to represent inductors. Figures 2-6A and 2-6B represent alternate, but equivalent, forms of the same thing; i.e., a fixed value, air-core inductor ("coil" in the vernacular). The other forms of inductor symbol shown in Fig. 2-6 are based on Fig. 2-6A but are just as valid if the "open-loop" form of Fig. 2-6B is used instead.

The form shown in Fig. 2-6C is a tapped fixed-value air-core inductor. By providing a tap on the coil, different values of fixed inductance are achieved. The inductance from one end of the coil to the tap is a fraction of the inductance available across the entire coil. By providing one or more taps, several different fixed values of inductance can be selected. Radio receivers and transmitters sometimes use the tap method, along with a bandswitch, to select different tuning ranges or "bands."

Variable inductors are shown in Figs. 2-6D and 2-6E. Both forms are used in schematic diagrams, although in some countries Fig. 2-6D implies a form of construction whereby a wiper or sliding electrical contact rides on the uninsulated turns of the coil. Figure 2-6E implies a construction where variable inductance is achieved by moving a magnetic core inside of the coil.

Figure 2-6F indicates a fixed value (or tapped, if desired) inductor with a powdered iron, ferrite or nonferrous (e.g., brass) core. The core will increase (ferrite or powdered iron) or decrease (brass) the inductance value relative to the same number of turns on an air core coil.

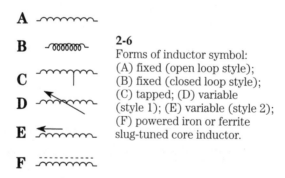

2-6
Forms of inductor symbol:
(A) fixed (open loop style);
(B) fixed (closed loop style);
(C) tapped; (D) variable
(style 1); (E) variable (style 2);
(F) powered iron or ferrite
slug-tuned core inductor.

Inductors in ac circuits

Impedance (Z) is the total opposition to the flow of ac in a circuit and as such it is analogous to resistance in dc circuits. Impedance is made up of a resistance component (R) and a component called reactance (X). Like resistance, reactance is measured in ohms. If the reactance is produced by an inductor, then it is called *inductive reactance* (X_L) and if by a capacitor it is called *capacitive reactance* (X_C). Inductive reactance is a function of the inductance and the frequency of the ac source:

$$X_L = 2\pi F L \qquad (2\text{-}13)$$

where

 X_L = inductive reactance in ohms (Ω)
 F = ac frequency in hertz (Hz)
 L = inductance in henrys (H).

In a purely resistive ac circuit (Fig. 2-7A), the current (I) and voltage (V) are said to be in phase with each other (i.e., they rise and fall at exactly the same times in the ac cycle). In vector notation (Fig. 2-7B), the current and voltage vectors are along the same axis, which is an indication of the zero-degree phase difference between the two.

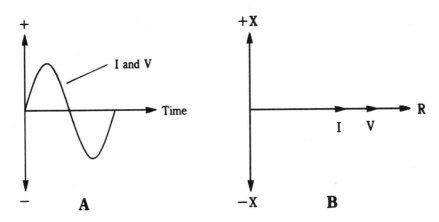

2-7 (A) Current and voltage in phase with each other (sine wave with vector notation); (B) vector relationship.

In an ac circuit that contains only an inductor (Fig. 2-8A), and is excited by a sine-wave ac source, the change in current is opposed by the inductance. As a result, the current in an inductive circuit lags behind the voltage by 90°. This is shown vectorially in Fig. 2-8B, and as a pair of sine waves in Fig. 2-8C.

The ac circuit that contains a resistance and an inductance (Fig. 2-9B) shows a phase shift (θ), shown vectorially in Fig. 2-9A, other than the 90° difference seen in purely inductive circuits. The phase shift is proportional to the voltage across the inductor and the current flowing through it. The impendance of this circuit is found by the Pythagorean rule described earlier, also called the root of the sum of squares method (see Fig. 2-9):

$$Z = \sqrt{R^2 + X_L^2}. \tag{2-14}$$

The coils used in radio receivers come in a variety of different forms and types, but all radios (except the very crudest untuned crystal sets) will have at least one coil.

Now let us turn our attention to the other member of the LC tuned circuit, the capacitor.

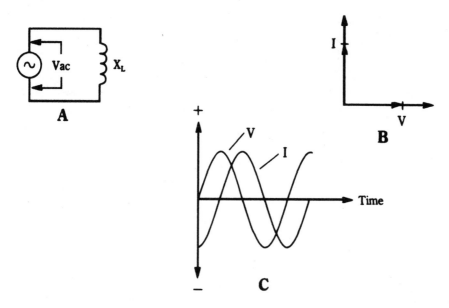

2-8 (A) Inductor ac circuit; (B) vector relationships; (C) sine-wave representation.

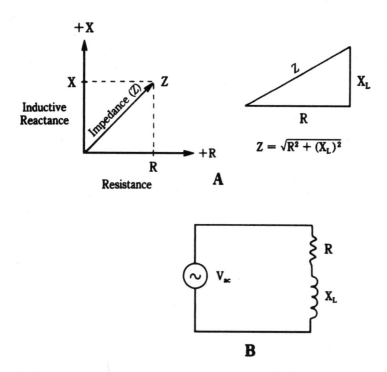

$$Z = \sqrt{R^2 + (X_L)^2}$$

2-9 (A) Resistor-inductor (RL) circuit; (B) vector relationships.

Air-core inductors

An air-core inductor actually has no core so it might also be called a *coreless coil*. Although it can be argued that the performance of an inductor differs between air and vacuum cores, the degree of difference is so negligible as to fall into the "decimal dust" category.

Three different forms of air-core inductor can be recognized. If the length (b) of a cylindrical coil is greater than, or equal to, the diameter (d) then the coil is said to be solenoid-wound. But if the length is much shorter than the diameter then the coil is said to be *loop-wound*. There is a gray area around the breakpoint between these inductors where the loop-wound coil seems to work somewhat like a solenoid-wound coil, but in the main most loop-wound coils are such that $b < d$. The principal uses of loop-wound coils is in making loop antennas for interference nulling and radio direction finding (RDF) applications.

Solenoid-wound air-core inductors

An example of the solenoid-wound air-core inductor was shown in Fig. 2-4. This form of coil is longer than its own diameter. The inductance of the solenoid-wound coil is given by the following:

$$L_{\mu H} = \frac{a^2 N^2}{9a + 10b},$$ (2-15)

where

$L_{\mu H}$ is the inductance in microhenrys (μH)
a is the coil radius in inches (in)
b is the coil length in inches (in)
N is the number of turns in the coil.

This equation will allow calculation of the inductance of a known coil, but we usually need to know the number of turns (N) required to achieve some specific inductance value determined by the application. For this purpose, we rearrange the equation in the form:

$$N = \frac{\sqrt{L(9a + 10b)}}{a}.$$ (2-16)

Several different forms of solenoid-wound air-core inductor are shown in Fig. 2-10. The version shown in Fig. 2-10A is homemade on a form of 25-mm PVC plumbing pipe, sawn to a length of about 80 mm and fitted at one end with solder lugs secured with small machine screws and hex nuts. The main inductor is shown in the dark-colored #24 AWG enamel-coated wire, while a small winding is made from #26 AWG insulated hook-up wire. The use of two coaxial inductors makes this assembly a transformer. The small primary winding can be connected between ground and an aerial, and the larger secondary winding can be resonated with a variable capacitor.

The air-core coil shown in Fig. 2-10B is part of an older radio transmitter, where it forms the anode tuning inductance.[1] Because several different frequency bands must be accommodated, there are actually several coils on the same form. The re-

---

1 Amateur radio operators might recognize this unit as the Heathkit DX-60B transmitter.

quired sections are switch-selected according to the band of operation. Another method for achieving tapped inductance is shown in Fig. 2-10C. This coil is a commercial air-core coil made by Barker & Williamson. Some models of this coil stock come with alternate windings indented to facilitate the connection of a tap. In any case, it is easy to press in the windings adjacent to the connection point. The variant shown in Fig. 6-10D was found, oddly enough, on a 1000-W HF motorcycle mobile. The whip antenna is resonated by using an alligator clip to select the number of turns from the coil.

A

B

2-10 (A) Solenoid wound inductor with transformer coupling link; (B) inductor used as a tank circuit in radio transmitter; (C) tapped inductor; (D) tapped inductor on mobile antenna; (E) roller inductor.

C

D

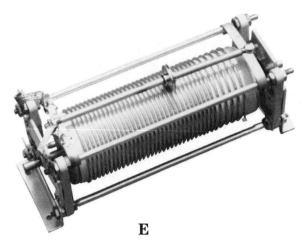

E

2-10 *Continued*

The air-core coil shown in Fig. 2-10E is a rotary inductor of the type found in some HF transmitters, antenna-tuning units, and other applications where continuous control of frequency is required. The inductor coil is mounted on a ceramic form that can be rotated using a shaft protruding from one end. As the form rotates, a movable shorting element rides along the turns of the coil to select the required inductance. Notice in Fig. 2-10E that the pitch (number of turns per unit length) is not constant along the length of the coil. This "pitch-winding" method is used to provide a nearly constant change of inductance for each revolution of the adjustment shaft.

Adjustable coils

There are several practical problems with the preceding standard fixed coil. For one thing, the inductance cannot easily be adjusted—either to tune the radio or to trim the tuning circuits to account for the tolerances in the circuit.

Air-core coils are difficult to adjust. They can be lengthened or shortened, the number of turns can be changed, or a tap or series of taps can be established on the coil in order to allow an external switch to select the number of turns that are allowed to be effective. None of these methods is terribly elegant—even though all have been used in one application or another.

The solution to the adjustable inductor problem that was developed relatively early in the history of mass-produced radios, and is still used today, is to insert a powdered iron or ferrite core (or "slug") inside the coil form (Fig. 2-11). The permeability of the core will increase or decrease the inductance according to how much of the core is inside the coil. If the core is made with either a hexagonal hole or screwdriver slot, then the inductance of the coil can be adjusted by moving the core in or out of the coil. These coils are called *slug-tuned inductors*.

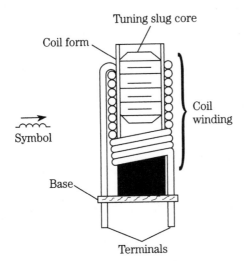

Tuning slug core

Coil form

Coil winding

Symbol

Base

Terminals

2-11
Slug-tuned ferrite or powdered iron-core inductor.

Capacitors and capacitance

Capacitors are the other component used in radio tuning circuits. Like the inductor, the capacitor is an energy-storage device. Although the inductor stores electrical energy in a magnetic field, the capacitor stores energy in an electrical (or "electrostatic") field. Electrical charge, Q, is stored in the capacitor. But more about that shortly.

The basic capacitor consists of a pair of metallic plates that face each other and are separated by an insulating material, called a *dielectric*. This arrangement is shown schematically in Fig. 2-12A and in a more physical sense in Fig. 2-12B. The fixed capacitor shown in Fig. 2-12B consists of a pair of square metal plates separated by a dielectric. Although this type of capacitor is not terribly practical, it was once used quite a bit in transmitters. Spark transmitters of the 1920s often used a glass and tinfoil capacitor that looked very much like Fig. 2-12B. Layers of glass and foil are sandwiched together to form a high-voltage capacitor. A 1-ft square capacitor made of $1/8$-in-thick glass and foil has a capacitance of about 2000 picofarads (pF).

Units of capacitance

Capacitance (C) is a measure of a capacitor's ability to store current or, more properly, electrical charge. The principal unit of capacitance is the farad (F) (named after physicist Michael Faraday). One farad is the capacitance that will store one coulomb of electrical charge (6.28×1018 electrons) at an electrical potential of 1 V, or, in math form:

$$1F = \frac{1Q}{1V},$$

(2-17)

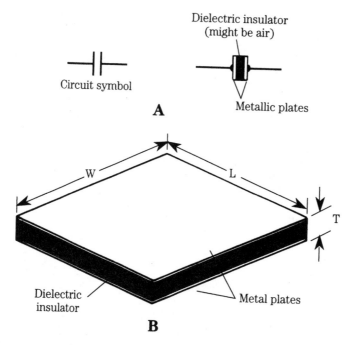

2-12 (A) Capacitors consist of a pair of conductors separated by a dielectric (insulator); (B) parallel plate capacitor.

where

C = capacitance in farads (F)

Q = electrical charge in coulombs (q)

V = electrical potential in volts (V)

The farad is far too large for practical electronics work, so subunits are used. The microfarad (μF) is 0.000001 farads (1F = 10^6 μF). The picofarad (pF) is 0.000001 μF = 10^{-12} farads. In older radio texts and schematics, the picofarad was called the *micromicrofarad* ($\mu\mu$F).

The capacitance of the capacitor is directly proportional to the area of the plates (in terms of Fig. 2-12B, $L \times W$), inversely proportional to the thickness (T) of the dielectric (or the spacing between the plates, if you prefer), and directly proportional to the dielectric constant (K) of the dielectric.

The dielectric constant is a property of the insulating material used for the dielectric. The dielectric constant is a measure of the material's ability to support electric flux and is thus analogous to the permeability of a magnetic material. The standard of reference for the dielectric constant is a perfect vacuum, which is said to have a value of K = 1.00000. Other materials are compared with the vacuum. The values of K for some common materials are as follows:

- vacuum: 1.0000
- dry air: 1.0006
- paraffin (wax) paper: 3.5

- glass: 5 to 10
- mica: 3 to 6
- rubber: 2.5 to 35
- dry wood: 2.5 to 8
- pure (distilled) water: 81

The value of capacitance in any given capacitor is found using the following:

$$C = \frac{0.0885\, KA(N - 1)}{T}, \qquad (2\text{-}18)$$

where

C = capacitance in picofarads
K = dielectric constant
A = area of one of the plates ($L \times W$), assuming that the two plates are identical
N = number of identical plates
T = thickness of the dielectric.

Breakdown voltage

The capacitor works by supporting an electrical field between two metal plates. This potential, however, can get too large. When the electrical potential (i.e., the voltage) gets too large, free electrons in the dielectric material (there are a few, but not too many, in any insulator) might flow. If a stream of electrons gets started, then the dielectric might break down and allow a current to pass between the plates. The capacitor is then shorted. The maximum breakdown voltage of the capacitor must not be exceeded. However, for practical purposes, there is a smaller voltage called the *dc working voltage (WVdc) rating,* which is the maximum safe voltage that can be applied to the capacitor. Typical values found in common electronic circuits are from 8 to 1000 WVdc, with values to many kilovolts less commonly available.

Circuit symbols for capacitors

The circuit symbols used to designate fixed-value capacitors are shown in Fig. 2-13A. Both types of symbols are common. In certain types of capacitors, the curved plate shown on the left in Fig. 2-13A is usually the outer plate (i.e., the one closest to the outside package of the capacitor). This end of the capacitor is often indicated by a colored band next to the lead attached to that plate.

The symbol for the variable capacitor is shown in Fig. 2-13B. The symbol for a variable capacitor is the fixed-value symbol with an arrow through the plates. Small trimmer and padder capacitors are often denoted by the symbol in Fig. 2-13C. The variable set of plates is designated by the arrow.

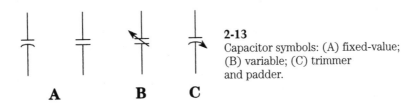

2-13
Capacitor symbols: (A) fixed-value;
(B) variable; (C) trimmer
and padder.

A   **B**   **C**

Fixed capacitors

Several types of fixed capacitors are used in typical electronic circuits. They are classified by dielectric type: paper, mylar, ceramic, mica, polyester, and others.

The construction of an old-fashioned paper capacitor is shown in Fig. 2-14. It consists of two strips of metal foil sandwiched on both sides of a strip of paraffin wax paper. The strip sandwich is then rolled up into a tight cylinder. This rolled-up cylinder is then packaged in either a hard plastic, bakelite, or paper-and-wax case. When the case is cracked, or the wax plugs are loose, replace the capacitor even though it tests as good—it won't be for long. Paper capacitors come in values from about 300 pF to about 4 μF. The breakdown voltages will be from 100 to 600 WVdc.

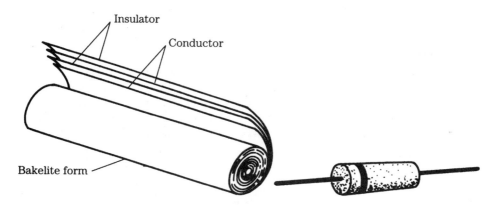

2-14 Paper capacitor construction.

The paper capacitor was used for a number of different applications in older circuits, such as bypassing, coupling, and dc blocking. Unfortunately, no component is perfect. The long rolls of foil used in the paper capacitor exhibit a significant amount of stray inductance. As a result, the paper capacitor is not used for high frequencies. Although they are found in some shortwave receiver circuits, they are rarely (if ever) used at VHF.

In modern applications, or when servicing older equipment containing paper capacitors, use a mylar dielectric capacitor in place of the paper capacitor. Select a unit with exactly the same capacitance rating and a WVdc rating that is equal to or greater than the original WVdc rating.

Several different forms of ceramic capacitors are shown in Fig. 2-15. These capacitors come in values from a few picofarads to 0.5 μF. The working voltages range from 400 WVdc to more than 30,000 WVdc. The common garden-variety disk ceramic capacitors are usually rated at either 600 to 1000 WVdc. Tubular ceramic capacitors are typically much smaller in value than disk or flat capacitors and are used extensively in VHF and UHF circuits for blocking, decoupling, bypassing, coupling, and tuning.

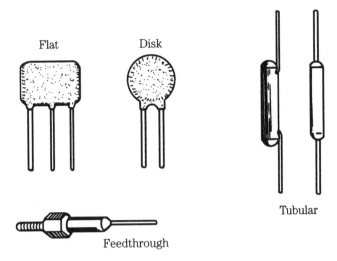

2-15 Ceramic capacitors.

The feedthrough type of ceramic capacitor is used to pass dc and low-frequency ac lines through a shielded panel. These capacitors are often used to filter or decouple lines that run between circuits separated by the shield for purposes of electromagnetic interference (EMI) reduction.

Ceramic capacitors are often rated as to the temperature coefficient. This specification is the change of capacitance per change in temperature in degrees Celsius. A "P" prefix indicates a positive temperature coefficient, an "N" indicates a negative temperature coefficient, and the letters "NPO" indicate a zero temperature coefficient (NPO stands for "negative positive zero"). Do not ad lib on these ratings when servicing a piece of electronic equipment. Use exactly the same temperature coefficient as the original manufacturer used. Nonzero temperature coefficients are often used in oscillator circuits to temperature compensate the oscillator's frequency drift.

Several different types of mica capacitors are shown in Fig. 2-16. The fixed mica capacitor consists of metal plates on either side of a sheet of mica or a sheet of mica silvered with a deposit of metal on either side. The range of values for mica capacitors tends to be 50 pF to 0.02 μF at voltages in the range of 400 to 1000 WVdc. The mica capacitor shown in Fig. 2-16C is called a *silver mica capacitor.* These capacitors have a low temperature coefficient, although for most applications, an NPO disk ceramic capacitor will provide better service than all but the best silver mica units (silver mica temperature coefficients have greater variation than NPO temperature coefficients). Mica capacitors are typically used for tuning and other uses in higher-frequency applications.

Today's equipment designer has a number of different dielectric capacitors available that were not commonly available (or available at all) a few years ago. Polycarbonate, polyester, and polyethylene capacitors are used in a wide variety of applications where the capacitors described in the previous paragraphs once ruled supreme. In digital circuits, tiny 100-WVdc capacitors carry ratings of 0.01 to 0.5 μF.

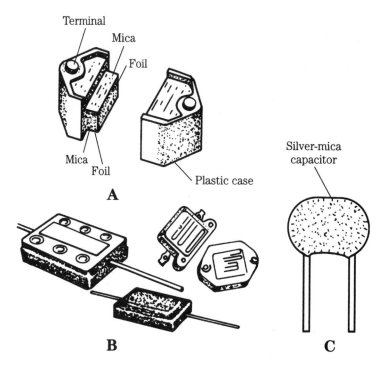

2-16 Mica capacitors.

These are used for decoupling the noise on the +5-Vdc power-supply line. In circuits, such as timers and op-amp Miller integrators, where the leakage resistance across the capacitor becomes terribly important, it might be better to use a polyethylene capacitor.

Check current catalogs for various old and new capacitor styles. The applications paragraph in the catalog will tell you what applications each can be used for as well as provide a guide to the type of antique capacitor it will replace.

Capacitors in ac circuits

When an electrical potential is applied across a capacitor, current will flow as charge is stored in the capacitor. As the charge in the capacitor increases, the voltage across the capacitor plates rises until it equals the applied potential. At this point, the capacitor is fully charged, and no further current will flow.

Figure 2-17 shows an analogy for the capacitor in an ac circuit. The actual circuit is shown in Fig. 2-17A. It consists of an ac source connected in parallel across the capacitor. The mechanical analogy is shown in Fig. 2-17B. The capacitor consists of a two-chambered cylinder in which the upper and lower chambers are separated by a flexible membrane or diaphragm. The wires are pipes to the ac source (which is a pump). As the pump moves up and down, pressure is applied to the first side of the diaphragm and then the other, alternately forcing fluid to flow into and out of the two chambers of the capacitor.

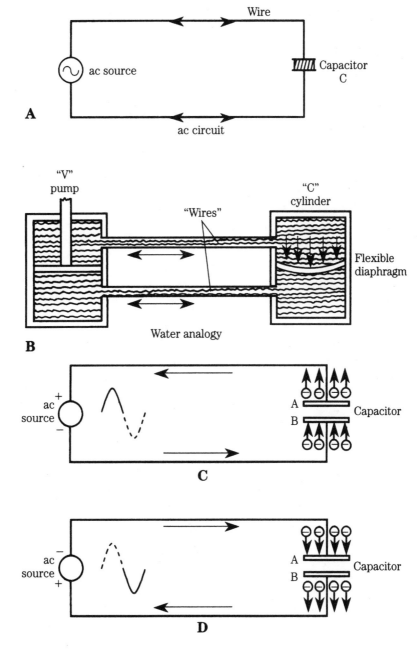

2-17 Capacitor circuit analogy and capacitor circuit action.

The ac circuit mechanical analogy is not perfect, but it works for these purposes. Apply these ideas to the electrical case. Figure 2-17 shows a capacitor connected across an ac (sine-wave) source. In Fig. 2-17A, the ac source is positive, so negatively charged electrons are attached from plate A to the ac source and electrons from the negative terminal of the source are repelled toward plate B of the capacitor. On the alternate half-cycle (Fig. 2-17B), the polarity is reversed. Electrons from the new negative pole of the source are repelled toward plate A of the capacitor and electrons from plate B are attracted toward the source. Thus, current will flow in and out of the capacitor on alternating half-cycles of the ac source.

Voltage and current in capacitor circuits

Consider the circuit in Fig. 2-18: an ac source connected in parallel with the capacitor. It is the nature of a capacitor to oppose these changes in the applied voltage (the inverse of the action of an inductor). As a result, the voltage lags behind the current by 90 degrees. These relationships are shown in terms of sine waves in Fig. 2-18B and in vector form in Fig. 2-18C.

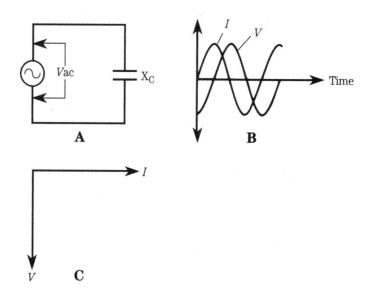

2-18 (A) Ac circuit using a capacitor; (B) sine-wave relationship between V and I; (C) vector representation.

Do you remember the differences in the actions of inductors and capacitors on voltage and current? In earlier texts, the letter E was used to denote voltage, so you could make a little mnemonic: "ELI the ICE man."

"ELI the ICE man" suggests that in the inductive (L) circuit, the voltage (E) comes before the current (I): ELI; in a capacitive (C) circuit, the current comes before the voltage: ICE.

The action of a circuit containing a resistance and capacitance is shown in Fig. 2-19A. As in the case of the inductive circuit, there is no phase shift across the resistor, so the R vector points in the east direction (Fig. 2-19B). The voltage across the capacitor, however, is phase-shifted 90° so its vector points south. The total resultant phase shift (ϕ) is found using the Pythagorean rule to calculate the angle between V_R and V_T.

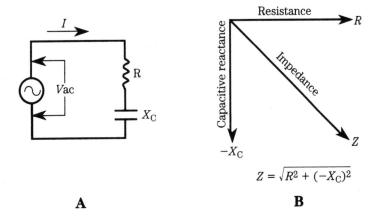

2-19 (A) Resistor-capacitor (RC) circuit; (B) vector representation of circuit action.

The impedance of an RC circuit is found in exactly the same manner as the impedance of an RL circuit; that is, as the root of the sum of the squares:

$$Z = \sqrt{R^2 + X_C^2}. \tag{2-19}$$

LC Resonant tank circuits

When you use an inductor and a capacitor together in the same circuit, the combination forms an LC resonant circuit, which is sometimes called a *tank circuit* or *resonant tank circuit*. These circuits can be used to tune a radio receiver as well as in other applications. There are two basic forms of LC resonant tank circuit: series (Fig. 2-20A) and parallel (Fig. 2-20B). These circuits have much in common and much that makes them fundamentally different from each other.

The condition of resonance occurs when the capacitive reactance and inductive reactance are equal. As a result, the resonant tank circuit shows up as purely resistive at the resonant frequency (see Fig. 2-20C) and as a complex impedance at other frequencies. The LC resonant tank circuit operates by an oscillatory exchange of energy between the magnetic field of the inductor and the electrostatic field of the capacitor, with a current between them carrying the charge.

Because the two reactances are both frequency-dependent, and because they are inverse to each other, the resonance occurs at only one frequency (F_R). You can calculate the standard resonance frequency by setting the two reactances equal to each other and solving for F. The result is as follows:

$$F = \frac{1}{2\pi\sqrt{LC}}, \tag{2-20}$$

where
    F is in hertz (Hz)
    L is in henrys (H)
    C is in farads (F).

Series-resonant circuits

The series-resonant circuit (Fig. 2-20A), like other series circuits, is arranged so that the terminal current (I) from the source (V) flows in both components equally.

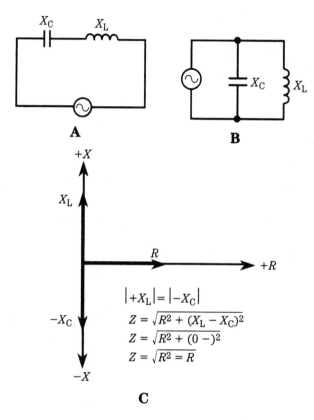

2-20 (A) Series inductor-capacitor (LC) circuit; (B) parallel LC circuit; (C) vector relationships.

The vector diagrams of Fig. 2-21A through 2-21C show the situation under three different conditions:

- Figure 2-21A: The inductive reactance is larger than the capacitive reactance, so the excitation frequency is greater than F_R. Notice that the voltage drop across the inductor is greater than that across the capacitor, so the total circuit looks like it contains a small inductive reactance.
- Figure 2-21B: The situation is reversed. The excitation frequency is less than the resonant frequency, so the circuit looks slightly capacitive to the outside world.
- Figure 2-21C: The excitation frequency is at the resonant frequency, so $X_C = X_L$ and the voltage drops across the two components are of equal but opposite phase.

In a circuit that contains a resistance, an inductive reactance, and a capacitive reactance, there are three vectors (X_L, X_C, and R) to consider (Fig. 2-22) plus a resultant vector. As in the other circuit, the north direction is X_L, the south direction is X_C, and the east direction is R. Using the parallelogram method, construct a resultant for R and X_C, which is shown as vector A. Next, construct the same kind of vector (B) for R and X_L. The resultant vector (C) is made using the parallelogram method on vectors A and B. Vector C represents the impedance of the circuit: The magnitude is represented by the length and the phase angle by the angle between vector C and R.

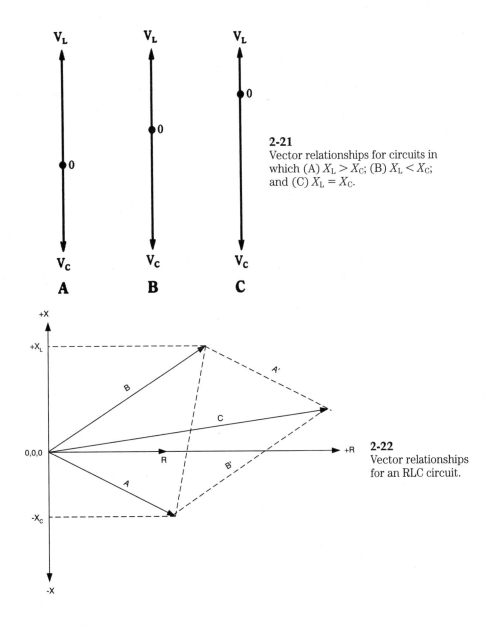

2-21
Vector relationships for circuits in which (A) $X_L > X_C$; (B) $X_L < X_C$; and (C) $X_L = X_C$.

2-22
Vector relationships for an RLC circuit.

Figure 2-23A shows a series-resonant LC tank circuit, and Fig. 2-23B shows the current and impedance as a function of frequency. Series-resonant circuits have a low impedance at the resonant frequency and a high impedance at all other frequencies. As a result, the line current from the source is maximum at the resonant frequency and the voltage across the source is minimum.

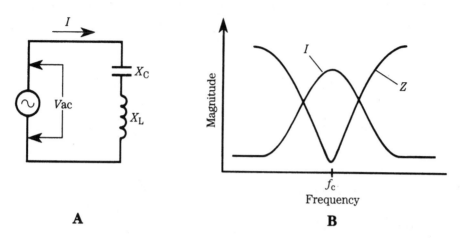

A   **B**

2-23 (A) Series-resonant LC circuit; (B) *I* and *Z* vs frequency response.

Parallel-resonant circuits

The parallel-resonant tank circuit (Fig. 2-24A) is the inverse of the series-resonant circuit. The line current from the source splits and flows separately in the inductor and capacitor.

The parallel-resonant circuit has its highest impedance at the resonant frequency and a low impedance at all other frequencies. Thus, the line current from the source is minimum at the resonant frequency (Fig. 2-24B) and the voltage across the LC tank circuit is maximum. This fact is important in radio tuning circuits, as you will see later.

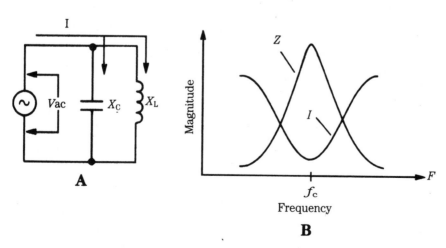

A   **B**

2-24 (A) Parallel-resonant LC circuit; (B) *I* and *Z* vs frequency response.

Choosing component values for LC-resonant tank circuits

In some cases, one of the two basic type of resonant tank circuit (either series or parallel) is clearly preferred over the other, but in other cases either could be used if it is used correctly. For example, in a wave trap (i.e., a circuit that prevents a particular frequency from passing) a parallel-resonant circuit in series with the signal line will block its resonant frequency while passing all frequencies removed from resonance; a series-resonant circuit shunted across the signal path will bypass its resonant frequency to common (or "ground") while allowing frequencies removed from resonance to pass.

LC-resonant tank circuits are used to tune radio receivers; it is these circuits that select the station to be received while rejecting others. A superheterodyne radio receiver (the most common type) is shown in simplified form in Fig. 2-25. According to the superhet principal, the *radio frequency being received* (F_{RF}) is converted to another frequency, called the *intermediate frequency* (F_{IF}) by being mixed with a *local oscillator signal* (F_{LO}) in a nonlinear mixer stage. The output of the untamed mixer would be a collection of frequencies defined by the following:

$$F_{IF} = mF_{RF} \pm nF_{LO}, \qquad (2\text{-}21)$$

where m and n are either integers or zero. For the simplified case that is the subject of this chapter only the first pair of products ($m = n = 1$) are considered, so the output spectrum will consist of $F_{RF}, F_{LO}, F_{RF} - F_{LO}$ (difference frequency), and $F_{RF} + F_{LO}$ (sum frequency). In older radios, for practical reasons, the difference frequency was selected for F_{IF} today either sum or difference frequencies can be selected, depending on the design of the radio.

Several LC tank circuits are present in this notional superhet radio. The antenna tank circuit (C_1/L_1) is found at the input of the RF amplifier stage, or if no RF amplifier is used it is at the input to the mixer stage. A second tank circuit (L_2/C_2), tuning the same range as L_1/C_1, is found at the output of the RF amplifier or at the input of the mixer. Another LC tank circuit (L_3/C_3) is used to tune the local oscillator; it is this tank circuit that sets the frequency that the radio will receive.

Additional tank circuits (only two shown) can be found in the IF amplifier section of the radio. These tank circuits will be fixed-tuned to the IF frequency, which in common AM broadcast-band (BCB) radio receivers is typically 450, 455, 460, or 470 kHz depending on the designer's choices (and sometimes country of origin). Other IF frequencies are also used, but these are most common. FM broadcast receivers typically use a 10.7-MHz IF, and shortwave receivers might use a 1.65-, 8.83-, or 9-MHz frequency or an IF frequency above 30 MHz.

The tracking problem

On a radio that tunes the front end with a single knob, which is the case in almost all receivers today, the three capacitors (C_1 through C_3 in Fig. 2-26) are typically ganged (i.e., mounted on a single rotor shaft). These three tank circuits must track each other; that is, when the RF amplifier is tuned to a certain radio signal frequency, the LO must produce a signal that is different from the RF frequency by the amount of the IF frequency. Perfect tracking is rare, but the fact that your single-knob-tuned radio works is testimony to the fact that the tracking is not too terrible.

The issue of tracking LC tank circuits for the AM BCB receiver has not been a major problem for many years: the band limits are fixed over most of the world and

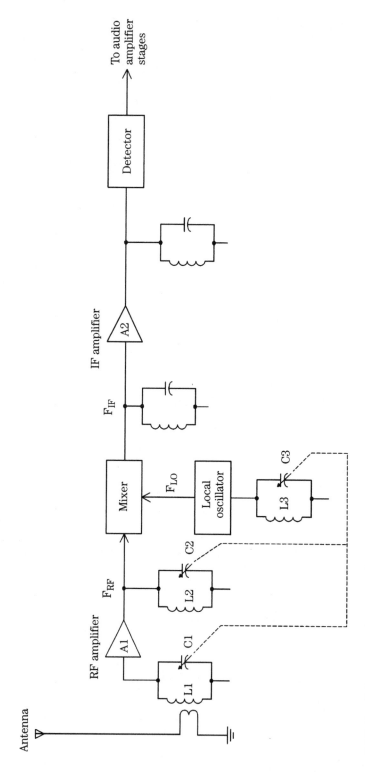

2-25 Parallel-resonant circuits used in a radio receiver.

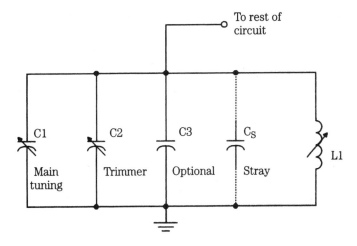

2-26 Capacitors used in a typical radio tuning circuit.

component manufacturers offer standard adjustable inductors and variable capacitors to tune the RF and LO frequencies. Indeed, some even offer three sets of coils: antenna, mixer input/RF amp output, and LO. The reason why the antenna and mixer/RF coils are not the same, despite tuning to the same frequency range, is that these locations see different distributed or "stray" capacitances. In the United States, it is standard to use a 10- to 365-pF capacitor and a 220-μH inductor for the 540- to 1700-kHz AM BCB. In other countries, slightly different combinations are used; for example, 320, 380, 440, and 500 pF and others are seen in catalogs.

Recently, however, two events coincided that caused me to examine the method of selecting capacitance and inductance values. First, I embarked on a design project to produce an AM DX-er's receiver that had outstanding performance characteristics. Second, the AM broadcast band was recently extended so that the upper limit is now 1700 rather than 1600 kHz. The new 540- to 1700-kHz band is not accommodated by the now-obsolete "standard" values of inductance and capacitance. So I calculated new candidate values.

The RF amplifier/antenna tuner problem

In a typical RF tank circuit, the inductance is kept fixed (except for a small adjustment range that is used for overcoming tolerance deviations) and the capacitance is varied across the range. The frequency changes as the square root of the capacitance changes. If F_1 is the minimum frequency in the range and F_2 is the maximum frequency then the relationship is as follows:

$$\frac{F_2}{F_1} = \sqrt{\frac{C_{max}}{C_{min}}} \qquad (2\text{-}22)$$

or, in a rearranged form that some find more applicable:

$$\left(\frac{F_2}{F_1}\right)^2 = \frac{C_{max}}{C_{min}}. \qquad (2\text{-}23)$$

In the case of the new AM receiver, I wanted an overlap of about 15 kHz at the bottom end of the band and 10 kHz at the upper end and so needed a resonant tank circuit that would tune from 525 to 1710 kHz. In addition, because variable capacitors are widely available in certain values based on the old standards, I wanted to use a "standard" AM BCB variable capacitor. A 10- to 380-pF unit from a U.S. vendor was selected.

The minimum required capacitance, C_{min}, can be calculated from the following:

$$\left(\frac{F_2}{F_1}\right)^2 C_{min} = C_{min} + \Delta C, \tag{2-24}$$

where
    F_1 is the minimum frequency tuned
    F_2 is the maximum frequency tuned
    C_{min} is the minimum required capacitance at F_2
    ΔC is the difference between C_{max} and C_{min}.

Example

Find the minimum capacitance needed to tune 1710 kHz when a 10- to 380-pF capacitor ($\Delta C = 380 - 10$ pF $= 370$ pF) is used and the minimum frequency is 525 kHz.

Solution

$$\left(\frac{F_2}{F_1}\right)^2 C_{min} = C_{min} + \Delta C$$

$$\left(\frac{1710 \text{ kHz}}{525 \text{ kHz}}\right)^2 C_{min} = C_{min} + 370 \text{ pF}$$

$$10.609 \, C_{min} = C_{min} + 370 \text{ pF}$$

$$C_{min} = 38.51 \text{ pF}.$$

The maximum capacitance must be $C_{min} + \Delta C$ or $38.51 + 370$ pF $= 408.51$ pF. Because the tuning capacitor (C_1 in Fig. 2-26) does not have exactly this range, external capacitors must be used and because the required value is higher than the normal value additional capacitors are added to the circuit in parallel to C_1. Indeed, because somewhat unpredictable "stray" capacitances also exist in the circuit, the tuning capacitor values should be a little less than the required values in order to accommodate strays plus tolerances in the actual (versus published) values of the capacitors. In Fig. 2-26, the main tuning capacitor is C_1 (10 to 380 pF), C2 is a small-value trimmer capacitor used to compensate for discrepancies, C_3 is an optional capacitor that might be needed to increase the total capacitance, and C_s is the stray capacitance in the circuit.

The value of the stray capacitance can be quite high—especially if other capacitors in the circuit are not directly used to select the frequency (e.g., in Colpitts and Clapp oscillators, the feedback capacitors affect the LC tank circuit). In the circuit that I was using, however, the LC tank circuit is not affected by other capacitors. Only the wiring strays and the input capacitance of the RF amplifier or mixer stage need be accounted. From experience, I apportioned 7 pF to C_s as a trial value.

The minimum capacitance calculated was 38.51; there is a nominal 7 pF of stray capacitance and the minimum available capacitance from C_1 is 10 pF. Therefore, the combined values of C_2 and C_3 must be 38.51 pF − 10 pF − 7 pF, or 21.5 pF. Because there is considerable reasonable doubt about the actual value of C_s, and because of tolerances in the manufacture of the main tuning variable capacitor (C_1), a wide range of capacitance for $C_2 + C_3$ is preferred. It is noted from several catalogs that 21.5 pF is near the center of the range of 45- and 50-pF trimmer capacitors. For example, one model lists its range as 6.8 to 50 pF; its center point is only slightly removed from the actual desired capacitance. Thus, a 6.8- to 50-pF trimmer was selected, and C_3 was not used.

Selecting the inductance is a matter of choosing the frequency and associated required capacitance at one end of the range and calculating from the standard resonance equation solved for L:

$$L_{\mu H} = \frac{10^6}{4\pi^2 F_{\text{low}}^2 C_{\text{max}}}$$

$$L_{\mu H} = \frac{10^6}{(4)(\pi^2)(525{,}000)^2 (4.085 \times 10^{-10})} = 224.97 \approx 225 \text{ pF.}$$

The RF amplifier input LC tank circuit and the RF amplifier output LC tank circuit are slightly different cases because the stray capacitances are somewhat different. In the example, I am assuming a JFET transistor RF amplifier, and it has an input capacitance of only a few picofarads. The output capacitance is not a critical issue in this specific case because I intend to use a 1-mH RF choke in order to prevent JFET oscillation. In the final receiver, the RF amplifier can be deleted altogether, and the LC tank circuit described will drive a mixer input through a link-coupling circuit.

The local oscillator (LO) problem

The local oscillator circuit must track the RF amplifier and must also tune a frequency range that is different from the RF range by the amount of the IF frequency (455 kHz). In keeping with common practice, I selected to place the LO frequency 455 kHz above the RF frequency. Thus, the LO must tune the range from 980 to 2165 kHz.

Three methods can be used to make the local oscillator track with the RF amplifier frequency when single-shaft tuning is desired: the trimmer capacitor method, the padder capacitor method, and the different-value cut-plate capacitor method.

Trimmer capacitor method The trimmer capacitor method is shown in Fig. 2-27 and is the same as the RF LC tank circuit. Using exactly the same method as before, a frequency ratio of 2165/980 yields a capacitance ratio of $(2165/980)^2 = 4.88{:}1$. The results are a minimum capacitance of 95.36 pF and a maximum capacitance of 465.36 pF. An inductance of 56.7 μH is needed to resonate these capacitances to the LO range.

There is always a problem associated with using the same identical capacitor for both RF and LO. It seems that there is just enough difference that tracking between

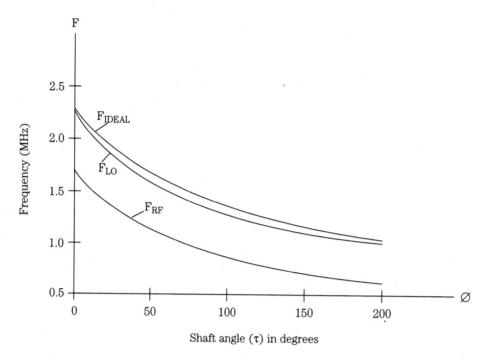

2-27 The tracking problem.

them is always a bit off. Figure 2-27 shows the ideal LO frequency and the calculated LO frequency. The difference between these two curves is the degree of nontracking. The curves overlap at the ends but are awful in the middle. There are two cures for this problem. First, use a padder capacitor in series with the main tuning capacitor (Fig. 2-28). Second, use a different-value cut-plate capacitor.

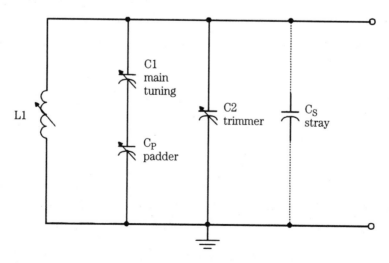

2-28 Oscillator-tuned circuit using padder capacitor.

Figure 2-28 shows the use of a padder capacitor (C_p) to change the range of the LO section of the variable capacitor. This method is used when both sections of the variable capacitor are identical. Once the reduced capacitance values of the C_1/C_p combination are determined, the procedure is identical to that above. But first, you must calculate the value of the padder capacitor and the resultant range of the C_1/C_p combination. The padder value is calculated from the following:

$$\frac{C_{1_{max}} C_p}{C_{1_{max}} + C_p} = \left(\frac{F_2}{F_1}\right)^2 \left(\frac{C_{1_{min}} C_p}{C_{1_{min}} + C_p}\right), \tag{2-25}$$

which solves for C_p. For the values of the selected main tuning capacitor and LO:

$$\frac{(380 \text{ pF})(C_p)}{(380 + C_p)\text{pF}} = (4.88)\left(\frac{(10 \text{ pF})(C_p)}{(10 + C_p \text{ pF})}\right). \tag{2-26}$$

Solving for C_p by the least common denominator method (crude, but it works) yields a padder capacitance of 44.52 pF. The series combination of 44.52 pF and a 10- to 380-pF variable yields a range of 8.2 to 39.85 pF. An inductance of 661.85 pF is needed for this capacitance to resonate over 980 to 2165 kHz.

Tuned RF/IF transformers

Most of the resonant circuits used in radio receivers are actually tuned RF transformers that couple signal from one stage to another. Figure 2-29 shows several popular forms of tuned, or coupled, RF/IF tank circuits. In Fig. 2-29A, one winding is tuned and the other is untuned. In the configurations shown, the untuned winding is the secondary of the transformer. This type of circuit is often used in transistor and other solid-state circuits or when the transformer has to drive either a crystal or mechanical bandpass filter circuit. In the reverse configuration (L_1 = output, L_2 = input), the same circuit is used for the antenna coupling network or as the interstage transformer between RF amplifiers in TRF radios.

The circuit in Fig. 2-29B is a parallel-resonant LC tank circuit that is equipped with a low-impedance tap. This type of circuit is used to drive a crystal detector or other low-impedance load. Another circuit for driving a low-impedance load is shown in Fig. 2-29C. This circuit splits the capacitance that resonates the coil into two series capacitors. As a result, it is a capacitive voltage divider. The circuit in Fig. 2-29D uses a tapped inductor for matching low-impedance sources (e.g., antenna circuits) and a tapped capacitive voltage divider for low-impedance loads. Finally, the circuit in Fig. 2-29E uses a tapped primary and tapped secondary winding in order to match two low-impedance loads while retaining the sharp bandpass characteristics of the tank circuit.

Construction of RF/IF transformers

The tuned RF/IF transformers built for radio receivers are typically wound on a common cylindrical form and surrounded by a metal shield that can prevent interaction of the fields of coils that are in close proximity to each other. Figure 2-30A shows the schematic for a typical RF/IF transformer, and the sectioned view (Fig. 2-30B) shows one form of construction. This method of building transformers

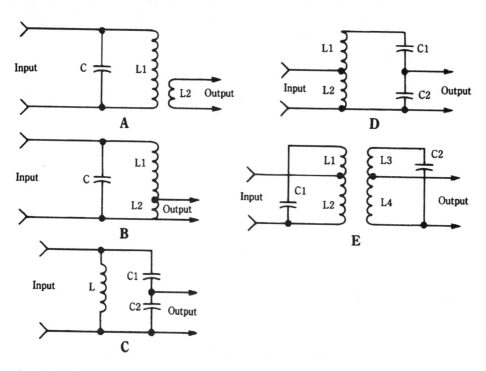

2-29 Coupling circuits used in radio receivers and transmitters.

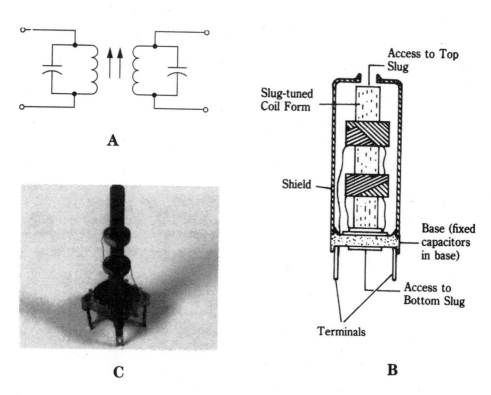

2-30 (A) Tuned RF or IF transformer; (B) IF/RF transformer construction; (C) actual transformer.

was common at the beginning of World War II and continued into the early transistor era. The capacitors were built into the base of the transformer, and the tuning slugs were accessed from holes in the top and bottom of the assembly. In general, expect to find the secondary at the bottom hole and the primary at the top holes.

The term *universal wound* refers to a cross-winding system that minimizes the interwinding capacitance of the inductor and raises the self-resonant frequency of the inductor (a good thing). An actual example of an RF/IF transformer is shown in Fig. 2-30C. The smaller type tends to be post-WWII, and the larger type is pre-WWII.

Bandwidth of RF/IF transformers

Figure 2-31A shows a parallel-resonant RF/IF transformer, and Fig. 2-31B shows the usual construction in which the two coils (L_1 and L_2) are wound at distance d apart on a common cylindrical form. The bandwidth of the RF/IF transformer is the difference between the frequencies where the signal voltage across the output winding falls off 6 dB from the value at the resonant frequency (F_R), as shown in Fig. 2-31C. If F_1 and F_2 are 6-dB (also called the *3-dB point* when signal power is measured instead of voltage) frequencies, then the bandwidth is $F_2 - F_1$. The shape of the frequency response curve in Fig. 2-31C is said to represent critical coupling.

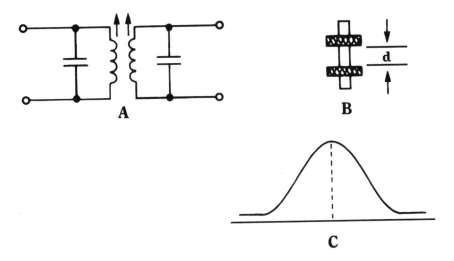

2-31 Moderate coupling IF transformer. (A) The circuit; (B) coil spacing; (C) bandpass response.

An example of a subcritical or undercoupled RF/IF transformer is shown in Fig. 2-32. As shown in Figs. 2-32A and 2-32B, the windings are farther apart than in the critically coupled case, so the bandwidth (Fig. 2-32C) is much narrower than in the critically coupled case. The subcritically coupled RF/IF transformer is often used in shortwave or communications receivers in order to allow the narrow bandwidth to discriminate between adjacent channels.

The overcritically coupled RF/IF transformer is shown in Fig. 2-33. Notice in Figs. 2-33A and 2-33B that the windings are closer together, so the bandwidth (Fig. 2-33C) is much broader. In some radio schematics and service manuals (not to

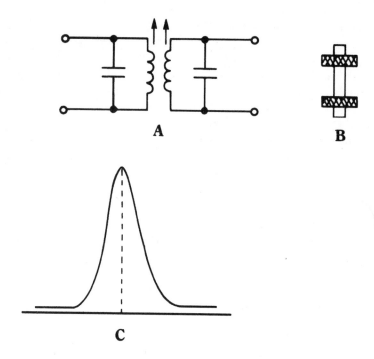

2-32 Loose coupling IF transformer. (A) The circuit; (B) coil spacing; (C) bandpass response.

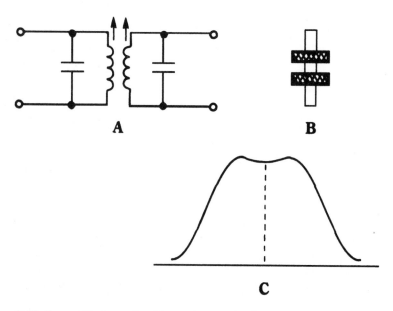

2-33 Over-critical coupling IF transformer. (A) The circuit; (B) coil spacing; (C) bandpass response.

mention early textbooks), this form of coupling was sometimes called *high-fidelity coupling* because it allowed more of the sidebands of the signal (which carry the audio modulation) to pass with less distortion of frequency response.

The bandwidth of the resonant tank circuit, or the RF/IF transformer, can be summarized in a figure of merit called Q. The Q of the circuit is the ratio of the bandwidth to the resonant frequency as follows: $Q = BW/FR$. An overcritically coupled circuit has a low Q, while a narrow bandwidth subcritically coupled circuit has a high Q.

A resistance in the LC tank circuit will cause it to broaden; that is, to lower its Q. The loaded Q (i.e., Q when a resistance is present, as in Fig. 2-34A) is always less than the unloaded Q. Some radios use a switched resistor (Fig. 2-34B) to allow the user to broaden or narrow the bandwidth. This switch might be labeled *fidelity* or *tone* or something similar on radio receivers.

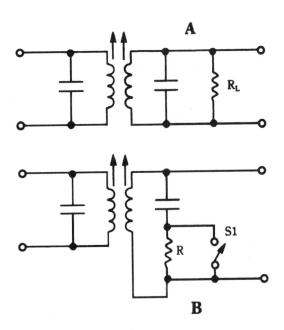

2-34 Resistor loading to broaden response. (A) Parallel method; (B) hi-fi switch.

Problems with RF/IF transformers

The IF and RF transformers represent a high potential for intermittent problems in radio receivers. There are two basic forms of problems with the IF transformer: intermittent operation and intermittent noise. The best cure for a bad IF or RF transformer is replacement, but because old IF and RF transformers are not always available today, more emphasis must be placed on repair of the transformer.

Figure 2-35A shows the basic circuit for a single-tuned IF transformer (others might have a tuned secondary winding). If one of the very-fine wires making up the coil breaks (Fig. 2-35B), then operation is interrupted. The problem can usually be diagnosed by lightly tapping on the shield can of the transformer or by using a signal

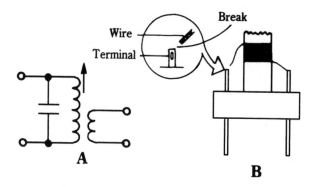

2-35 (A) IF transformer circuit; (B) typical defect.

tracer or signal generator to find where the signal is interrupted. In some cases, the plate voltage of the IF amplifier is interrupted when the transformer opens. This can be spotted using a dc voltmeter.

Repairing the IF transformer is a delicate operation. Examine the shield to determine how it is assembled. If it is secured with a screw or nut, simply remove the screw (or nut). If the IF transformer is sealed by metal tabs, then you must very carefully pry the tabs open (do not break them or bend them too far). Slide the base and coil assembly out of the shield. If there is enough slack on the broken wire, simply solder the wire back onto the terminal. The heat of the soldering iron tip will burn away the enamel insulation. If the wire does not have enough slack, then add a little length to the terminal by soldering a short piece of solid wire to the terminal and then solder the IF transformer wire to the added wire. In some cases, it is possible to remove a portion of one turn of the transformer winding in order to gain extra length, but this procedure will change the tuning.

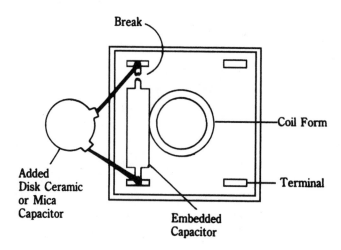

2-36 Replacement of the embedded capacitor with a disk ceramic or silvered mica fixed capacitor.

A noisy IF transformer needs replacement, but the problem is the lack of original or replacement components. The capacitor can be replaced on many IF transformers. If the capacitor is a discrete component soldered to the base, it is simple to replace it with another similar (if not identical) capacitor. But if the IF transformer uses an embedded mica capacitor (Fig. 2-36), then the job becomes more complex.

If the capacitor element is visible (as in Fig. 2-36), then clip one end of the capacitor where it is attached to the terminal. Bridge a disk ceramic or mica capacitor between the two terminals. The value of the capacitor must be found experimentally—it is the value that will resonate the coil to the IF.

If the capacitor is embedded and not visible, it might become necessary to install a new terminal and solder the added capacitor to that terminal (with appropriate external circuit modifications).

3
CHAPTER

Variable capacitors in RF circuits

Like all capacitors, variable capacitors are made by placing two sets of metal plates parallel to each other (Fig. 3-1A) separated by a dielectric of air, mica, ceramic, or a vacuum. The difference between variable and fixed capacitors is that, in variable capacitors, the plates are constructed in such a way that the capacitance can be changed. There are two principal ways to vary the capacitance: either the spacing between the plates is varied or the cross-sectional area of the plates that face each other is varied. Figure 3-1B shows the construction of a typical variable capacitor used for the main tuning control in radio receivers. The capacitor consists of two sets of parallel plates. The stator plates are fixed in their position and are attached to the frame of the capacitor. The rotor plates are attached to the shaft that is used to adjust the capacitance.

Another form of variable capacitor used in radio receivers is the compression capacitor shown in Fig. 3-1C. It consists of metal plates separated by sheets of mica dielectric. In order to increase the capacitance, the manufacturer may increase the area of the plates and mica or the number of layers (alternating mica/metal) in the assembly. The entire capacitor will be mounted on a ceramic or other form of holder. If mounting screws or holes are provided then they will be part of the holder assembly.

Still another form of variable capacitor is the piston capacitor shown in Fig. 3-1D. This type of capacitor consists of an inner cylinder of metal coaxial to, and inside of, an outer cylinder of metal. An air, vacuum, or (as shown) ceramic dielectric separates the two cylinders. The capacitance is increased by inserting the inner cylinder further into the outer cylinder.

The small-compression or piston-style variable capacitors are sometimes combined with air variable capacitors. Although not exactly correct word usage, the smaller capacitor used in conjunction with the larger air variable is called a trimmer capacitor. These capacitors are often mounted directly on the air variable frame or very close by in the circuit. In many cases, the "trimmer" is actually part of the air variable capacitor.

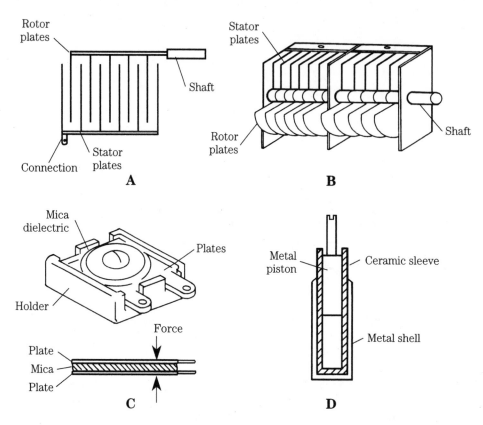

3-1 Air dielectric variable capacitors consists of interleaved stator and rotor plates. (A) Schematic view; (B) construction; (C) mica compression variable capacitor; (D) piston variable capacitor.

There are actually two uses for small variable capacitors in conjunction with the main tuning capacitor in radios. First, there is the true "trimmer," i.e., a small-valued variable capacitor in parallel with the main capacitor (Fig. 3-2A). These trimmer capacitors (C_2) are used to trim the exact value of the main capacitor (C_1). The other form of small capacitor is the padder capacitor (Fig. 3-2B), which is connected in series with the main capacitor. This error in terminology is calling both series and parallel capacitors "trimmers," when only the parallel connected capacitor is properly so-called.

The capacitance of an air variable capacitor at any given setting is a function of how much of the rotor plate set is shaded by the stator plates. In Fig. 3-3A, the rotor plates are completely outside of the stator plate area. Because the shading is zero, the capacitance is minimum. In Fig. 3-3B, however, the rotor plate set has been slightly meshed with the stator plate, so some of its area is shaded by the stator. The capacitance in this position is at an intermediate-value. Finally, in Fig. 3-3C, the rotor is completely meshed with the stator so the cross-sectional area of the rotor that is shaded by the stator is maximum. Therefore, the capacitance is also maximum. Remember these two rules:

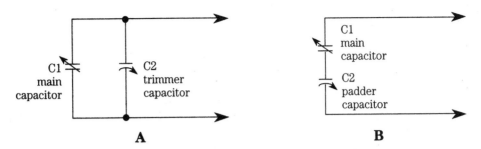

3-2 (A) Trimmer capacitor connected in parallel with the main tuning capacitor; (B) padder capacitor is connected in series with the main tuning capacitor.

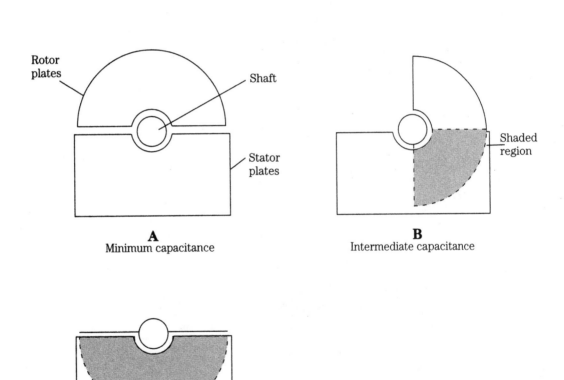

3-3 The capacitance of the air variable capacitor is determined by how much of the rotor plate is shaded by the stator plates. (A) Minimum capacitance; (B) intermediate capacitance; (C) maximum capacitance.

1. Minimum capacitance is found when the rotor plates are completely unmeshed with the stator plates; and
2. Maximum capacitance is found when the rotor plates are completely meshed with the stator plates.

Figure 3-4 shows a typical single-section variable capacitor. The stator plates are attached to the frame of the capacitor, which in most radio circuits is grounded. Front and rear mounts have bearing surfaces to ease the rotor's action. The ganged variable capacitor (Fig. 3-5) was invented to provide tracking between two related LC-tuned circuits, as in a radio receiver. Such capacitors are basically two (in the case of Fig. 3-5) or more variable capacitors mechanically ganged on the same rotor shaft.

3-4
Small air variable capacitor.

In Fig. 3-5, both sections of the variable capacitor have the same capacitance, so they are identical to each other. If this capacitor is used in a superheterodyne radio, the section used for the local oscillator (LO) tuning must be padded with a series capacitance in order to reduce the overall capacitance. This trick is done to permit the higher-frequency LO to track with the RF amplifiers on the dial.

3-5
Dual section air variable capacitor.

In many superheterodyne radios, you will find variable tuning capacitors in which one section (usually the front section) has fewer plates than the other section.

One section tunes the RF amplifier of the radio, and the other tunes the local oscillator. These capacitors are sometimes called *cut-plate capacitors* because the LO section plates are cut to permit tracking of the LO with the RF.

Straight-line capacitance vs straight-line frequency capacitors

The variable capacitor shown in Fig. 3-5 has the rotor shaft in the geometric center of the rotor plate half-circle. The capacitance of this type of variable capacitor varies directly with the rotor shaft angle. As a result, this type of capacitor is called a *straight-line capacitance model*. Unfortunately, as you will see in a later section, the frequency of a tuned circuit based on inductors and capacitors is not a linear (straight line) function of capacitance. If a straight line capacitance unit is used for the tuner, then the frequency units on the dial will be cramped at one end and spread out at the other (you've probably seen such radios). But some capacitors have an offset rotor shaft (Fig. 3-6A) that compensates for the nonlinearity of the tuning circuit. The shape of the plates and the location of the rotor shaft are designed to produce a linear relationship between the shaft angle and the resonant frequency of the tuned circuit in which the capacitor is used. A comparison between straight-line capacitance and straight-line frequency capacitors is shown in Fig. 3-6B.

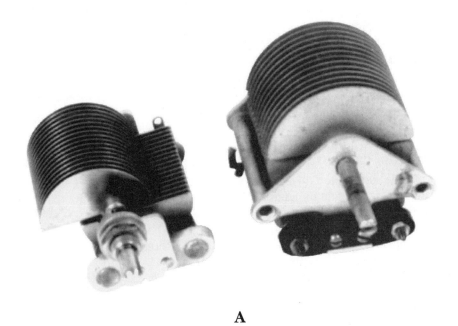

A

3-6 (A) Straight-line frequency capacitor; (B) comparison of straight-line capacitance with straight-line frequency capacitor.

3-6
Continued

B

Special variable capacitors

In the preceding sections, the standard forms of variable capacitor were covered. These capacitors are largely used for tuning radio receivers, oscillators, signal generators, and other variable-frequency LC oscillators. This section covers some special forms of variable capacitor.

Split-stator capacitors

The split-stator capacitor is one in which two variable capacitors are mounted on the same shaft. The split-stator capacitor normally uses a pair of identical capacitors, each the same value, turned by the same shaft. The rotor is common to both capacitors. Thus, the capacitor will tune either two tuned circuits at the same time or both halves of a balanced-tuned circuit (i.e., one in which the inductor is center-tapped and grounded).

Differential capacitors

Although some differential capacitors are often mistaken for split-stator capacitors, they are actually quite different. The split-stator capacitor is tuned in tandem, i.e., both capacitor sections have the same value at any given shaft setting. The differential capacitor, on the other hand, is arranged so that one capacitor section increases in capacitance and the other section decreases in exactly the same proportion.

Differential capacitors are used in impedance bridges, RF resistance bridges, and other such instruments. If you buy or build a high-quality RF impedance bridge for antenna measurements, for example, it is likely that it will have a differential capacitor as the main adjustment control. The two capacitors are used in two arms of a Wheatstone bridge circuit. Be careful of planning to build such a bridge, however. I recently bought the differential capacitor for such an instrument, and it cost nearly $50!

"Transmitting" variable capacitors

The one requirement of transmitting variable capacitors (and certain antenna tuning capacitors) is the ability to withstand high voltages. The high-power ham radio or AM broadcast transmitter will have a dc potential of 1500 to 7500 V on the RF amplifier anode, depending on the type of tube used. If amplitude-modulated, the potential can double. Also, if certain antenna defects arise, then the RF voltages in the circuit can rise quite high. As a result, the variable capacitor used in the final amplifier anode circuit must be able to withstand these potentials.

Two forms of transmitting variables are typically used in RF power amplifiers and antenna tuners. Figure 3-7 shows a transmitting air variable capacitor. The shaft of this particular capacitor is nylon, so it can be mounted either with the frame grounded or with the frame floating at high voltage. The other form of transmitting variable is the vacuum variable. This type of capacitor is a variation of the piston capacitor, but it has a vacuum dielectric (*K* factor = 1.0000). The model shown in Fig. 3-8 is a 18- to 1000-pF model that is driven from a 12-Vdc electric motor. Other vacuum variables are manually driven.

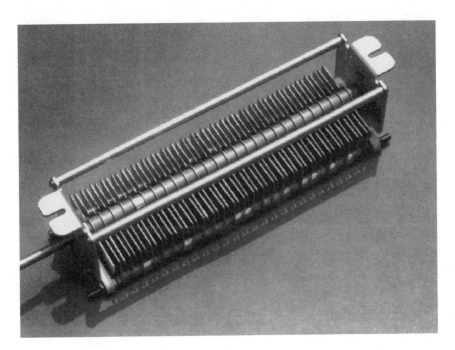

3-7 Transmitting air variable capacitor.

Solid-state capacitors

One of the problems with variable capacitors is that they are large, bulky things (look at all the photos) that must be mechanically operated. Modern electronic circuits, including most radios today, are electrically tuned using a varicap diode for the capacitor function. These "capacitors" operate because the junction capacitance

3-8 Vacuum variable capacitor.

(C_t) of a PN junction diode is a function of the reverse bias voltage applied across the diode. The varicap (a.k.a. "varactor") is therefore a variable capacitor in which the capacitor is a function of an applied voltage. Maximum capacitances run from 15 to 500 pF, depending on the type.

Figure 3-9 shows the usual circuit for a varicap diode. D_1 is the varactor, and capacitor C_1 is a dc-blocking capacitor. Normally, the value of C_1 is set many times higher than the capacitance of the diode. The total capacitance is as follows:

$$C = \frac{C_1 C_t}{C_1 + C_t}.$$ (3-1)

Capacitor C_1 will affect the total capacitance only negligibly if $C_1 > C_t$.

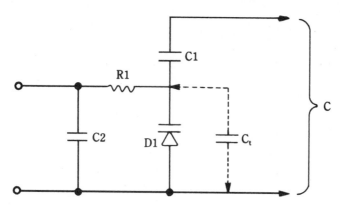

3-9 Variable-capacitance diode (varactor) in a typical circuit.

The control circuit for the varactor is series current-limiting resistor R_1. This resistor is typically 10 to 470 kΩ. The shunt capacitor (C_2) is used to decouple RF from the circuit from getting to other circuits and noise signals from other circuits from affecting the capacitor.

Varactors come in several different standard diode packages, including the two-terminal "similar to 182" package shown in Fig. 3-10. Some variants bevel the edge of the package to denote which is the cathode. In other cases, the package style will be like other forms of diode. Varactors are used in almost every form of diode package, up to and including the package used for 50- to 100-A stud-mounted rectifier diodes.

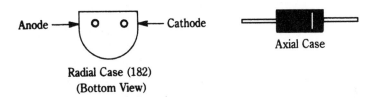

Anode → | Cathode

Radial Case (182)
(Bottom View)

Axial Case

3-10 Typical varactor cases.

How do varactors work?

Varactors are specially made PN junction diodes that are designed to enhance the control of the PN junction capacitance with a reverse bias voltage. Figure 3-11 shows how this capacitance is formed. A PN junction consists of P- and N-type semi-

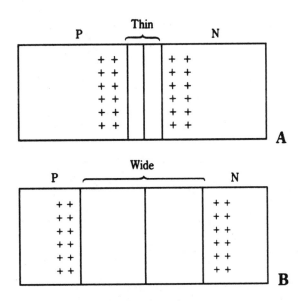

3-11 Varactor diode operation. (A) Thin depletion zone means maximum capacitance; (B) wider depletion zone means less capacitance.

conductor material placed in juxtaposition with each other, as shown in Fig. 3-11. When the diode is forward-biased, the charge carriers (electrons and holes) are forced to the junction interface, where positively charged holes and negatively charged electrons annihilate each other (causing a current to flow). But under reverse bias situations (e.g., those shown in Fig. 3-11), the charges are drawn away from the junction interface.

Figure 3-11A shows the situation where the reverse bias is low. The charge carriers are drawn only a little way from the junction, creating a thin insulating depletion zone. This zone is an insulator between the two charge-carrying P- and N-regions, and this situation fulfills the criterion for a capacitor: two conductors separated by an insulator. Figure 3-11B shows the situation where the reverse bias is increased. The depletion zone is increased, which is analogous to increasing the separation between plates.

The varactor is not an ideal capacitor (but then again, neither are "real" capacitors). Figure 3-12 shows the equivalent circuit for a varactor. Figure 3-12A shows the actual model circuit, and Fig. 3-12B shows one that is simplified, but nonetheless is valuable to understanding the varactor's operation. The equivalent circuit of Fig. 3-12B assumes that certain parameter shown in Fig. 3-12A are negligible.

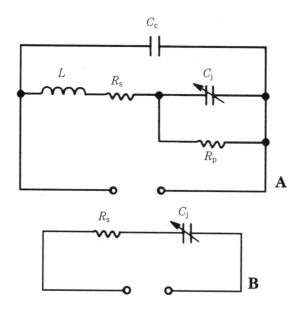

3-12 (A) Equivalent circuit of a varactor diode; (B) simplified circuit.

Figure 3-13 shows a typical test circuit for the varactor. A variable dc voltage is applied as a reverse bias across the diode. A series resistor serves both to limit the current should the voltage exceed the avalanche or zener points (which could destroy the diode) and also to isolate the diode from the rest of the circuitry. Without a high-value resistor (10 to 470 kΩ is the normal range; 100 kΩ is typical) in series with the dc supply, stray circuit capacitances and the power-supply output capaci-

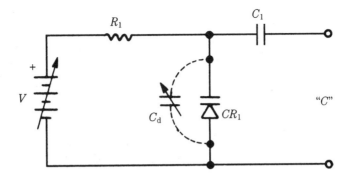

3-13 Varactor diode-control circuit.

tance would swamp the typically low value of varactor capacitance. The capacitor at the output (C_1) is used to block the dc from affecting other circuits or the dc in other circuits from affecting the diode. The value of this capacitor must be very large in order to prevent it from affecting the diode capacitance (C_d).

Varactor-tuning voltage sources The capacitance of a varactor is a function of the applied reverse bias potential. Because of this, it is essential that a stable, noise-free source of bias is provided. If the diode is used to tune an oscillator, for example, frequency drift will result if the dc potential is not stable. Besides ordinary dc drift, noise affects the operation of varactors. Anything that varies the dc voltage applied to the varactor will cause a capacitance shift.

Electronic servicers should be especially wary of varactor-tuned circuits in which the tuning voltage is derived from the main regulated power supply without an intervening regulator that serves only the tuning voltage input of the oscillator. Dynamic shifts in the regulator's load, variations in the regulator voltage, and other problems can create local oscillator drift problems that are actually power-supply problems and have nothing at all to do with the tuner (despite the apparent symptoms).

The specifications for any given varactor are given in two ways. First is the nominal capacitance taken at a standard voltage (usually 4 Vdc, but 1 and 2 Vdc are also used). The other is a capacitance ratio expected when the dc reverse bias voltage is varied from 2 to 30 Vdc (whatever the maximum permitted applied potential is for that diode). Typical is the NTE replacement line type 614. According to the *NTE Service Replacement Guide and Cross-Reference*, the 614 has a nominal capacitance of 33 pF at 4 Vdc reverse bias potential and a "C_2/C_{30}" capacitance ratio of 3:1.

Varactor applications

Varactors are electronically variable capacitors. In other words, they exhibit a variable capacitance that is a function of a reverse bias potential. This phenomenon leads to several common applications in which capacitance is a consideration. Figure 3-14 shows a typical varactor-tuned LC tank circuit. The link-coupled inductor (L_2) is used to input RF to the tank when the circuit is used for RF amplifiers (etc.). The

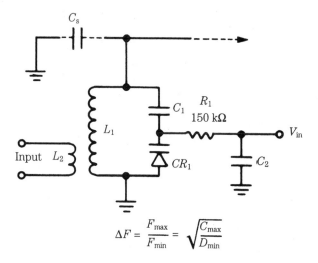

$$\Delta F = \frac{F_{max}}{F_{min}} = \sqrt{\frac{C_{max}}{D_{min}}}$$

3-14 Varactor diode-tuned circuit.

principal LC tank circuit consists of the main inductor (L_1) and a capacitance made up from the series equivalent of C_1 and varactor CR_1. In addition, you must also take into account the stray capacitance (C_s) that exists in all electronic circuits. The blocking capacitor and series-resistor functions were covered in the preceding paragraphs. Capacitor C_2 is used to filter the tuning voltage, V_{in}.

Because the resonant frequency of an LC-tuned tank circuit is a function of the square root of the inductance-capacitance product, we find that the maximum/minimum frequency of the varactor-tuned tank circuit varies as the square root of the capacitance ratio of the varactor diode. This value is the ratio of the capacitance at minimum reverse bias over capacitance at maximum reverse bias. A consequence of this is that the tuning characteristic curve (voltage vs frequency) is basically a parabolic function.

Note and Warning!

Cleaning Variable Capacitors

The main tuning capacitors in old radios are often full of crud, grease, and dust. Similarly, ham radio operators working the hamfest circuit looking for linear amplifier and antenna tuner parts often find just what they need, but the thing is all gooped-up. Several things can be done about it. First, try using dry compressed air. It will remove dust, but not grease. Aerosol cans of compressed air can be bought from a lot of sources, including automobile parts stores and photography stores.

Another method, if you have the hardware, is to ultrasonically clean the capacitor. The ultrasonic cleaner, however, is expensive; unless you have one, do not rush out to lay down the bucks.

Still another way is to use a product, such as Birchwood Casey Gun Scrubber. This product is used to clean firearms and is available in most gun shops. Firearms goop-up because gun grease, oil, unburned powder, and burned powder residue combine to create a crusty mess that is every bit as hard to remove as capacitor gunk. A related product is the degunking compound used by auto mechanics.

At one time, carbon tetrachloride was used for this purpose . . . and you will see it listed in old radio books. However, carbon tet is now well-recognized as a health hazard. DO NOT USE CARBON TETRACHLORIDE for cleaning, despite the advice to the contrary found in old radio books.

4
CHAPTER

Winding your own coils

Inductors and capacitors are the principal components used in RF tuning circuits. The resonant frequency of a tank circuit is the frequency to which the inductor–capacitor combination is tuned and is found from:

$$F = \frac{1}{2\pi\sqrt{LC}} \qquad (4\text{-}1)$$

or, if either the inductance (L) or capacitance (C) is known or preselected then the other can be found by solving Eq. (4-1) for the unknown, or:

$$C = \frac{1}{39.5\,F^2 L} \qquad (4\text{-}2)$$

and

$$L = \frac{1}{39.5\,F^2 C}. \qquad (4\text{-}3)$$

In all three equations, L is in henrys, C is in farads, and F is in hertz (don't forget to convert values to microhenrys and picofarads after calculations are made).

Capacitors are easily obtained in a wide variety of values. But tuning inductors are either unavailable or are available in other people's ideas of what you need. As a result, it is often difficult to find the kinds of parts that you need. This chapter will look at how to make your own slug-tuned adjustable inductors, RF transformers, and IF transformers (yes, you can build your own IF transformers).

Tuning inductors can be either air-, ferrite-, or powdered-iron-core coils. The air-core coils are not adjustable unless either an expensive roller-contact mechanism or clumsy taps on the winding of the coil are provided. However, the ferrite- and powdered-iron slug-tuned core coils are adjustable.

Figure 4-1 shows one form of slug-tuned adjustable coil. The form is made of plastic, phenolic, fiberglass, nylon, or ceramic materials and is internally threaded. The windings of the coil (or coils in the case of RF/IF transformers) are wound onto the form. The equation for calculating the inductance of a single-layer air-core coil

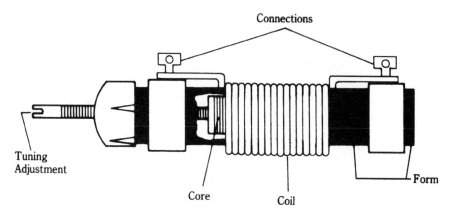

Connections

Tuning
Adjustment

Core

Coil

Form

4-1 Variable inductor on open coil form.

was discussed in Chapter 2. For non-air-core coils, the inductance is multiplied by a factor that is determined by the properties of the core material.

The tuning slug is a ferrite- or powdered-iron-core coil that mates with the internal threads in the coil form. A screwdriver slot or hex hole in either (or both) end allows you to adjust it. The inductance of the coil depends on how much of the core is inside the coil windings.

Amidon Associates coil system

It was once difficult to obtain coil forms to make your own project inductors. Amidon Associates Inc. makes a series of slug-tuned inductor forms that can be used to make any-value coil that you are likely to need. Figure 4-2A shows an Amidon form, and Fig. 4-2B shows an exploded view.

L–57 COIL FORM - - typical assembly

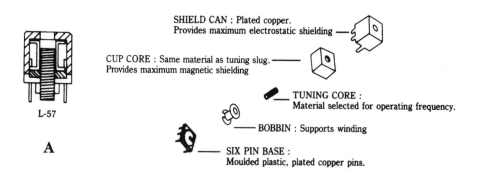

SHIELD CAN : Plated copper.
Provides maximum electrostatic shielding

CUP CORE : Same material as tuning slug.
Provides maximum magnetic shielding

TUNING CORE :
Material selected for operating frequency.

BOBBIN : Supports winding

SIX PIN BASE :
Moulded plastic, plated copper pins.

L-57

A

B

4-2 Shielded slug-tuned variable inductor.

Table 4-1 gives the type numbers, frequency ranges (in megaHertz), and other specifications for the coil forms made by Amidon. Three sizes of coil form are offered. The L-33s are 0.31" square and 0.40" high, the L-43s are 0.44" square and 0.50" high, and the L-57s are 0.56" square and 0.50" high. The last number (e.g., −1, −6, −10) in each type number indicates the type of material, which in turn translates to the operating frequency range (see Table 4-1). Now, see how the coil forms are used.

Table 4-1. Amidon coil form specifications

Part number	Frequency range (MHz)	A_L value	Ratio	Q_{max}
L-33-1	0.30–1.0	76	1.7:1	80
L-33-2	1.00–10	68	1.5:1	90
L-33-3	0.01–0.5	80	1.8:1	70
L-33-6	10–50	60	1.5:1	100
L-33-10	25–100	54	1.4:1	120
L-33-17	50–200	48	1.3:1	130
L-43-1	0.30–1.00	115	1.6:1	110
L-43-2	1.00–10	98	1.6:1	120
L-43-3	0.01–0.5	133	1.8:1	90
L-43-6	10–50	85	1.4:1	130
L-43-10	25–100	72	1.3:1	150
L-43-17	50–200	56	1.2:1	200
L-57-1	0.30–1.00	175	3:1	*
L-57-2	1.00–10	125	2:1	*
L-57-3	0.01–0.5	204	3:1	*
L-57-6	10–50	115	2:1	*
L-57-10	25–100	100	2:1	*
L-57-17	50–200	67	1.5:1	*

Determine the required inductance from Eq. (4-3). For an experiment to see how this coil system works, I decided to build a 15-MHz WWV converter that reduced the WWV radio station frequency to an 80- to 75-m ham-band frequency. Thus, I needed a circuit that would tune 15 MHz. It is generally a good idea to have a high capacitance-to-inductance ratio in order to maintain a high Q factor. I selected a 56-pF NPO capacitor for the tuned circuit because it is in the right range, and a dozen or so were in my junk box. According to Eq. (4-3), therefore, I needed a 2-μH inductor.

To calculate the number of turns (N) required to make any specific inductance, use the following equation:

$$N = 100\sqrt{\frac{L}{0.9A_L}}, \tag{4-4}$$

where

L = inductance in microhenrys (μH)

N = the number of turns.

The A_L factor is a function of the properties of the core materials and is found in Table 4-1; the units are microhenrys per 100 turns (μH/100 turns). In my case, I selected an L-57-6, which covers the correct frequency range and has an A_L value of 115 μH/100 turns. According to Eq. (4-4), therefore, I need 14 turns of wire.

The coil is wound from no. 26 to no. 32 wire. Ideally, Litz wire is used, but that is both hard to find and difficult to solder. For most projects ordinary enamel-coated magnet wire will suffice. A razor knife (such as X-acto) and soldering iron tip can be used to remove the enamel from the ends of the wire. Because the forms are small, I recommend using the no. 32 size.

Winding the coil can be a bit tricky if your vision needs augmentation as much as mine. But, using tweezers, needlenose pliers, and a magnifying glass on a stand made it relatively easy. Figure 4-3 shows the method for winding a coil with a tapped winding. Anchor one end of the wire with solder on one of the end posts and use this as the reference point. In my case, I wanted a 3-turn tap on the 14-turn coil, so I wound 3 turns then looped the wire around the center post. After this point was soldered, the rest of the coil was wound and then anchored at the remaining end post. A dab of glue or clear fingernail polish will keep the coil windings from moving.

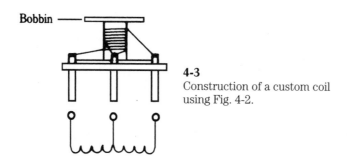

4-3
Construction of a custom coil using Fig. 4-2.

If you make an RF/IF transformer, then there will be two windings. Try to separate the primary and secondary windings if both are tuned. If one winding is not tuned, then simply wind it over the "cold" (i.e., ground) end of the tuned winding—no separation is needed.

The Amidon coil forms are tight, but they do have sufficient space for very small disk ceramic capacitors inside. The 56-pF capacitors that I selected fit nicely inside the shielded can of the coil, so I elected to place it there. Thus, I've basically made a 15-MHz RF/IF transformer.

After constructing the 15-MHz RF coil, I tested it and found that the slug tuned the coil to 15 MHz with a nice tolerance on either side of the design resonant frequency. It worked!

Although slug-tuned inductors are sometimes considered a bit beyond the hobbyist or ham, that is not actually true. The Amidon L-series coil forms can easily be used to make almost any inductor that you are likely to need.

Making your own toroid-core inductors and RF transformers

A lot of construction projects intended for electronic hobbyists and amateur radio operators call for inductors or radio-frequency (RF) transformers wound on toroidal cores. A toroid is a doughnut-shaped object, i.e., a short cylinder (often with rounded edges) that has a hole in the center (see Fig. 4-4). The toroidal shape is desirable for inductors because it permits a relatively high inductance value with few turns of wire, and, perhaps most important, the geometry of the core makes it self-shielding. That latter attribute makes the toroid inductor easier to use in practical RF circuits. Regular solenoid-wound cylindrical inductors have a magnetic field that goes outside the immediate vicinity of the windings and can thus intersect nearby inductors and other objects. Unintentional inductive coupling can cause a lot of serious problems in RF electronic circuits so they should be avoided wherever possible. The use of a toroidal shape factor, with its limited external magnetic field, makes it possible to mount the inductor close to other inductors (and other components) without too much undesired interaction.

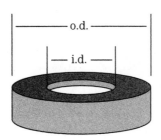

4-4
(A) Trifilar wound transformer circuit; (B) actual windings; (C) glue or silicone seal is used to hold the ends of the windings.

Materials used in toroidal cores

Toroid cores are available in a variety of materials that are usually grouped into two general classes: powdered iron and ferrite. These groups are further subdivided.

Powdered-iron materials

Powdered-iron cores are available in two basic formulations: carbonyl irons and hydrogen-reduced irons. The carbonyl materials are well-regarded for their temperature stability; they have permeability (μ) values that range from 1μ to about 35μ. The carbonyls offer very good Q values to frequencies of 200 MHz. Carbonyls are used in high-power applications as well as in variable-frequency oscillators and wherever temperature stability becomes important. However, notice that no powdered-iron material or ferrite is totally free of temperature variation, so oscillators using these cores must be temperature compensated for proper operation. The hydrogen-reduced iron devices offer permeabilities up to 90μ, but are lower Q than carbonyl devices. They are most used in electromagnetic interference (EMI) filters. The powdered-iron materials are the subject of Table 4-2.

Table 4-2. Powdered-iron core materials

Material	Permeability (μ)	Comments
0	1	Used up to 200 MHz; inductance varies with method of winding
1	20	Made of carbonyl C; similar to mixture no. 3 but is more stable and has a higher volume resistivity
2	10	Made of carbonyl E; high Q and good volume resistivity over range of 1 to 30 MHz
3	35	Made of carbonyl HP; very good stability and good Q over range of 0.05 to 0.50 MHz
6	8	Made of carbonyl SF; similar to mixture no. 2 but has higher Q over range 20 to 50 MHz
10	6	Type W powdered iron; good Q and high stability from 40 to 100 MHz
12	3	Made of a synthetic oxide material; good Q but only moderate stability over the range 50 to 100 MHz
15	25	Made of carbonyl GS6; excellent stability and good Q over range 0.1 to 2 MHz; recommended for AM BCB and VLF applications
17	3	Carbonyl material similar to mixture no. 12 but has greater temperature stability but lower Q than no. 12
26	75	Made of hydrogen reduced iron; has very high permeability; used in EMI filters and DC chokes

Ferrite materials

The name *ferrite* implies that the materials are iron-based (they are not), but ferrites are actually grouped into nickel-zinc and manganese-zinc types. The nickel-zinc material has a high-volume resistivity and high Q over the range 0.50 to 100 MHz. The temperature stability is only moderate, however. The permeabilities of nickel-zinc materials are found in the range 125 to 850μ. The manganese-zinc materials have higher permeabilities than nickel-zinc and are on the order of 850 to 5000μ. Manganese-zinc materials offer high Q over 0.001 to 1 MHz. They have low volume resistivity and moderate saturation flux density. These materials are used in switching power supplies from 20 to 100 kHz and for EMI attenuation in the range 20 to 400 MHz. See Table 4-3 for information on ferrite materials.

Toroid-core nomenclature

Although there are several different ways to designate toroidal cores, the one used by Amidon Associates is perhaps that most commonly found in electronic hobbyist and amateur radio published projects.

Although the units of measure are the English system, which is used in the United States and Canada and formerly in the UK, rather than SI units, their use with respect to toroids seems widespread. The type number for any given core will consist of three elements: xx–yy–zz. The "xx" is a one- or two-letter designation of the

Table 4-3. Ferrite materials

Material	Permeability (μ)	Remarks[1]
33	850	M–Z; used over 0.001 to 1 MHz for loopstick antenna rods; low volume resistivity
43	850	N–Z; medium-wave inductors and wideband transformers to 50 MHz; high attenuation over 30 to 400 MHz; high volume resistivity
61	125	N–Z; high Q over 0.2 to 15 MHz; moderate temperature stability; used for wideband transformers to 200 MHz
63	40	High Q over 15 to 25 MHz; low permeability and high volume resistivity
67	40	N–Z; high Q operation over 10 to 80 MHz; relatively high flux density and good temperature stability; similar to Type 63, but has lower volume resistivity; used in wideband transformers to 200 MHz
68	20	N–Z; excellent temperature stability and high Q over 80 to 180 MHz; high volume resistivity
72	2000	High Q to 0.50 MHz but used in EMI filters from 0.50 to 50 MHz; low volume resistivity
J/75	5000	Used in pulse and wideband transformers from 0.001 to 1 MHz and in EMI filters from 0.50 to 20 MHz; low volume resistivity and low core losses
77	2000	0.001 to 1 MHz; used in wideband transformers and power converters and in EMI and noise filters from 0.5 to 50 MHz
F	3000	Is similar to Type 77 above but offers a higher volume resistivity, higher initial permeability, and higher flux saturation density; used for power converters and in EMI/noise filters from 0.50 to 50 MHz

[1] N–Z: nickel–zinc;
M–Z: manganese–zinc.

general class of material, i.e., powdered iron (xx = "T") or ferrite (xx = "TF"). The "yy" is a rounded-off approximation of the outside diameter (*o.d.* in Fig. 4-4) of the core in inches; "37" indicates a 0.375" (9.53 mm) core, while "50" indicates a 0.50" (12.7 mm) core. The "zz" indicates the type (mixture) of material. A mixture no. 2 powdered-iron core of 0.50" diameter would be listed as a T-50-2 core. The cores are color-coded to assist in identification.

Inductance of toroidal coils

The inductance of the toroidal core inductor is a function of the permeability of the core material, the number of turns, the inside diameter (*i.d.*) of the core, the outside diameter (*o.d.*) of the core, and the height (*h*) (see Fig. 4-1) and can be approximated by:

$$L = 0.011684 \, h \, N^2 \, \mu_r \, \log_{10}\left(\frac{o.d.}{i.d.}\right) H. \qquad (4\text{-}5)$$

This equation is rarely used directly, however, because toroid manufacturers provide a parameter called the A_L *value*, which relates inductance per 100 or 100 turns of wire. Tables 4-4 and 4-5 show the A_L values of common ferrite and powdered-iron cores.

Table 4-4. Common powdered-iron A_L values

Core size	Core material type (mix)								
	26	3	15	1	2	6	10	12	0
12	—	60	50	48	20	17	12	7	3
16	—	61	55	44	22	19	13	8	3
20	—	90	65	52	27	22	16	10	3.5
37	275	120	90	80	40	30	25	15	4.9
50	320	175	135	100	49	40	31	18	6.4
68	420	195	180	115	57	47	32	21	7.5
94	590	248	200	160	84	70	58	32	10.6
130	785	350	250	200	110	96	—	—	15
200	895	425	—	250	120	100	—	—	—

Table 4-5. Common ferrite-core A_L values

Core size[1]	Material type					
	43	61	63	72	75	77
23	188	24.8	7.9	396	990	356
37	420	55.3	17.7	884	2210	796
50	523	68	22	1100	2750	990
50A	570	75	24	1200	2990	1080
50B	1140	150	48	2400	—	2160
82	557	73.3	22.8	1170	3020	1060
114	603	79.3	25.4	1270	3170	1140
114A	—	146	—	2340	—	—
240	1249	173	53	3130	6845	3130

[1] Core type no. prefix: TF-yy-zz.

Winding toroid inductors

There are two basic ways to wind a toroidal core inductor: distributed (Fig. 4-5A) and close-spaced (Fig. 4-5B). In distributed toroidal inductors, the turns of wire that are wound on the toroidal core are spaced evenly around the circumference of the core, with the exception of a gap of at least 30° between the ends (see Fig. 4-5A). The gap ensures that stray capacitance is kept to a minimum. The winding covers 270° of the core. In close winding (Fig. 4-5B), the turns are made so that adjacent turns of wire touch each other. This pratice raises the stray capacitance of the winding, which affects the resonant frequency, but can be done in many cases with little or no ill effect (especially where the capacitance and resonant point shift are negli-

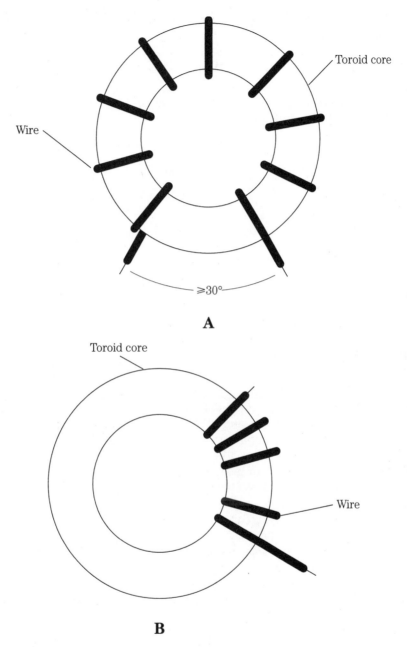

4-5 Toroid winding styles: (A) distributed; (B) close wound.

gible). In general, close winding is used for inductors in narrowband-tuned circuits, and distributed winding is used for broadband situations, such as conventional and balun RF transformers. The method of winding has a small effect on the final inductance of the coil. Although this makes calculating the final inductance less predictable, it also provides a means of final adjustment of actual inductance in the circuit as-built.

Calculating the number of turns

As in all inductors, the number of turns of wire determines the inductance of the finished coil. In powdered-iron cores, the A_L rating of the core is used with fair confidence to predict the number of turns needed.

For powdered-iron cores:

$$N = 100 \frac{\sqrt{L_{\mu H}}}{A_L} \qquad (4\text{-}6)$$

For ferrite cores:

$$N = 1000 \sqrt{\frac{L_{mH}}{A_L}}, \qquad (4\text{-}7)$$

where

N = the number of turns
$L_{\mu H}$ = inductance in microhenrys (μH)
L_{mH} = inductance in millihenrys (mH)
A_L is a property of the core material.

Building the toroidal device

The toroid core or transformer is usually wound with enameled or formvar-insulated wire. For low-powered applications (receivers, VFOs, etc.) the wire will usually be no. 22 through no. 36 (with no. 26 being very common) AWG. For high-power applications, such as transmitters and RF power amplifiers, a heavier grade of wire is needed. For high-power RF applications, no. 14 or no. 12 wire is usually specified, although wire as large as no. 6 has been used in some commercial applications. Again, the wire is enameled or formvar-covered insulated wire.

In the high-power case, it is likely that high voltages will exist. In high-powered RF amplifiers, such as used by amateur radio operators in many countries, the potentials present across a 50-Ω circuit can reach hundreds of volts. In those cases, it is common practice to wrap the core with a glass-based tape, such as Scotch 27.

High-powered applications also require a large-area toroid rather than the small toroids that are practical at lower power levels. Cores in the FT-150-zz to FT-240-zz or T-130-zz to T-500-zz are typically used. In some high-powered cases, several identical toroids are stacked together and wrapped with tape to increase the power-handling capacity. This method is used quite commonly in RF power amplifier and antenna tuner projects.

Binding the wires

It sometimes happens that the wires making up the toroidal inductor or transformer become loose. Some builders prefer to fasten the wire to the core using one of the two methods shown in Fig. 4-6. Figure 4-6A shows a dab of glue, silicone adhesive, or the high-voltage sealant Glyptol (sometimes used in television receiver high-voltage circuits) to anchor the end of the wire to the toroid core.

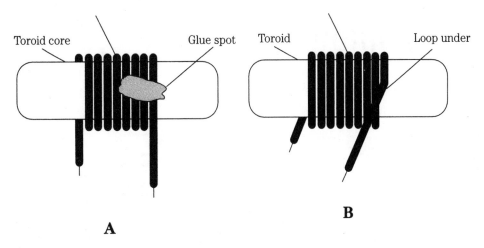

4-6 Methods for fastening the wire on a toroid winding: (A) glue spot and (B) "tuck under" method.

Other builders prefer the method shown in Fig. 4-6B. In this method, the end of the wire is looped underneath the first full turn and pulled taut. This method will effectively anchor the wire, but some say it creates an anomaly in the magnetic situation that might provoke interactions with nearby components. In my experience, that situation is not terribly likely, and I use the method regularly with no observed problems thus far.

When the final coil is ready, and both the turns count and spacing are adjusted to yield the required inductance, the turns can be anchored to the coil placed in service. A final sealant method is to coat the coil with a thin layer of clear lacquer, or "Q-dope" (this product is intended by its manufacturer as an inductor sealant).

Mounting the toroidal core

Toroids are a bit more difficult to mount than solenoid-wound coils (cylindrical coils), but the rules that one must follow are not as strict. The reason for loosening of the mounting rules is that the toroid, when built correctly, is essentially self-shielding, so less attention (not *no* attention!) can be paid to the components that surround the inductor. In the solenoid-wound coil, for example, the distance between adjacent coils and their orientation is important. Adjacent coils, unless well shielded, must be placed at right angles to each other to lessen the mutual coupling between the coils. However, toroidal inductors can be closer together and either coplanar or adjacent planar can be placed with respect to each other. Although some spacing must be maintained between toroidal cores (the winding and core manufacture not being perfect), the required average distance can be less than for solenoid-wound cores.

Mechanical stability of the mounting is always a consideration for any coil (indeed, any electronic component). For most benign environments, the core can be mounted directly to a printed wiring board (PWB) in the manner of Figs. 4-7A and 4-7B. In Fig. 4-7A, the toroidal inductor is mounted flat against the board; its leads

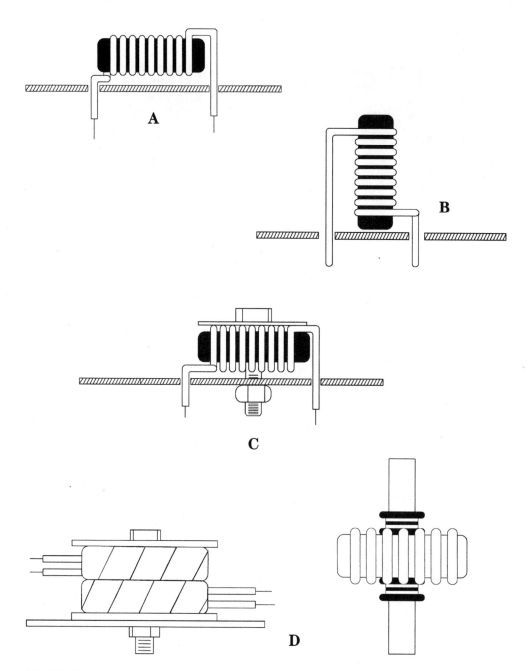

4-7 (A) Flat mounting; (B) on-end mounting; (C) secured mounting (use nylon machine screws); (D) mounting high-power or high-voltage toroidal inductors or transformers; (E) suspending toroid inductors on a dowel; (F) mounting method for a "single-turn primary" transformer in RF watt-meters or VSWR meters.

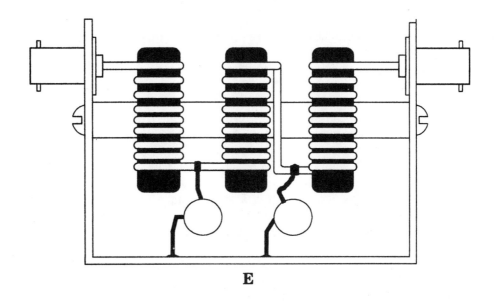

E

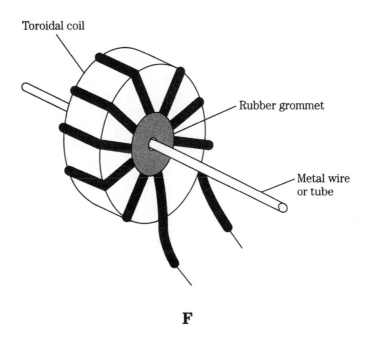

F

4-7 *Continued*

are passed through holes in the board to solder pads underneath. The method of Fig. 4-7B places the toroid at right angles to the board, but it still uses the leads soldered to copper pads on the PWB to anchor the coil. It is wise to use a small amount of RTV silicone sealant or glue to held the coil to the board once it is found to work satisfactorily.

If the environment is less benign with respect to vibration levels, then a method similar to Fig. 4-7C can be used. Here, the toroid is fastened to the PWB with a set of nylon machine screw and nut hardware and a nylon or fiber washer. In high-powered antenna tuning units, it is common to see an arrangement similar to that in Fig. 4-7D. In this configuration, several toroidal cores are individually wrapped in glass tape, then the entire assembly is wrapped as a unit with the same tape. This assembly is mounted between two insulators, such as plastics, ceramic, or fiberboard, which are held together as a "sandwich" by a nylon bolt and hex nut.

Figure 4-7E shows a method for suspending toroidal cores in a shielded enclosure. I've used this method to make five-element low-pass filters for use in my basement laboratory. The toroidal inductors are mounted on a dowel, which is made of some insulating material, such as wood, plastic, plexiglass, Lexan, or other synthetic. If the dowel is sized correctly, then the inductors will be a tight slip fit and there will be no need for further anchoring. Otherwise, a small amount of glue or RTV silicone sealant can be used to stabilize the position of the inductor.

Some people use a pair of undersized rubber grommets over the dowel, one pressed against either side of the inductor (see inset to Fig. 4-7E). If the grommets are taut enough, then no further action is needed. Otherwise, they can be glued to the rod.

A related mounting method is used to make current transformers in homemade RF power meters (Fig. 4-7F). In this case, a rubber grommet is fitted into the center of the toroid, and a small brass or copper rod is passed through the center hole of the grommet. The metal rod serves as a one-turn primary winding. A sample of the RF current flowing in the metal rod is magnetically coupled to the secondary winding on the toroid, where it can be fed to an oscilloscope for display or rectified, filtered, and displayed on a dc current meter that is calibrated in watts or VSWR units.

Torodial RF transformers

Both narrowband-tuned and broadband RF transformers can be accommodated by toroidal powdered-iron and ferrite cores. The schematic symbols used for transformers are shown in Fig. 4-8. These symbols are largely interchangeable and are all seen from time to time. In Fig. 4-8A, the two windings are shown adjacent to each other, but the core is shown along only one of them. This method is used to keep the drawing simple and does not imply in any way that the core does not affect one of the windings. The core can be represented either by one or more straight lines, as shown, or by dotted lines. The method shown in Fig. 4-8B is like the conventional transformer representation in which the windings are juxtaposed opposite each other with the core between them. In Fig. 4-8C, the core is extended and the two windings are along one side of the core bars.

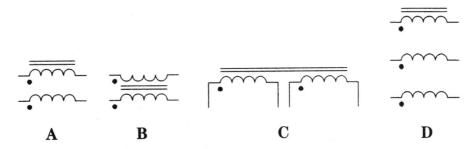

4-8 Transformer schematic styles.

In each of the transformer representations of Fig. 4-8, dots are on the windings. These dots show the "sense" of the winding and represent the same end of the coils. Thus, the wires from two dotted ends are brought to the same location, and the two coils are wound in the same direction. Another way of looking at it is that if a third winding were used to excite the core from an RF source, the phase of the signals at the dot end will be the same; the phase of the signal at the undotted ends will also be the same, but will be opposite that of the dotted ends.

I was once questioned by a reader concerning the winding protocol for toroidal transformers, as shown in books and magazine articles. My correspondent included a partial circuit (Fig. 4-9A) as typical of the dilemma. The question was How do you wind it? and a couple of alternative methods were proposed. At first I thought it was a silly question because the answer was obvious, and then I realized that perhaps I was wrong; to many people, the answer was not that obvious. The answer to this question is that all windings are wound together in a multifilar manner.

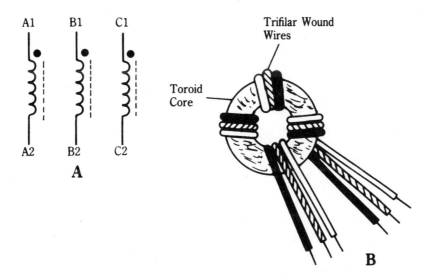

4-9 Trifilar wound transformer: (A) circuit symbol; (B) winding details; (C) using parallel wires; (D) using twisted wires; and (E) glue spot for securing windings.

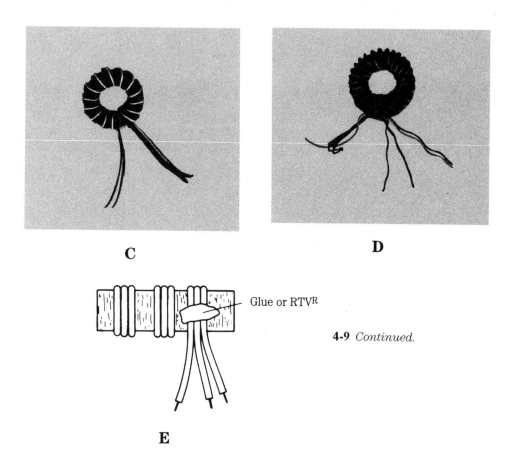

C

D

Glue or RTV^R

4-9 Continued.

E

Because there are three windings, in this case, we are talking about trifilar windings. Figure 4-9B shows the trifilar winding method. For the sake of clarity, I have colored all the three wires differently so that you can follow it. This practice is also a good idea for practical situations. Because most of my projects use nos. 26, 28, or 30 enameled wire to wind coils, I keep three colors of each size on hand and wind each winding with a different color. That makes it a whole lot easier to identify which ends go with each other.

The dots in the schematic and on the pictorial are provided to identify one end of the coil windings. Thus, the dot and no-dot ends are different from each other; in circuit operation, it usually makes a difference which way the ends are connected into the circuit (the issue is signal phasing).

Figures 4-9C and 4-9D shows two accepted methods for winding a multifilar coil on a toroidal core. Figure 4-9C is the same method as in Fig. 4-9B, but on an actual toroid instead of a pictorial representation. The wires are laid down parallel to each other as shown previously. The method in Fig. 4-9D uses twisted wires. The three wires are chucked up in a drill and twisted together before being wound on the core.

With one end of the three wires secured in the drill chuck, anchor the other end of the three wires in something that will hold it taut. I use a bench vise for this purpose. Turn the drill on the slow speed and allow the wires to twist together until the desired pitch is achieved.

Be very careful when performing this operation. If you don't have a variable-speed electric drill (so that it can be run at very low speeds), then use an old-fashioned manual hand drill. If you use an electric drill, wear eye protection. If the wire breaks or gets loose from its mooring, it will whip around wildly until the drill stops. That whipping wire can cause painful welts on your skin, and it can easily cause permanent eye damage.

Of the two methods for winding toroids, the method shown in Figs. 4-9B and 4-9C is preferred. When winding toroids, at least those of relatively few windings, pass the wire through the doughnut hole until the toroid is about in the middle of the length of wire. Then, loop the wire over the outside surface of the toroid and pass it through the hole again. Repeat this process until the correct number of turns are wound onto the core. Be sure to press the wire against the toroid form and keep it taut as you wind the coils.

Enameled wire is usually used for toroid transformers and inductors, and that type of wire can lead to a problem. The enamel can chip and cause the copper conductor to contact the core. On larger cores, such as those used for matching transformers and baluns used at kilowatt power levels, the practical solution is to wrap the bare toroid core in a layer of fiberglass packing tape. Wrap the tape exactly as if it were wire, but overlap the turns slightly to ensure that the entire circumference of the core is covered.

On some projects, especially those in which the coils and transformers use very fine wire (e.g., no. 30), I have experienced a tendency for the wire windings to unravel after the winding is completed. This problem is also easily curable. At the ends of the windings place a tiny dab of rubber cement or RTV silicone sealer (see Fig. 4-5E).

Conventional RF transformers

One of the principal uses of transformers in RF circuits is impedance transformation. When the secondary winding of a transformer is connected to a load impedance, the impedance seen "looking into" the primary will be a function of the load impedance and the turns ratio of the transformer (see Fig. 4-10A). The relationship is:

$$\frac{N_p}{N_s} = \sqrt{\frac{Z_p}{Z_s}}. \tag{4-8}$$

With the relationship of Eq. (4-8), you can match source and load impedances in RF circuits.

Example

Assume that you have a 3- to 30-MHz transistor RF amplifier with a base input impedance of 4 Ω (Z_s), and that transistor amplifier has to be matched to a 50-Ω

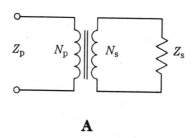

A

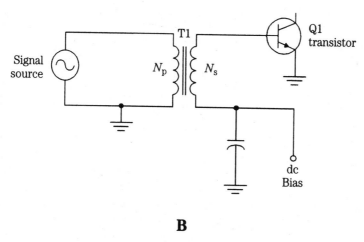

B

4-10 (A) A transformer with primary and secondary labels; (B) typical transformer circuit with transistor load.

source impedance (Z_p), as shown in Fig. 4-10B. What turns ratio is needed to effect the impedance match? Let's calculate:

$$\frac{N_p}{N_s} = \sqrt{\frac{Z_p}{Z_s}}$$

$$\frac{N_p}{N_s} = \sqrt{\frac{50\,\Omega}{4\,\Omega}}$$

(4-9)

$$\frac{N_p}{N_s} = \sqrt{12.5} = 3.53{:}1.$$

A general design rule for the value of inductance used in transformers is that the inductive reactance at the lowest frequency must be 4 times (4×) the impedance connected to that winding. In the case of the 50-Ω primary of this transformer, the inductive reactance of the primary winding should be $4 \times 50 = 200\ \Omega$. The inductance should be:

$$L = \frac{200\,\Omega\,10^6}{2\pi F}$$

$$L = \frac{200\,\Omega 10^6}{(2)(3.1415)(3{,}000{,}000\,\text{Hz})} = 10.6\mu\text{H}.$$

Now that you know that a 10.6-μH inductance is needed, you can select a toroidal core and calculate the number of turns needed. The T-50-2 (RED) core covers the correct frequency range and is of a size that is congenial to easy construction. The T-50-2 (RED) core has an A_L value of 49, so the number of turns required is as follows:

$$N = 100\sqrt{\frac{10.6\,\mu\text{H}}{49}} = 46.5 \text{ turns} \approx 47 \text{ turns}.$$

The number of turns in the secondary must be such that the 3.53 : 1 ratio is preserved when 47 turns are used in the primary:

$$N_s = \frac{47}{3.53} = 13.3 \text{ turns}.$$

If you wind the primary with 47 turns and the secondary with 13 turns, then you can convert the 4-Ω transistor base impedance to the 50-Ω systems impedance.

Example

A beverage antenna can be constructed for the AM broadcast band (530 to 1700 kHz). By virtue of its construction and installation, it exhibits a characteristic impedance, Z_o, of 600 Ω. What is the turns ratio required of a transformer at the feed end (Fig. 4-11) to match a 50-Ω receiver input impedance?

$$\frac{N_s}{N_p} = \sqrt{\frac{600\,\Omega}{50\,\Omega}} = 3.46{:}1.$$

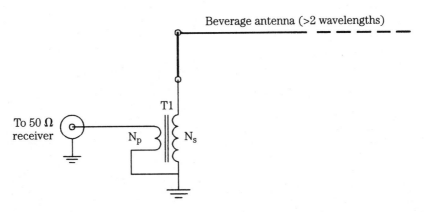

4-11 Coupling transformer used to feed a Beverage antenna.

The secondary requires an inductive reactance of 4 × 600 Ω = 2400 Ω. To obtain this inductive reactance at the lowest frequency of operation requires an inductance of:

$$L = \frac{(2400\,\Omega)(10^6)}{(2)(\pi)(530{,}000\text{ Hz})} = 721\,\mu\text{H}.$$

Checking a table of powdered-iron toroid cores, you will find that the −15 (RED/WHT) mixture will operate over 0.1 to 2 MHz (see Table 4-6.). Selecting a T-106-15 (Red/White) core provides an A_L value of 345. The number of turns required to create an inductance of 721 μH is:

$$N = 100\sqrt{\frac{721\,\mu\text{H}}{345}} = 145\text{ turns.}$$

The primary winding must have:

$$N_p = \frac{145}{3.46}\text{ turns} = 42\text{ turns.}$$

Winding this transformer, as designed, should provide adequate service.

Table 4-6. Properties of powdered-iron core types

Material type	Color code	Mu (μ)	Frequency (MHz)
41	Green	75	—
3	Grey	35	0.05–0.5
15	Red/white	25	0.1–2
1	Blue	20	0.5–5
2	Red	10	1–30
6	Yellow	8	10–90
10	Black	6	60–150
12	Green/white	3	100–200
0	Tan	1	150–300

Ferrite and powdered-iron rods

Ferrite rods (Fig. 4-12A) are used in the low-frequency medium wave, AM broadcast band, and LF/VLF receivers to form "loopstick" antennas. These antennas are often used in radio-direction finders because they possess a "figure-8" reception pattern that counterposes two deep nulls with a pair of main lobes. Another application of ferrite rods is to make balun transformers (Fig. 4-12B) or filament transformers in vacuum-tube high-power RF amplifiers. Ferrite rod inductors are also used in any application, up to about 10 MHz, where a high inductance is needed.

Two permeability figures are associated with the ferrite rod, but only one of them is easily available (see Table 4-7): the initial permeability. This figure is used in the equations for inductance. The effective permeability is a little harder to pin down and is dependent on such factors as (a) the length/diameter ratio of the rod, (b) location of the coil on the rod (centered is most predictable), (c) the spacing between turns of wire, and (d) the amount of air space between the wire and the rod. In general, maximizing effective A_L and inductance requires placing the coil in the center of the rod. The best Q, on the other hand, is achieved when the coil runs nearly the entire length of the rod.

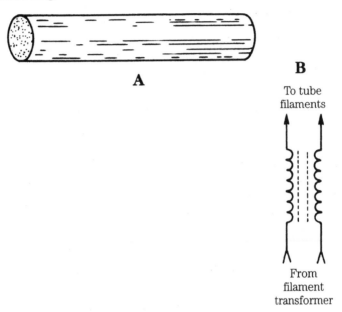

B

To tube
filaments

A

From
filament
transformer

4-12 (A) Ferrite rod; (B) filament transformer RF choke.

Table 4-7. Properties of some common ferrite rods

Part no. turns	Permeability	Approx. A_L[1]	Ampere
R61-025-400	125	26	110
R61-033-400	125	32	185
R61-050-400	125	43	575
R61-050-750	125	49	260
R33-037-400	800	62	290
R33-050-200	800	51	465
R33-050-400	800	59	300
R33-050-750	800	70	200

[1] Approximate value for coil centered on rod, covering nearly the entire length, made of no. 22 wire. Actual A_L may vary with situation.

Several common ferrite rods available from Amidon Associates are shown in Table 4-7. The "R" indicates "rod" while the number associated with the "R" (e.g., R61) indicates the type of ferrite material used in the rod. The following numbers (e.g., 025) denote the diameter (025 = 0.25″, 033 = 0.33″, and 050 = 0.50″); the length of the rod is given by the last three digits (200 = 2″, 400 = 4″, and 750 = 7.5″. The Type 61 material is used from 0.2 to 10 MHz, and the Type 33 material is used in VLF applications.

In some cases, the A_L rating of the rod will be known, and in others, only the permeability (μ) is known. If either is known, you can calculate the inductance produced by any given number of turns.

For the case where the A_L is known:

$$L_{\mu H} = N_p^2 A_L \times 10^{-4} \tag{4-10}$$

For the case where μ is known:

$$L_{\mu H} = (4 \times 10^{-9}) \pi N_p^2 \mu \left(\frac{A_e \, (\text{cm}^2)}{l_e \, (\text{cm})} \right), \tag{4-11}$$

where
  N_p is the number of turns of wire
  μ is the core permeability
  A_e is the cross-sectional area of the core (cm^2)
  l_e is the length of the rod's flux path (cm)

Example

Find the number of turns required on an R61-050-750 ferrite rod to make a 220-μH inductor for use in the AM receiver with a 10- to 365-pF variable capacitor.

Solution

Solving Eq. (4-11) for N:

$$N = \sqrt{\frac{L_{\mu H}}{A_L \times 10^{-4}}}$$

$$N = \sqrt{\frac{220 \, \mu H}{(43)(10^{-4})}} = 226.$$

Figure 4-13 shows several popular ways for ferrite rod inductors to be wound for service as loopstick antennas in radio receivers. Figure 4-13A shows a transformer circuit in which the main tuned winding is broken into two halves, A and B. These windings are at the end regions of the rod but are connected together in series at the center. The main coils are resonated by a dual capacitor, C_{1A} and C_{1B}. A coupling winding, of fewer turns, is placed at the center of the rod and is connected to the coaxial cable going to the receiver. In some cases, the small coil is also tuned but usually with a series capacitor (see Fig. 4-13B). Capacitor C_2 is usually a much larger value than C_1 because the coupling winding has so many fewer turns than the main tuning windings. The antenna in Fig. 4-13C is a little different. The two halves of the

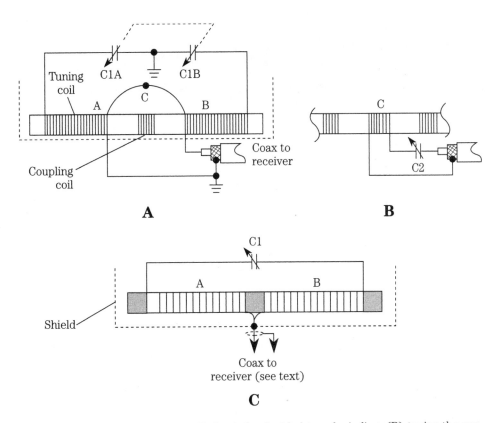

4-13 Loopstick antenna: (A) overall circuit for doubled-tuned winding; (B) tuning the coupling loop; and (C) singled tuned variety.

tuning winding are connected together at the center and connected directly to the center conductor of the coaxial cable or to a single downlead. A single capacitor is used to resonate the entire winding (both halves).

Project 4-1

A radio direction-finding antenna can be used for a number of purposes, only one of which is finding the direction from which a radio signal arrives. Another use is in suppressing cochannel and adjacent-channel interference. This becomes possible when the desired station is in a direction close to right angles from the line between the receiver and the desired transmitter. Reduction of the signal strength of the interfering signal is possible because the loopstick antenna has nulls off both ends.

Figure 4-14 shows a loopstick antenna mounted in a shielded compartment for radio direction finding. The shield is used to prevent electrical field coupling from nearby sources, such as power lines and other stations, yet doesn't affect the reception of the magnetic field of radio stations. The aluminum can be one-half of an electronic hobbyist's utility box, of appropriate dimensions, or can be built custom from

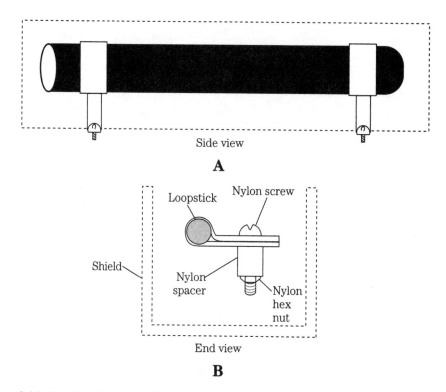

Side view

A

B

End view

4-14 Mounting the loopstick antenna in a shielded enclosure.

Harry & Harriet Homeowner Do-It-Yourself hardware stores. The loopstick antenna is mounted by nonmetallic cable ties to nylon spacers that are, in turn, fastened to the aluminum surface with nylon hardware.

The number of turns required for the winding can be found experimentally, but starting with the number called for by the preceding formula. The actual number of turns depends in part on the frequency of the band being received and the value of the capacitors used to resonate the loopstick antenna.

Noncylindrical air-core inductors

Most inductors used in radio and other RF circuits are either toroidal or solenoid-wound cylindrical. There is, however, a class of inductors that are neither solenoidal nor toroidal. Many loop antennas are actually inductors fashioned into either triangle, square, hexagon, or octagon shapes. Of these, the most common is the square-wound loop coil (Fig. 4-15). The two basic forms are: flat wound (Fig. 4-15A) and depth wound (Fig. 4-15B). The equation for the inductance of these shaped coils is a bit difficult to calculate, but the equation provided by F. W. Grover of the U.S. National Bureau of Standards in 1946 is workable:

$$L_{\mu H} = K_1 N^2 A \left(Ln\left(\frac{K_2 AN}{(N+1)B}\right) + K_3 + \left(\frac{K_4(N+1)B}{AN}\right)\right), \qquad (4\text{-}12)$$

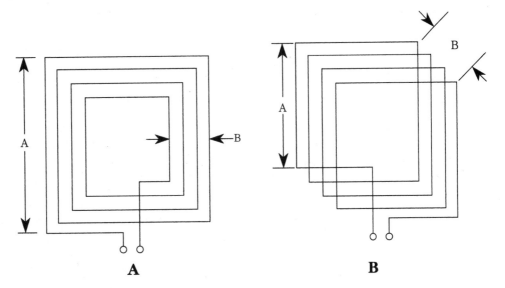

4-15 Inductive-loop direction-finding antenna.

where

  A is the length of each side in centimeters (cm)

  B is the width of the coil in centimeters (cm)

  N is the number of turns in the coil (close wound)

  K_1, K_2, K_3, and K_4 are 0.008, 1.4142, 0.37942, and 0.3333 respectively.

Whenever conductors are placed side by side, as in the case of the loop-wound coil, there is a capacitance between the conductors (even if formed by a single loop of wire, as in a coil). This capacitance can be significant when dealing with radio circuits. The estimate of distributed loop capacitance (in picofarads) for square loops is given by Bramslev as about 60A, where A is expressed in meters (m). The distributed capacitance must be accounted for in making calculations of resonance. Subtract the distributed capacitance from the total capacitance required in order to find the value of the capacitor required to resonate the loop-wound coil.

Binocular- ("balun" or "bazooka") core inductors

The toroidal core has a certain charm because it is easy to use, is predictable, and is inherently self-shielding because of its geometry. But there is another core shape that offers very high inductance values in a small volume. The binocular core (Fig. 4-16) offers very high A_L values in small packages, so you can create very high inductance values without them being excessively large. A binocular core that uses type 43 ferrite, and is about the same weight and size as the T-50-43 ($A_L = 523$), has an A_L value of 2890. Only 8.8 turns are required to achieve 225 μH on this core.

There are actually two different types of binocular core in Fig. 4-16. The Type 1 binocular core is shown in Fig. 4-16A. It is larger than the Type 2 (Fig. 4-16B) and has larger holes. It can, therefore, be used for larger-value inductors and transformers. The Type 2 core can be considered as a two-hole ferrite bead.

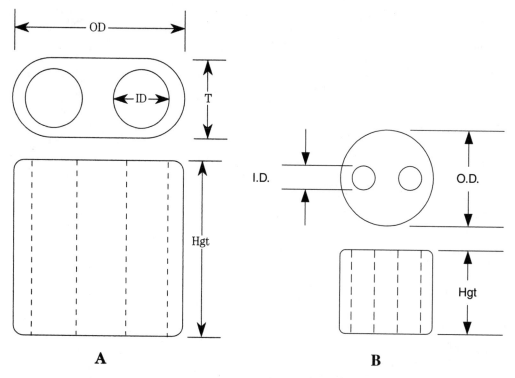

4-16 Bazooka balun cores: (A) Type 1 and (B) Type 2.

Table 4-8 shows several popular-sized binocular cores and their associated A_L values. The center two digits of each part number is the type of ferrite material used to make the core (e.g., BN-xx-202), and the last digits refer to the size and style of the core.

Three different ferrite materials are commonly used in binocular cores. The type 43 material is a nickel-zinc ferrite and has a permeability of 850. It is used for wideband transformers up to 50 MHz and has high attenuation, from 30 to 400 MHz. It can be used in tuned RF circuits from 10 to 1000 kHz. The type 61 material is also nickel-

Table 4-8. Binocular cores

Part no.	Material	A_L value	Size	Style
BN-43-202	43	2890	A	1
BN-43-2302	43	680	B	1
BN-43-2402	43	1277	C	1
BN-43-3312	43	5400	D	1
BN-43-7051	43	6000	E	1
BN-61-202	61	425	A	1
BN-61-2302	61	100	B	1

zinc and has a permeability of 125. It offers moderate to good thermal stability and a high Q over the range 200 kHz to 15 MHz. It can be used for wideband transformers up to 200 MHz. The type 73 material has a permeability of 2500 and offers high attenuation from 500 kHz to 50 MHz.

The binocular core can be used for a variety of different RF inductor devices. Besides the single, fixed inductor, it is also possible to wind conventional transformers and balun transformers of various types on the core.

Figure 4-17 shows Type 1 binocular cores wound in various ways. The normal manner of winding the turns of the inductor is shown in Fig. 4-17A; the wire is passed from hole to hole around the central wall between the holes. The published A_L values for each core are based on this style of winding, and it is the most commonly used.

An edge-wound coil is shown in Fig. 4-17B. In this coil, the turns are wound around the outside of the binocular core. To check the difference, I wound a pair of BN-43-202 cores with 10 turns of no. 26 wire; one in the center (Fig. 4-17A) and one around the edge (Fig. 4-17B). The center-wound version produced 326 μH of inductance, and the edge-wound version produced 276 μH with the same number of turns.

Counting the turns on a binocular core is a little different than you might expect. A single U-shaped loop that enters and exits the core on the same side (Fig. 4-17C) counts as one turn. When the wire is looped back through a second time (Fig. 4-17D), there are two turns.

Winding the binocular core

Some people think that it is easier to use these cores than toroids, and after spending a rainy weekend winding LF and AM BCB coils (after pumping groundwater out of the basement workshop!), I am inclined to agree partially. The "partially" means that they are easier to work than toroids if you do it correctly. It took me some experimenting to figure out a better way than holding the core in one hand and the existing wires already on the core in another hand and then winding the remaining coils with a third hand. Not being a Martian, I don't have three hands, and my "Third Hand" bench tool didn't seem to offer much help. Its alligator clip jaws were too coarse for the no. 36 enameled wire that I was using for the windings. So, enter a little "mother of invention" ingenuity (it's amazing how breaking a few wires can focus one's attention on the problem). In fact, I came up with two related methods between gurgles of my portable sump pump.

The first method is shown in Fig. 4-18. The binocular core is temporarily affixed to a stiff piece of cardboard stock such as a 5 × 7" card or a piece cut from the stiffener used in men's shirts at the laundry. The cardboard is taped to the work surface, and the core is taped to the cardboard. One end of the wire that will be used for the winding is taped to the cardboard with enough leader to permit working the end of the coil once it is finished (2 to 3"). Pass the wire through the holes enough times to make the coil needed and then anchor the free end to the cardboard with tape. If the device has more than one winding, make each one in this manner, keeping the ends taped down as you go. Once all of the windings are in place, seal the assembly

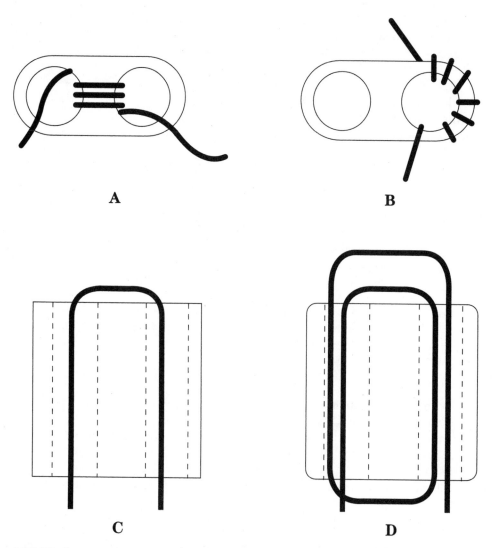

4-17 Winding style for Bazooka balun cores: (A) through the center; (B) around the edge (less-predictable inductance); (C) single turn winding (note: no doubling back); and (D) two-turn winding.

with Q-Dope or some other sealant (RTV silicone, rubber cement, etc.). Q-Dope is intended for inductors and can be purchased from G-C dealers or by mail from Ocean State Electronics [P.O. Box 1458, 6 Industrial Drive, Westerly, RI, 02891; phone 1-800-866-6626 (orders only), 401-596-3080 (voice) or 401-596-3590 (fax)].

The second method involves making a header for the binocular core. This header can be permanent and can be installed into the circuit just like any other coil with a header. When built correctly, the header will be spaced on 0.100" centers, so it is compatible with DIP printed circuit boards and perforated wiring board. I used

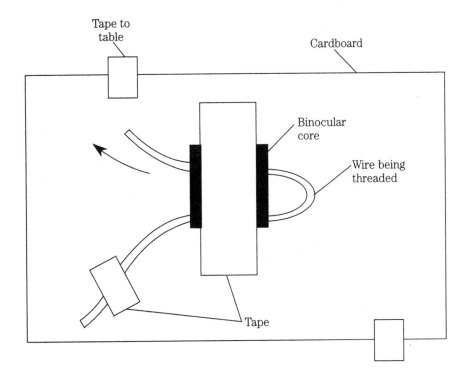

4-18 Winding "jig" for Bazooka balun core.

perforated wiring board of the sort that has printed circuit pads (none of which connected to each other) at each hole.

Figure 4-19A shows the basic configuration for my homebrew header (a DIP header can also be used). These connectors are intended to connect wiring or other components to a DIP printed circuit board designed for digital integrated circuits. I found that a small segment of printed perfboard, 0.100″ centers on the holes, that contained five rows by nine columns of holes (see inset to Fig. 4-19B) was sufficient for the 0.525 × 0.550″ BN-xx-202. Larger or smaller hole matrices can be cut for larger or smaller binocular cores.

The connections to the header are perfboard push terminals (available any place that perfboard and printed-circuit making supplies are sold). I used the type of perfboard that had solder terminals so that the push terminals can be held to the board with solder. Otherwise, they have a distinct tendency to back out of the board with handling.

When the header is finished, the binocular core is fastened to the top surface of the header with tape, then the pins of the header are pushed into a large piece of perfboard. This step is done to stabilize the assembly on the work surface. It might be a good idea to stabilize the perfboard to the table with tape to keep it from moving as you wind the coils.

Once the header and core are prepared, then it is time to make the windings. Scrape the insulation off one end of the wire for about 1/4″. An X-acto knife, scalpel,

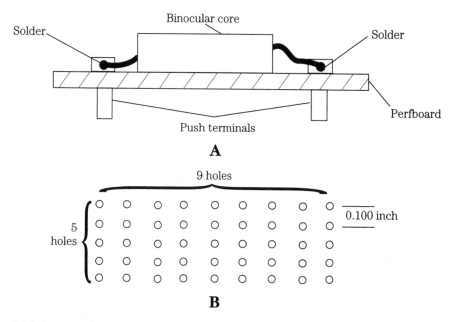

4-19 Construction of a perfboard carrier or header for Bazooka balun cores.

or similar tool can be used to do this job. Turn the wire over several times to make sure that the enamel insulation is scraped around the entire circumference. Some people prefer to burn the insulation off with a soldering iron, which tins the end of the wire as it burns the insulation away. I've found that method to be successful when the smaller gauges are used, but when good-quality no. 26 or larger wire is used, the scraping method seems to work better. If the scraping method is used, then follow the scraping by tinning the exposed end of the wire with solder. Each winding of the transformer can be made by threading the wire through the core as needed. As each winding is finished, the loose end is cleaned, tinned, and soldered to its push terminal. After all windings are completed, then seal the assembly with Q-Dope or equivalent.

Homebrew binocular cores

You can build custom "binocular cores" from toroidal cores. The toroids are easily obtained from many sources and in many different mixtures of both powdered iron and ferrite. You can also make larger binocular cores using toroids because of the wide range of toroid sizes. Actual binocular cores are available in a limited range of mixtures and sizes. Figure 4-20 shows the common way to make your own binocular core: Stack a number of toroid cores in the manner shown. It is common practice to wrap each stack in tape, then place the two stacks together and wrap the assembly together. Although four toroids are shown on each side, any number can be used.

A variation on the theme is shown in Fig. 4-21. This binocular core is designed to have a single-turn winding consisting of a pair of brass tubes passed through the

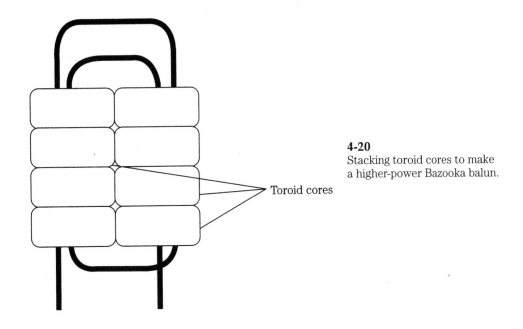

4-20
Stacking toroid cores to make
a higher-power Bazooka balun.

Toroid cores

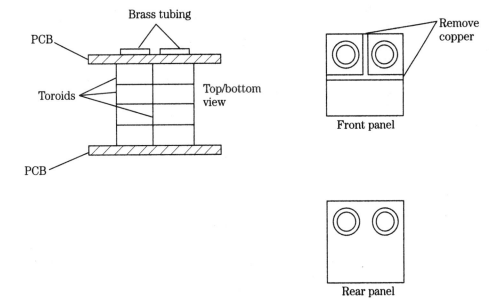

4-21 Construction of the high-power Bazooka balun core with stacked toroid cores and copper (PCB material) end plates.

center holes of the toroid stacks. The ends of the stacks are held together with a pair of copper-clad PC boards. The read panel has no copper removed, and the front panel is etched to isolate the two brass tubes. The pads around the brass tubes at the front end are used to make connections to the tubing (which serves as a single-turn winding). The other winding of the transformer is made of ordinary insulated wire, which is passed through the brass tubes the correct number of turns to achieve the desired turns ratio. This type of binocular core was once popular with ham operators who built their own solid-state RF power amplifiers. The higher-power transformers needed to match the impedances of the base and collector terminals of the RF transistors were not easily available on the market, so they had to "roll their own."

The binocular core is not as well-known as the toroid core, but for many applications, it is the core of choice. This is especially true when low frequencies are used or when large inductances are needed in a small package . . . and you don't want to work your arm off hand-winding the large number of turns that would be required on a solenoid-wound or toroidal core.

References

F. W. Grover, *Inductance Calculation: Working Formulas and Tables,* D. VanNostrand Co., Inc. (New York, 1946).

G. Bramslev, "Loop Aerial Reception," *Wireless World,* Nov. 1952, pp. 469–472.

5
CHAPTER

Radio receivers: theory and projects

One of the things that fascinated people about radio from its very earliest days is that signals arrive seemingly by magic through the air from long distances. It was very quickly discovered that radio signals are electromagnetic waves—exactly like light and infrared, except for frequency and wavelength. Radiowaves have a much lower frequency than light; therefore, the wavelengths are much longer. The wavelengths of radio signals range from around 25,000 m at the low-VLF range down to millimeters in the upper microwave spectrum.

Consider three different forms of signals: continuous wave (CW), amplitude modulation (AM), and frequency modulation (FM). The CW signal is only considered very briefly because it is largely irrelevant to most readers of this book. The CW signal (Fig. 5-1) is made up of sinusoidal oscillations at the transmitter frequency. For example, a 500-kHz maritime signal oscillates 500,000 times per second. The critical point about the CW signal is that it has a constant amplitude while it is on. If the signal is turned on and off, to form dots and dashes of the Morse code, it is possible to send messages with the CW signal. This type of signal was transmitted by ships at sea during the early days of wireless . . . and indeed maritime radiotelegraph CW is still heard on the air today (even though less and less—The U.S. Coast Guard has stopped monitoring 500 kHz).

Modulation is the act of adding information to an unmodulated radio signal (called the *carrier*). That unmodulated signal is, by the way, the same as an unkeyed CW signal. The modulating signal is usually audio from speech or music sources. In radio, the technical term for the modulating signal is *intelligence*, although considering the content of certain talk shows, the Citizen's Band, and other radio transmissions, it is difficult to keep a straight face when talking about the modulating signal as "intelligence." Figure 5-2 shows the relationship between the modulating signal (Fig. 5-2A), the carrier (Fig. 5-2B), and the resultant amplitude-modulated signal (Fig. 5-2C).

The transmitter modulator stage superimposes the audio signal onto the carrier, resulting in the characteristic signal shown in Fig. 5-2C. This signal is received at the

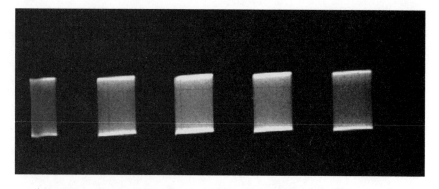

5-1 CW signal.

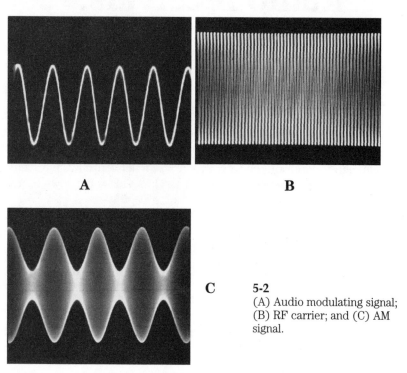

A

B

C

5-2
(A) Audio modulating signal;
(B) RF carrier; and (C) AM
signal.

radio set. Notice that the peaks of the AM signal vary in step with the audio modulating signal. The frequency of the AM carrier remains constant, but its intensity or amplitude varies with the audio signal. In this case, the frequency or phase of the carrier varies in step with the audio signal, and the amplitude remains constant.

The tuner

The airspace surrounding the radio is filled with a cacophony of electromagnetic signals that are ready for the plucking. Very early in the radio business, it was found that radios not only worked better with tuning units but would separate signals on different wavelengths (frequencies). The standard radio tuner consists of an inductor and a capacitor. Figure 5-3 shows an RF tuner circuit. An inductor stores energy in a surrounding magnetic field, and the capacitor stores energy in an electrostatic field between the two plates.

Inductors and capacitors both provide some opposition to the flow of ac current (including radio frequency or "RF" frequencies). This opposition is called *reactance* and for inductors it is *inductive reactance* (X_L); for capacitors, it is called *capacitive reactance* (X_C). At a certain frequency, called resonance, these two reactances are equal ($X_L = X_C$). When this occurs (in the LC "tank" circuit), the electrostatic field of the capacitor and the magnetic field oscillate so that energy is swapped back and forth between the fields of the inductor and capacitor. The tuner will accept this frequency but reject others. The resonant frequency of the LC "tank" circuit of Fig. 5-3 is given by:

$$F = \frac{1}{2\pi\sqrt{LC}},$$ (5-1)

where

f is the frequency in hertz (Hz)
L is the inductance in henrys (H)
C is the capacitance in farads (F).

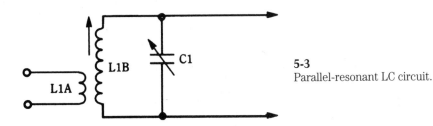

5-3
Parallel-resonant LC circuit.

Radio tuners don't normally use the simple LC resonant tank circuit of Fig. 5-3 but rather depend on the RF transformer principle shown in Fig. 5-4. The primary winding of the RF transformer (L_{1A}) consists of a few turns of wire wound on the lower end of the same coil form as the main tuning coil, L_{1B}.

If either the inductor or capacitor is made variable, then an operator can select from a number of resonant frequencies by varying the frequency of the tuned circuit. In most radios, the tuner control is a variable capacitor because of the ease with which these components can be made.

When the radiowave is captured by the antenna, a small current (I_1) at the same frequency as the radiowave is set up in the antenna–ground path. One-half of the radiowave causes the current to flow toward the ground (Fig. 5-4B). Because the primary (L_{1A}) and secondary (L_{1B}) are magnetically coupled, current I_1 induces current I_2, which flows in the LC tank circuit producing a signal voltage (V) across the output terminals. Similarly, when the radiowave reverses polarity, the antenna–ground current also reverses direction (Fig. 5-4C). The current in the secondary is also

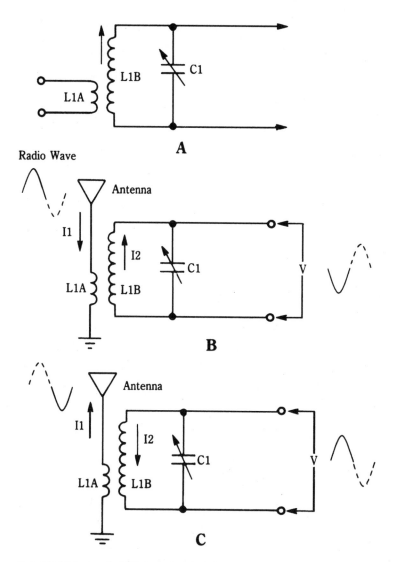

5-4 (A) Link-coupled RF transformer LC-tuned circuit; (B) antenna coupler circuit using Fig. 4-2A circuit on one half of incoming signal; and (C) same circuit on alternate half-cycle (note current direction reversal).

therefore reversed, creating the other half of the output signal voltage (V). As the radiowave oscillates between its peaks at the radio transmitter frequency, the currents (I_1 and I_2) and signal voltage (V) also oscillate at the same frequency. Only the resonant frequency oscillates efficiently in the LC tank circuit, so only it will appear at the output terminals—even though many waves set up currents in the antenna–ground circuit.

As a point of interest, even though most radio tuners use a fixed inductor and a variable capacitor (which is ganged to the tuning knob on the radio front panel), a few models out there used a fixed capacitor and a variable inductor (Fig. 5-5). The coil has a hollow coil form and is tuned with a powdered-iron or ferrite core that is connected to the tuning dial through a dial cord. These radios were produced during World War II, when the metal needed to make a variable capacitor was far too precious to the war effort for use in consumer radios. Also, some very high-quality commercial and military receivers used permeability (i.e., inductor) tuning mechanisms; car radios also typically used PTM assemblies in the days before all-electronic tuning.

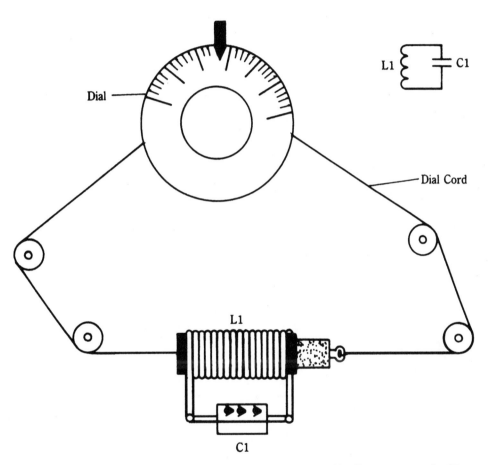

5-5 Inductor-tuned radio tuning circuit was common in World War II consumer and military "morale" radio receivers.

The selectivity of the radio is its ability to separate stations adjacent to each other on the dial. The number of LC-tuned circuits in the signal path determines the selectivity. Figure 5-6 shows the curves to expect with one tuned circuit (A), two tuned circuits (B), and three tuned circuits (C). Notice that in each case, the LC tuned circuit does not pick out only one single frequency, but rather it picks up a small band of frequencies that surround the resonant frequency.

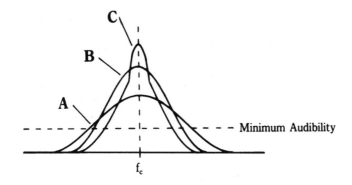

5-6 Bandwidth effects of three different coupling schemes.

The selectivity of the receiver is measured in terms of the bandwidth of the tuned circuit. The standard method for specifying selectivity is to measure the frequency response of the circuit to find the points where the power drops −3 dB or the voltage drops −6 dB (same point, but specified differently). Figure 5-7 shows the response of a moderately selective LC tank circuit. The bandwidth is the frequency difference $F_2 - F_1$, where F_1 and F_2 are the frequencies at which the response drops −6 dB (voltage), or −3 dB (power), from the level found at resonant frequency, F_R.

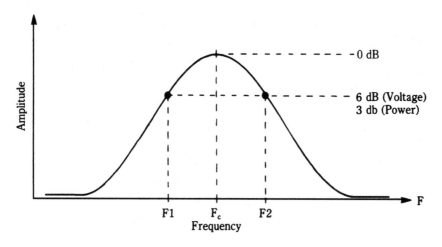

5-7 Bandpass characteristics.

Tuned radio-frequency (TRF) receivers

The tuned radio frequency (TRF) radio uses one or more radio frequency (RF) amplifiers to boost the weak signal received from the antenna. The simplest form of TRF radio (Fig. 5-8) consists of a single RF amplifier stage, with frequency selection (LC tank) circuits at the input and output. The signal at the antenna input of the RF

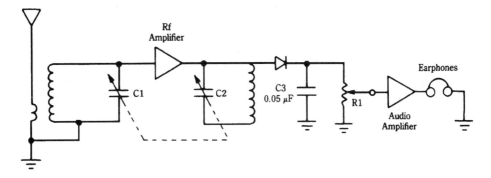

5-8 Simple TRF radio receiver.

amplifier is very weak, but because of the amplification of the tube in the amplifier, it is boosted to a stronger level at the output.

The RF amplifier is followed by a detector, which is, in turn, followed by an audio amplifier and a reproducer (e.g., loudspeaker). The detector demodulates the RF signal to recover the audio signal that was impressed on the RF carrier at the transmitter. This audio signal is boosted to a higher power level in the audio frequency (AF) amplifier and is then fed to a loudspeaker, earphones, or other audio reproducer.

More complex TRF radios will use two or more RF amplifiers in cascade (Fig. 5-9) prior to the detector and AF amplifier. The extra amplification boosts the signal level even higher than is possible in the single-stage version. Also, the additional tuned circuits tend to sharpen the selectivity quite a bit. If a single tuning shaft operates all of the variable capacitors, i.e., the capacitors are ganged together (as shown by the dotted line in Fig. 5-9), then "single-knob" operation is possible.

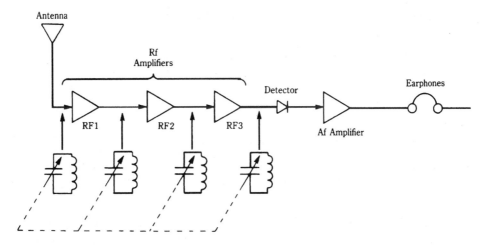

5-9 Multistage TRF radio receiver.

Superheterodyne receivers

The superheterodyne radio receiver (Fig. 5-10) was invented in the early 1920s, but only a very few sources could supply them because of patent restrictions. Later, however, patents were pooled as the radio industry grew and eventually the patents expired. The superheterodyne design was so superior that within a decade it took over all but a very few radios and is still today the basic design of all AM and FM radio receivers.

The block diagram to a superheterodyne receiver is shown in Fig. 5-10. The basic idea of the superheterodyne is to convert all the RF carrier signals from the radio waves to a fixed frequency, where it can be amplified and otherwise processed. The stages of the basic superheterodyne receiver consists of a mixer, a local oscillator (LO), an intermediate frequency (IF) amplifier, a detector, and an AF amplifier. The latter two stages are also used in TRF radios and serve exactly the same function in superheterodynes. Better-quality superhet radios also include an RF amplifier.

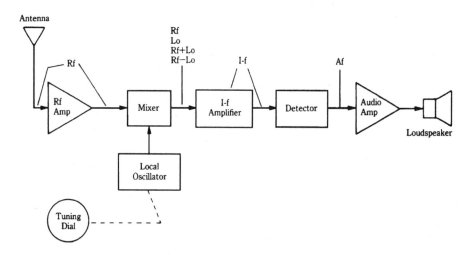

5-10 Block diagram to a superheterodyne receiver.

The RF amplifier boosts the weak signal from the antenna and provides the radio with some additional selectivity. It also prevents the signal from the LO from being coupled to the antenna, where it could be radiated into space. The RF amplifier is sometimes called the *preselector* in radio schematics.

The output of the RF amplifier is applied to the input of the mixer stage. Also the input to the mixer is the local oscillator (LO) signal. The two are mixed together in a nonlinear circuit; two produces at least four output frequencies: RF, LO, RF − LO (difference), and RF + LO (sum). Either the sum or difference frequency can be selected for the intermediate frequency (IF), but in the vast majority of antique or classic radios, it was the difference frequency that was used. For AM radios, the IF frequency is 455 or 460 kHz in home radios and 260 or 262.5 kHz in auto radios; FM broadcast receivers and most VHF/UHF monitors use 10.7 MHz as the IF. In many receivers, the LO and mixer are combined into a single stage called a *converter*. Another name for the mixer or converter is *first detector*.

The IF amplifier provides the radio with the largest amount of signal gain and the tightest selectivity. It is the fact that the IF operates on only one frequency (e.g., 455 kHz) which permits the very high gain to be achieved without oscillation and other difficulties.

The detector operates at the IF frequency and will demodulate the IF frequency to recover the audio. From the detector, it is passed on to the AF amplifier and the reproducer (speaker).

Using the NE-602/NE-612 chips

The Signetics NE-602A and NE-612 are monolithic integrated circuits containing a double-balanced mixer (DBM), an oscillator, and an internal voltage regulator in a single eight-pin package (Fig. 5-11). The DBM section operates to 500 MHz, and the internal oscillator section works to 200 MHz. The primary uses of the NE-602 are in HF and VHF receivers, frequency converters, and frequency translators. The device can also be used as a signal generator in many popular inductor-capacitor (LC) variable-frequency oscillator (VFO), piezoelectric-crystal (XTAL), or swept-frequency configurations. This chapter explores the various configurations for the dc power supply, the RF input, the local oscillator, and the output circuits. Certain applications of the device are also covered.

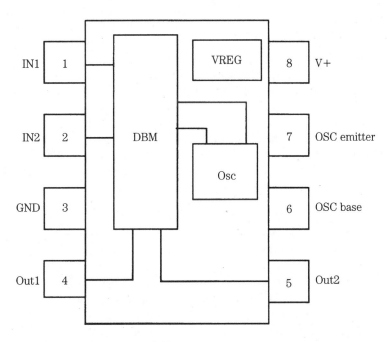

5-11 Internal circuit of NE-602.

The NE-602 version of the device operates over a temperature range of 0 to +70°C, and the SA-602 operates over the extended temperature range of −40 to +85°C. The most common form of the device is probably the NE-602N, which is an eight-pin mini-DIP package. Eight-lead SO surface-mount ("D-suffix") packages are also available. In this chapter, the NE-602N is featured, although the circuits also work with the other packages and configurations. The improved NE-602AN and NE-602AD are now available.

Because the NE-602 contains both a mixer and a local oscillator, it can operate as a radio receiver "front-end" circuit. It features good noise rejection and reasonable third-order intermodulation performance. The noise figure is typically 5 dB at a frequency of 45 MHz. The NE-602 has a third-order intercept point on the order of −15 dBm, referenced to a matched input, although it is recommended that a maximum signal level of −25 dBm ($\approx$3.16 mW) be observed. This signal level corresponds to about 12.6 mV into a 50-Ω load, or 68 mV into the 1500-Ω input impedance of the NE-602. The NE-602 is capable of providing 0.2 μV sensitivity in receiver circuits without external RF amplification. One criticism of the NE-602 is that it appears to sacrifice some dynamic range for high sensitivity—a problem said to be solved in the "A" series (e.g., NE-602AN and in the NE-612).

Frequency conversion or translation

The process of frequency conversion is called *heterodyning*. When two signals of different frequencies (F_1 and F_2) are mixed in a nonlinear circuit, a collection of different frequencies will appear in the output of the circuit. These are characterized as F_1, F_2, and ($nF_1 \pm mF_2$), where n and m are integers. In most practical situations, n and m are 1, so the total output spectrum will consist at least of F_1, F_2, $F_1 + F_2$, and $F_1 - F_2$. Of course, if the two input circuits contain harmonics, then additional products are found in the output. In superheterodyne radio receivers, either the sum or difference frequency is selected as the intermediate frequency (IF). In order to make the frequency conversion possible, a circuit needs a local oscillator and a mixer circuit (both of which are provided by the NE-602).

The local oscillator (LO) consists of a VHF NPN transistor with the base connected to pin 6 of the NE-602, and the emitter is connected to pin 7; the collector of the oscillator transistor is not available on an external pin. There is also an internal buffer amplifier, which connects the oscillator transistor to the DBM circuit. Any of the standard oscillator circuit configurations can be used with the internal oscillator, provided that access to the collector terminal is not required. Thus, Colpitts, Clapp, Hartley, Butler, and other oscillator circuits can be used with the NE-602 device, and the Pierce and Miller oscillator circuits are not.

The double-balanced mixer (DBM) circuit is shown in Fig. 5-12; it consists of a pair of cross-connected differential amplifiers (Q_1/Q_2 with Q_5 as a current source; similarly Q_3/Q_4 with Q_6 working as a current source). This configuration is called a *Gilbert transconductance cell*. The cross-coupled collectors form a push–pull output (pins 4 and 5) in which each output pin is connected to the $V+$ power-supply terminal through 1500-Ω resistances. The input is also push–pull and is cross-coupled between the two halves of the cell. The local oscillator signal is injected into each cell-half at the base of one of the transistors.

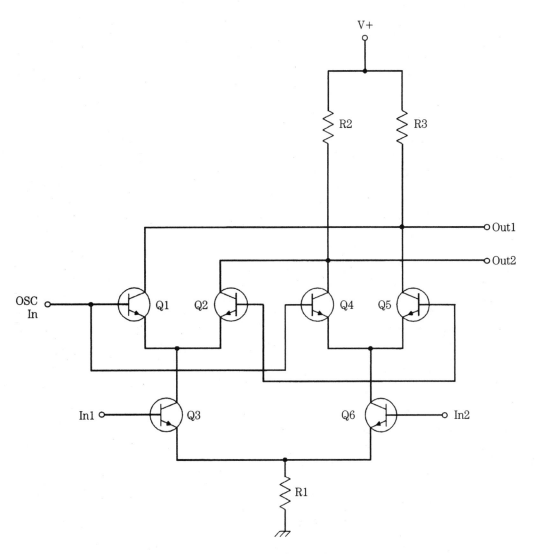

5-12 Transconductance cell double-balanced mixer used in NE-602.

Because the mixer is "double-balanced," it has a key attribute that makes it ideal for use as a frequency converter or receiver front-end: suppression of the LO and RF input signals in the outputs. In the NE-602 chip, the output signals are $F_1 + F_2$, and $F_1 - F_2$; neither LO nor RF signals appear in the output in any great amplitude. Although some harmonic products appear, many are also suppressed because of the DBM action.

dc power-supply connections on the NE-602
The $V+$ power-supply terminal of the NE-602 is pin 8, and the ground connection is pin 3; both must be used for the dc power connections. The dc power supply range is +4.5 to +8 Vdc, with a current drain ranging from 2.4 to 2.8 mA.

It is highly recommended that the $V+$ power-supply terminal (pin 8) be bypassed to ground with a capacitor of 0.01 to 0.1 μF. The capacitor should be mounted as close to the body of the NE-602 as is practical; short leads are required in radio-frequency (RF) circuits.

Figure 5-13A shows the recommended power-supply configuration for situations where the supply voltage is +4.5 to +8 V. For best results, the supply voltage should be voltage-regulated. Otherwise, the local oscillator frequency might not be stable, which leads to problems. A series resistor ($\approx$100 to 180 Ω) is placed between the $V+$ power supply and the $V+$ terminal on the NE-602. If the power-supply voltage is raised to +9 V, then increase the value of the series resistance an order of magnitude to 1000 to 1500 Ω (Fig. 5-13B).

If the dc power-supply voltage is either unstable, or is above +9 V, then it is highly recommended that a means of voltage regulation be provided. In Fig. 5-13C, a zener diode is used to regulate the NE-602 $V+$ voltage to 6.8 Vdc—even though the supply voltage ranges from +9 to +18 V (a situation found in automotive applications). An alternative voltage regulator circuit is shown in Fig. 5-13D. This circuit uses a three-terminal IC voltage regulator to provide $V+$ voltage to the NE-602. Because the NE-602 is a very low current drain device, the lower power versions of the regulators (e.g., 78Lxx) can be used. The low-power versions also permit the NE-602 to have its own regulated power supply—even though the rest of the radio receiver uses a common dc power supply. Input voltages of +9 to more than +28 Vdc, depending on the regulator device selected, can be used for this purpose. The version of Fig. 5-13D uses a 78L09 to provide +9 V to the NE-602, although 78L05 and 78L06 can also be used to good effect.

NE-602 input circuits

The RF input port of the NE-602 uses pins 1 and 2 to form a balanced input. As is often the case in differential amplifier RF mixers, the RF input signals are applied to the base terminals of the two current sources (Q5 and Q6 in Fig. 5-12). The input impedance of the NE-602 is 1500 Ω, shunted by 3 pF at lower frequencies, although in the VHF region, the impedance drops to about 1000 Ω.

Several different RF input configurations are shown in Fig. 5-14; both single-ended (unbalanced) and differential (balanced) input circuits can be used with the NE-602. In Fig. 5-14A, a capacitor-coupled, untuned, unbalanced input scheme is shown. The signal is applied to pin 1 (although pin 2 could have been used instead) through a capacitor, C_1, that has a low impedance at the operating frequency. The signal level should be less than −25 dBm, or about 68 mV rms (180 mV peak to peak). Whichever input is used, the alternate input is unused and should be bypassed to ground through a low-value capacitor (0.001 to 0.1 μF, depending on the frequency).

A wideband transformer coupled RF input circuit is shown in Fig. 5-14B. In this configuration, a wideband RF transformer is connected so that the secondary is applied across pins 1 and 2 of the NE-602, with the primary of the transformer connected to the signal source or antenna. The turns ratio of the transformer can be used to transform the source impedance to 1500 Ω (the NE-602 input impedance). Either conventional or toroid-core transformers can be used for T_1. As in the previous circuit, one input is bypassed to ground through a low-reactance capacitor.

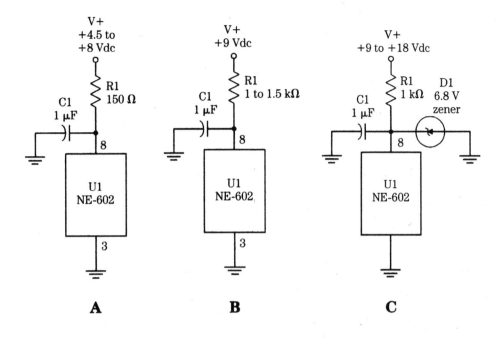

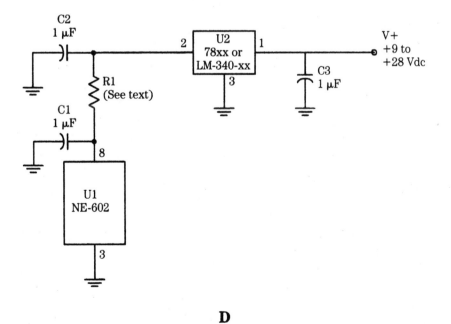

5-13 Dc power supply connections for the NE-602.

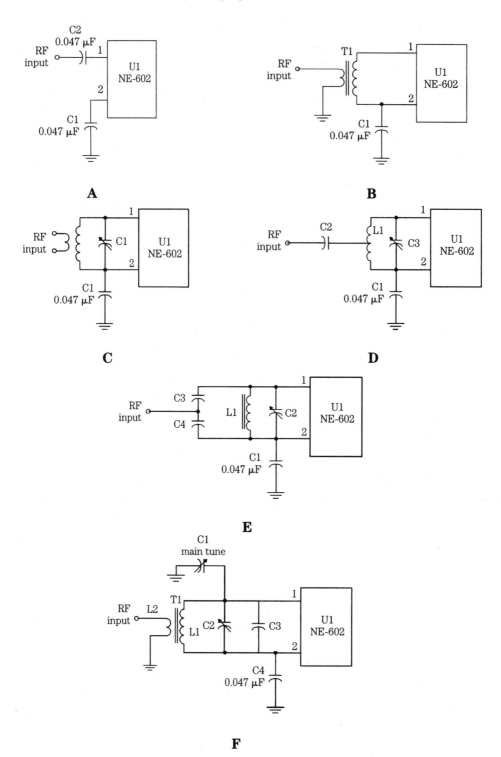

5-14 Input circuits for the NE-602.

Tuned RF input circuits are shown in Figs. 5-14C–14E and 5-15. Each of these circuits performs two functions: (a) it selects the desired RF frequency while rejecting others and (b) it matches the 1.5-kΩ input impedance of the NE-602 to the source or antenna system impedance (e.g., 50 Ω). The circuit shown in Fig. 5-14D uses an inductor (L_1) and capacitor (C_1) tuned to the input frequency, as do the other circuits, but the impedance matching function is done by tapping the inductor; a dc-blocking capacitor is used between the antenna connection and the coil. A third capacitor, C_3, is used to bypass one of the inputs (pin 2) to ground.

Another version of the circuit is shown in Fig. 5-14E. It is similar in concept to the previous circuit but uses a tapped capacitor voltage divider (C_3/C_4) for the impedance-matching function. Resonance with the inductor is established by the combination of C_1, the main tuning capacitor, in parallel with the series combination of C_2 and C_3:

$$C_{\text{tune}} = C_1 + \frac{C_2 C_3}{C_2 + C_3}. \tag{5-2}$$

The previous two circuits are designed for use when the source or antenna system impedance is less than the 1.5-Ω input impedance of the NE-602. The circuit of Fig. 5-14F can be used in all three situations: input impedance lower than, higher than, or equal to the NE-602 input impedance—depending on the ratio of the number of turns in the primary winding (L_2) to the number of turns in the secondary winding (L_1). The situation shown schematically in Fig. 5-14F is for the case where the source impedance is less than the input impedance of the NE-602.

The secondary of the RF transformer (L_1) resonates with a capacitance made up of C_1 (main tuning), C_2 (trimmer tuning or bandspread), and a fixed capacitor, C_3. An advantage of this circuit is that the frame of the main tuning capacitor is grounded. This feature is an advantage because most tuning capacitors are designed for grounded-frame operation, so construction is easier. In addition, most of the variable-frequency oscillator circuits used with the NE-602 also use a grounded-frame capacitor. The input circuit of Fig. 5-14F can therefore use a single dual-section capacitor for single-knob tuning of both RF input and local oscillator.

Figure 5-15 shows a tuned-input circuit that relies, at least in part, on a voltage variable-capacitance ("varactor") diode for the tuning function. The total tuning capacitance that resonates transformer secondary L_2 is the parallel combination of C_1 (trimmer), C_2 (a fixed capacitor), and the junction capacitance of varactor diode, D1. The value of capacitor C_3 is normally set to be large compared with the diode capacitance so that it will have little effect on the total capacitance of series combination C_3/C_{D1}. In other cases, however, the capacitance of C_3 is set close to the capacitance of the diode, so it becomes part of the resonant circuit capacitance.

A varactor diode is tuned by varying the reverse bias voltage applied to the diode. Tuning voltage V_t is set by a resistor voltage divider consisting of R_1, R_2, and R_3. The main tuning potentiometer (R_1) can be a single-turn model, but for best resolution of the tuning control use a 10- or 15-turn potentiometer. The fine-tuning potentiometer can be a panel-mounted model for use as a bandspread control or a trimmer model for use as a fine adjustment of the tuning circuit (a function also shared by trimmer capacitor, C_1).

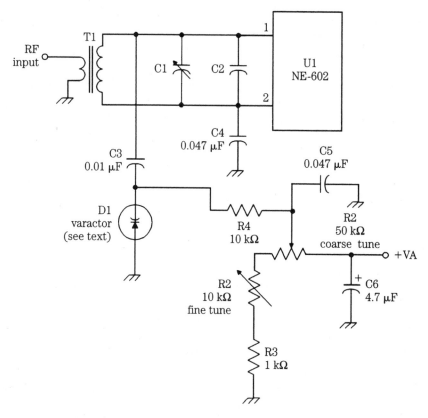

5-15 Varactor-tuned input circuit for NE-602.

The voltage used for the tuning circuit (V_A) must be well-regulated, or the tuning will shift with variations of the voltage. Some designers use a separate three-terminal IC regulator for V_A, but that is not strictly necessary. A more common situation is to use a single low-power 9-V three-terminal IC voltage regulator for both the NE-602 and the tuning network. However, it will only work when the diode needs no more than +9 Vdc for correct tuning of the desired frequency range. Unfortunately, many varactor diodes require a voltage range of about +1 to +37 V to cover the entire range of available capacitance.

In due course, the oscillator circuits will also show a version of the Fig. 5-15 circuit that is tuned by a sawtooth waveform (for swept frequency operation) or a digital-to-analog converter (for computer-controlled frequency selection).

NE-602 output circuits

The NE-602 output circuit consists of the cross-coupled collectors of the two halves of the Gilbert transconductance cell (Fig. 5-12) and are available on pins 4 and 5. In general, it doesn't matter which of these pins is used for the output; in single-ended output configurations, only one terminal is used and the alternate output terminal is ignored. Each output terminal is connected internally to the NE-602 to $V+$ through separate 1.5-kΩ resistors.

Figure 5-16A shows the wideband, high-impedance (1.5 kΩ) output configuration. Either pin 4 or 5 (or both) can be used. A capacitor is used to provide dc blocking. This capacitor should have a low reactance at the frequency of operation, so values between 0.001 and 0.1 μF are generally selected.

Transformer output coupling is shown in Fig. 5-16B. In this circuit, the primary of a transformer is connected between pins 4 and 5 of the NE-602. For frequency converter or translator applications, the transformer could be a broadband RF transformer wound on either a conventional slug-tuned form or a toroid form. For direct-conversion autodyne receivers, the transformer would be an audio transformer. The standard 1:1 transformers used for audio coupling can be used. These transformers are sometimes marked as impedance ratio rather than turns ratio (e.g., 600 Ω:600 Ω, or 1.5K:1.5K).

Frequency converters and translators are the same thing, except that the *converter* terminology generally refers to a stage in a superhet receiver and *translator* is more generic. For these circuits, the broadband transformer will work, but it is probably better to use a tuned RF/IF transformer for the output of the NE-602. The resonant circuit will reject all but the desired frequency product; e.g., the sum or difference IF frequency. Figure 5-16C shows a common form of resonant output circuit for the NE-602. The tuned primary of the transformer is connected across pins 4 and 5 of the NE-602, and a secondary winding (which could be tuned or untuned) is used to couple the signal to the following stages.

A single-ended RF tuned transformer output network for the NE-602 is shown in Fig. 5-16D. In this coupling scheme, the output terminal of the IC is coupled to the $V+$ dc power-supply rail through a tuned transformer. Perhaps a better solution to the single-ended problem is the circuit of Fig. 5-16E. In this circuit, the transformer primary is tapped for a low impedance and the tap is connected to the NE-602 output terminal through a dc-blocking capacitor. These transformers are easily available in either 455-kHz or 10.7-MHz versions and can also be made relatively easily.

Still another single-ended tuned output circuit is shown in Fig. 5-16F. In this circuit, one of the outputs is grounded for RF frequencies through a capacitor. Tuning is a function of the inductance of L_1 and the combined-series capacitance of C_1, C_2, and C_3. By tapping the capacitance of the resonant circuit at the junction of C_2–C_3, it is possible to match a lower impedance (e.g., 50 Ω) to the 1.5-kΩ output impedance of the NE-602.

The single-ended output network of Fig. 5-16G uses a low-pass filter as the frequency-selective element. This type of circuit can be used for applications, such as a heterodyne signal generator in which the local oscillator frequency of the NE-602 is heterodyned with the signal from another source applied to the RF input pins of the IC. The difference frequency is selected at the output when the low-pass filter is designed so that its cut-off frequency is between the sum and difference frequencies.

In Fig. 5-16H, an IF filter is used to select the desired output frequency. These filters are available in a variety of different frequencies and configurations, including the Collins mechanical filters that were once used extensively in high-grade communications receivers (262-, 455-, and 500-kHz center frequencies). Current high-

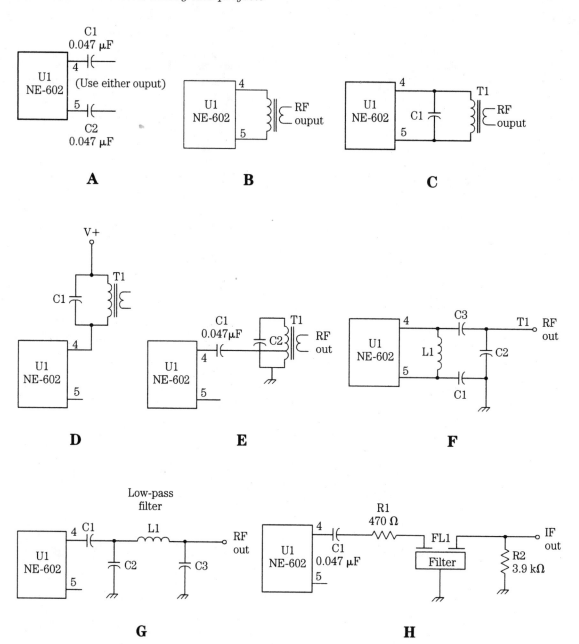

5-16 Output circuits for NE-602.

grade communications receivers typically used crystal IF filters centered on 8.83, 9, or 10.7 MHz or 455 kHz (with bandwidths of 100 Hz to 30 kHz). Even some broadcast radio receivers use IF filters. Such filters are made of piezoceramic material and are usually centered on 260 or 262.5 kHz (AM auto radios), 455 or 460 kHz (other AM radios), or 10.7 MHz (FM radios). The lower-frequency versions are typically made with 4-, 6-, or 12-kHz bandwidths, and the 10.7-MHz versions have bandwidths of 150 to 300 kHz (200 kHz is most common).

In the circuit, Fig. 5-16H it is assumed that the low-cost ceramic AM or FM filters are used (for other types, compatible resistances or capacitances are needed to make the filter work properly). The input side of the filter (FL1) in Fig. 5-16H is connected to the NE-602 through a 470-Ω resistor and an optional dc blocking capacitor (C_1). The output of the filter is terminated in a 3.9-kΩ resistor. The difference IF frequency resulting from the conversion process appears at this point.

One of the delights of the NE-602 chip is that it contains an internal oscillator circuit that is already coupled to the internal double-balanced mixer. The base and emitter connections to the oscillator transistor inside the NE-602 are available through pins 6 and 7, respectively. The internal oscillator can be operated at frequencies up to 200 MHz. The internal mixer works to 500 MHz. If higher oscillator frequencies are needed, then use an external local oscillator. An external signal can be coupled to the NE-602 through pin 6, but must be limited to less than about −13.8 dBm, or 250 mV, across 1500 Ω.

The next section shows some of the practical local oscillator (LO) circuits that can be successfully used with the NE-602, including one that allows digital or computer control of the frequency. Oscillator circuits are covered in greater detail in Chapter 13.

NE-602 local oscillator circuits

There are two general methods for controlling the frequency of the LO in any oscillator circuit: inductor-capacitor (LC) resonant tank circuits or piezoelectric crystal resonators. Both forms are considered, but first, the crystal oscillators.

Figure 5-17A shows the basic Colpitts crystal oscillator. It will operate with fundamental-mode crystals on frequencies up to about 20 MHz. The feedback network consists of a capacitor voltage divider (C_1/C_2). The values of these capacitors are critical and should be approximately:

$$C_1 = \frac{100}{\sqrt{F_{\text{MHz}}}} \text{ pF} \tag{5-3}$$

$$C_2 = \frac{1000}{F_{\text{MHz}}} \text{ pF.} \tag{5-4}$$

The values predicted by these equations are approximate, but work well under circumstances where external stray capacitance does not dominate the total. However, the practical truth is that capacitors come in standard values and these might not be exactly the values required by Eqs. (5-3) and (5-4).

When the capacitor values are correct, the oscillation will be consistent. If you pull the crystal out, then reinsert it, the oscillation will restart immediately. Alternatively, if the power is turned off and back on again, the oscillator will always restart.

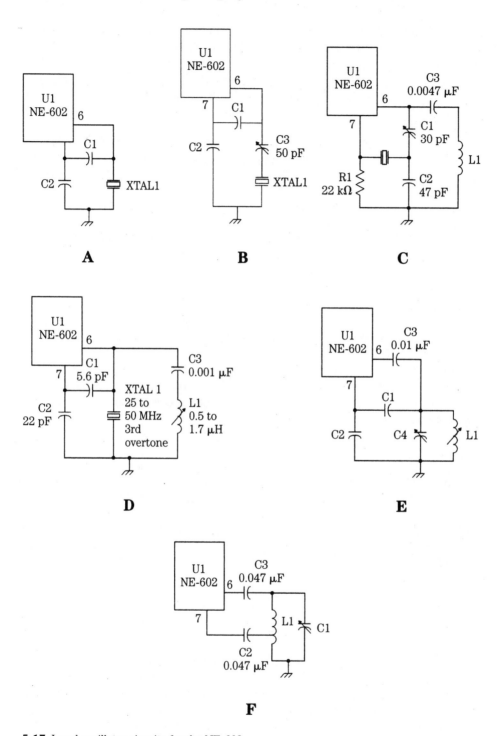

5-17 Local oscillator circuits for the NE-602.

If the capacitor values are incorrect, the oscillator will either fail to run or will operate intermittently. Generally, an increase in the capacitances will suffice to make operation consistent.

A problem with the circuit of Fig. 5-17A is that the crystal frequency is not controllable. The actual operating frequency of any crystal depends, in part, on the circuit capacitance seen by the crystal. The calibrated frequency is typically valid when the load capacitance is 20 or 32 pF, but this can be specified to the crystal manufacturer at the time of ordering. In Fig. 5-17B, a variable capacitor is placed in series with the crystal to set the frequency. This trimmer capacitor can be adjusted to set the oscillation frequency to the desired frequency.

The two previous crystal oscillators operate in the fundamental mode of crystal oscillation. The resonant frequency in the fundamental mode is set by the dimensions of the slab of quartz used for the crystal; the thinner the slab, the higher the frequency. Fundamental-mode crystals work reliably up to about 20 MHz, but above 20 MHz, the slabs become too thin for safe operation. Above about 20 MHz, the thinness of the slabs of fundamental-mode crystal causes them to fracture easily. An alternative is to use overtone mode crystals. The overtone frequency of a crystal is not necessarily an exact harmonic of the fundamental mode but is close to it. The overtones tend to be close to odd-integer multiples of the fundamental (3rd, 5th, and 7th). Overtone crystals are marked with the appropriate overtone frequency rather than the fundamental.

Figures 5-17C and 5-17E are overtone-mode crystal-oscillator circuits. The circuit in Fig. 5-17C is the Butler oscillator. The overtone crystal is connected between the oscillator emitter of the NE-602 (pin 7) and a capacitive voltage divider that is connected between the oscillator base (pin 6) and ground. There is also an inductor in the circuit (L_1); this inductor must resonate with C_1 to the overtone frequency of crystal Y_1. Figure 5-17C can use either 3rd- or 5th-overtone crystals up to about 80 MHz. The circuit in Fig. 5-17D is a 3rd-overtone crystal oscillator that works from 25 to about 50 MHz and is simpler than Fig. 5-17C.

A pair of variable-frequency oscillator (VFO) circuits are shown in Figs. 5-17E and 5-17F. The circuit in Fig. 5-17E is the Colpitts oscillator version, and Fig. 5-17F is the Hartley oscillator version. In both oscillators, the resonating element is an inductor/capacitor (LC) tuned resonant circuit. In Fig. 5-17E, however, the feedback network is a tapped capacitor voltage divider; in Fig. 5-17F, it is a tap on the resonating inductor. In both cases, a dc-blocking capacitor to pin 6 is needed in order to prevent the oscillator from being dc-grounded through the resistance of the inductor.

Voltage-tuned NE-602 oscillator circuits

Figure 5-18 shows a pair of VFO circuits in which the capacitor element of the tuned circuit is a voltage variable-capacitance diode, or varactor (D_1 in Figs. 5-18A and 5-18B). These diodes exhibit a junction capacitance that is a function of the reverse-bias potential applied across the diode. Thus, the oscillating frequency of these circuits is a function of tuning voltage V_t. The version shown in Fig. 5-18A is the parallel-resonant Colpitts oscillator, and that in Fig. 5-18B is the series-tuned Clapp oscillator.

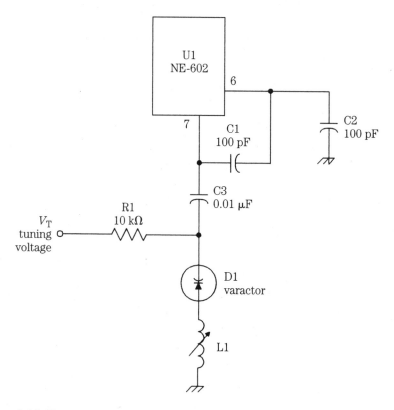

5-18 Varactor-tuned oscillator.

The NE-602 is a well-behaved RF chip that will function in a variety of applications from receivers, to converters, to oscillators, to signal generators.

Receiver circuits you can build

One of the joys of experimenting with electronic circuits is building a project that not only works well but is useful and provides a certain amount of either enjoyment or utility later on. Radio receivers fall into that category but are generally considered advanced projects that cannot be accomplished by the average experimenter. Many people erroneously believe that only kit-built receivers are candidate projects for them because of the alleged complexity of radio receivers. However, modern integrated circuit electronics make it a lot easier to design and build your own receiver today than ever before. This section shows a few of the types of circuits that make up a radio receiver and offers some suggestions to custom tailor your own radio in the VLF, AM broadcast, or shortwave bands.

Direct-conversion receivers, popular with homebrewers (also see Chapter 6 for detailed information on direct-conversion receivers), also use a frequency-translation process, but it is entirely different in its nature from the superhet principle. The straight superhet will not demodulate CW and SSB signals unless a product detector is used in place of the envelope detector that "decodes" AM. But the direct-conversion receiver is basically a product detector that operates at the RF frequency to demodulate that signal directly to audio. The local oscillator in a direct-conversion receiver is tuned to the same frequency as the RF signal coming into the antenna. The difference frequency, when the LO is adjusted correctly, is the modulating audio frequency in the case of SSB and a selected sidetone frequency for CW (i.e., the frequency of the tone that you hear in the output).

The output of the superhet converter stage in a direct-conversion receiver is audio, but it might contain some residual elements of the RF and LO signals or other mixer products than the difference frequency. In order to smooth this output signal, therefore, a low-pass filter that passes only the audio signals is typically used downstream from the converter.

The direct-conversion receiver is not covered in detail here, but you can infer the design principles from the superhet paragraphs that follow. To cover the circuits that you can use to make your own home-designed receiver, let's start at the front end and work toward the detector and audio amplifier stages.

Front-end circuits

The front end of the radio receiver consists of the RF amplifier (if used) and the converter or mixer/LO stages. The basis for the designs are the Signetics NE-602 balanced-mixer integrated circuit. This device has limited dynamic range but is sufficient because it compensates with a better-than-average noise figure and sufficient conversion gain to make it possible to not use an RF amplifier in most projects.

The pinouts for the NE-602 IC converter are shown in Fig. 5-19. The two inputs together form a balanced pair, along with two outputs. In some cases, both outputs are used, but in others only a single output (pin 4) is used. Either an internal or external oscillator can be used with the NE-602, and both versions are presented here. In the selected circuits, the supply voltage for the NE-602 will be +5 Vdc to +9 Vdc.

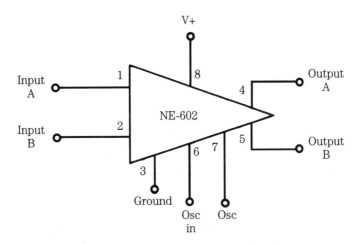

5-19 Another view of the NE-602 integrated circuit.

Figure 5-20 shows a simple superheterodyne converter circuit based on the NE-602 IC. The input circuit is broadbanded by using an RF transformer built on a toroid powdered-iron form. The turns ratio L_2/L_1 is typically 10:1 to 12:1; i.e., there are 10 to 12 turns on L_2 for every turn on L_1. Experiments and published data show good starting numbers are 20 to 24 turns on L_2, with 2 to 3 turns in L_1 for frequencies in the upper shortwave region. As the frequency is decreased, the number of turns is increased to about 34 to 40 turns on L_2 at the AM broadcast band.

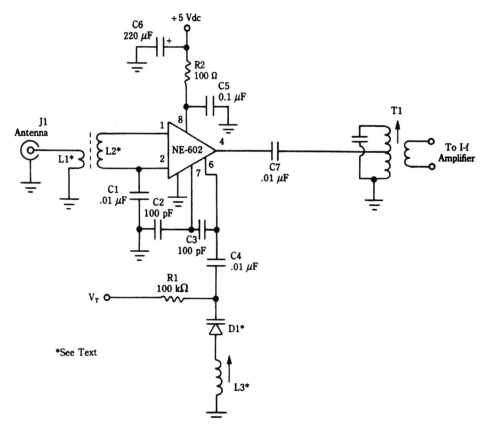

5-20 NE-602 receiver circuit.

The capacitors and the 100-Ω resistor in the $V+$ circuit (connected to pin 8 of the NE-602) are used for isolation and decoupling. These components prevent RF in the NE-602 circuit from traveling to other stages in the radio via the dc power line or, alternatively, signals from other stages from modulating the converter stage (or possibly causing oscillations).

The oscillator circuit in Fig. 5-20 consists of the components attached to pins 6 and 7 of the NE-602 IC. The actual tuning is set by a variable-capacitance diode (varactor), D_1. This diode is essentially a voltage-tuned capacitor. The oscillating frequency should be set 455 kHz above the desired RF frequency using Eq. (5-1). If you know (or have on hand) either the capacitor or the inductor:

$$L = \frac{1}{39.5 \, F^2 C} \tag{5-5}$$

or

$$C = \frac{1}{39.5 \, F^2 L}. \tag{5-6}$$

In all three expressions, the frequency is in hertz, inductance in henrys, and the capacitance in farads (don't forget to convert units from microhenrys, megahertz or kilohertz, and picofarads).

Example
Find the value of capacitance needed to resonate a 3.4-µH inductor to 9 MHz. Note: 9 MHz = 9,000,000 Hz and 3.4 µH = 0.0000034 H.

Solution

$$C = \frac{1}{(39.5)(9,000,000 \text{ Hz})^2 (0.0000034 \text{ H})}$$

$$C = \frac{1}{1.088 \times 10^{10}}$$

$$C = 9.19 \times 10^{-11} \text{ farads} = 91.9 \text{ pF}$$

When making calculations, allow about 10 pF for stray-circuit capacitances. Also, the tuning circuit can consist of the diode plus a parallel capacitance. I've used a 22-pF varactor, 27-pF disk ceramic, and a 6- to 70-pF trimmer capacitor to tune HF receiver projects. Standard tuning diodes are available from replacement-parts vendors, such as NTE. Part numbers of use in the HF region are the NTE-613 (22 pF) and NTE-614 (33 pF). For a wider tuning range, select the NTE-618 (440 pF).

The output of the NE-602 must be tuned to the IF frequency, i.e., the difference between the received RF signal and the local oscillator signal, by a tuned IF transformer (T1 in Fig. 5-20). For most receiver projects, this difference frequency should be 455-kHz because of the easy availability of the coils. Several sources offer these coils, but perhaps the easiest to obtain are the Bell/J. W. Miller (19070 Reyes Avenue, Rancho Dominguez, CA 90224) and the Toko coils marketed by Digi-Key (P.O. Box 677, Thief River Falls, MN 56701-0677; 1-800-344-4539).

IF amplifier circuits

The IF amplifier provides most of the gain in a superhet receiver, and it also supplies the narrowest bandwidth in the signal chain. Thus, the IF amplifier is chiefly responsible for both the sensitivity and selectivity of the radio receiver. In fancy radio receivers, the IF amplifier can be quite complex, as witnessed by the designs used in modern shortwave and ham radio receivers in the kilobuck range. This one is a considerably simpler design, but is nonetheless effective. For those who need a more complex design, I recommend various editions of Bill Orr's *The Radio Handbook* or *The ARRL Radio Amateur's Handbook.*

Figure 5-21 shows the basic circuit for an IF amplifier building block. For receivers requiring moderate gain, for example, an AM receiver intended primarily for local reception, then only one stage needs to be used. In shortwave receivers, or wherever superior sensitivity is sought, it is permissible to cascade two or three such stages to boost amplification. The gain of this stage is on the order of 50 dB, and some projects might need as much as 80 to 110 dB.

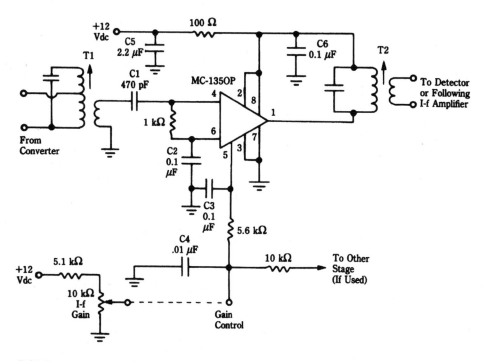

5-21 Integrated circuit IF amplifier.

The basis for the IF amplifier building block is the MC-1350P gain block chip (this chip is also available as NTE replacement part number NTE-746).

The input circuit is tuned by either a single- or double-tuned transformer of the same sort used at the output of the NE-602 stage covered previously. The secondary impedance of this transformer should be on the order of 1000 Ω in order to properly match the input impedance of the MC-1350P. The output of the MC-1350P is also tuned by a similar IF transformer.

The gain in the MC-1350P device is controlled by the voltage applied to pin 5. This voltage must be varied between +3 and +9 Vdc in order to set the overall gain of the stage. The gain-control voltage supplied to pin 5 can be derived from a manual gain control (i.e., potentiometer voltage divider), automatic gain control (AGC) circuit, or a combination of both.

Figure 5-21 shows a sample manual gain control circuit based on a 10-kΩ potentiometer and a 5.1-kΩ fixed resistor. If a fixed gain is desired, then a fixed-ratio voltage-divider chain can be used instead. Also, it is not strictly necessary to use the same gain control for all stages in the combined IF amplifier. You can, for example, make one stage a variable-gain circuit and the other(s) fixed-gain.

Detector and audio circuits

Once the IF amplifier builds the signal up to a point where it can be successfully demodulated with good volume, you will want a detector circuit to recover the modulating audio. The simplest AM detector is the envelope detector shown in Fig. 5-22. The envelope detector is a single-diode rectifier at the output of the last IF transformer. The diode should be a germanium type, such as 1N34, 1N60, 1N270, ECG-109, or NTE-109 devices. A volume-control potentiometer serves as the load for the diode, and a 0.01-μF capacitor filters out a residual IF signal that remains in the circuit following demodulation.

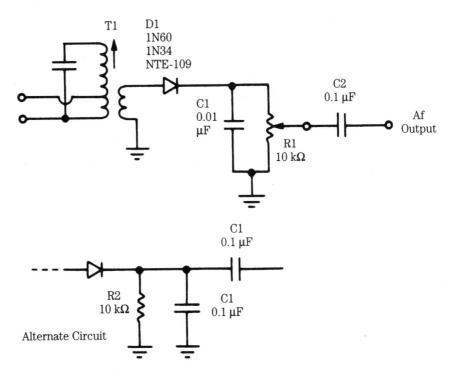

5-22 AM envelope detector circuit.

The detector stage must be capacitor-coupled to protect the following stage because the demodulation process produces a small dc offset. The capacitor strips off the dc and makes the signal once again ac. In some cases, the audio stage following the detector contains the volume control, so an alternate circuit shown in Fig. 5-22 is used at the detector.

The basis for the audio stage is the easily obtained LM-386 IC audio subsystem chip (Fig. 5-23A). This chip is easily obtained in replacement semiconductor lines, from RadioShack and from a number of mail-order sources. Although the low power level leaves something to be desired, the LM-386 is both better behaved and more easily obtained than certain higher-powered audio subsystem chips.

Figure 5-23B shows the basic circuit for the LM-386 chip when it is used as an audio stage for a receiver project. Notice that the circuit is extremely simple in that it basically has just input, output, ground, and *V+* connections. The circuit is able to

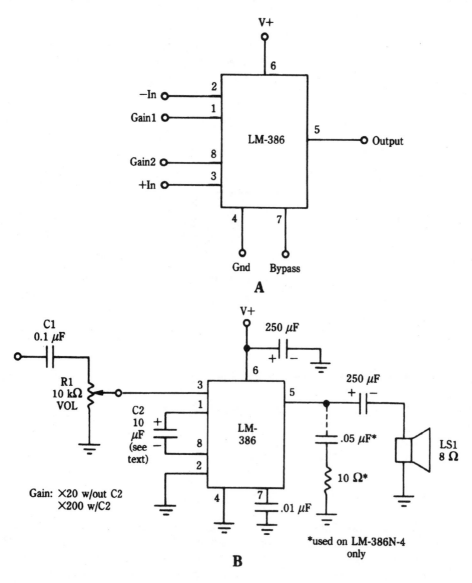

5-23 (A) LM-386 audio amplifier integrated circuit; (B) LM-386 standard circuit.

provide two levels of gain. If capacitor C_2 is used, then the gain is $\times 200$; if the capacitor is deleted, the gain is $\times 20$. Normally, the gain will be set to $\times 20$ in receiver projects unless the design uses little gain ahead of the detector (which might be the case in the simplest superhets or in direct-conversion designs).

The LM-386 is relatively well-behaved in projects, which means that it takes only ordinary care in layout to achieve good results. It is possible, however, that because of the high gain, this IC will oscillate mercilessly if proper layout is not followed. Be sure to keep the inputs and outputs physically separated and don't allow the ground connections to wind all over the wiring board. Also, make sure that the $V+$ bypass capacitor is used . . . and has sufficient value (as shown).

Direct conversion

Figure 5-24 shows the NE-602 chip used as a direct-conversion receiver. In this type of circuit, the local oscillator (LO) operates on the same frequency as the received RF signal. As a result, when the two are zero beat, the difference IF frequency is the audio modulating the incoming carrier. In CW reception, the LO is tuned to a frequency a few hundred hertz from the incoming RF. In that case, the difference frequency will be a beat note that is interpreted as a CW signal.

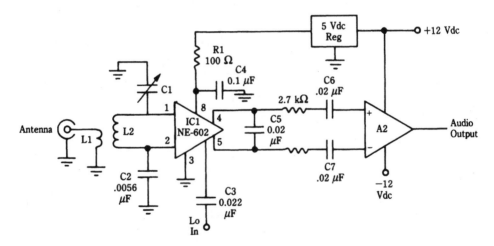

5-24 Simple direct conversion receiver using the NE-602 "single IC radio."

This section was not intended to design a radio receiver for you but rather to empower you to design your own radio receivers in the VLF, AM, and shortwave bands. The circuits are well-behaved and are easily built using readily available components.

Mechanical-filter IF amplifier

Selectivity is one of the "holy grails" of receiver designers and in many instances is more important than sensitivity or certain other specifications. The selectivity of the receiver is a measure of its ability to reject unwanted signals. The selectivity of a receiver is usually specified as the bandwidth between the points on the bandpass characteristic, where the gain falls off -6 dB (voltage) from the midband response. Also important is the skirt factor, which is a measure of the shape of the IF bandpass curve. The skirt factor is the ratio of the bandwidth of the passband at -60 dB down to the bandwidth at -6 dB down from the midband response.

The IF amplifier provides most of the selectivity characteristic of the modern superhet receiver. But there is only so much that can be done with simple LC circuits. If you want to achieve a very narrow selectivity characteristic, then it is necessary to use either a crystal filter or a mechanical filter. The mechanical filter provides the best tradeoff between selectivity and skirt factor. Unfortunately, new Collins mechanical filters cost a great deal of money. Surplus dealers, however, offer former military mechanical filters for a much lower cost. Fair Radio Sales of Lima, OH sells the filters from receivers, such as the R-390, R-392, and so on, at a reasonable price. These filters will have an IF frequency of 260, 455, or 500 kHz depending on the model. There are also several bandwidths available: 500 Hz and 1.8, 2.9, 4, 8, and 16 kHz.

Figure 5-25A shows the internal works of a typical mechanical filter, and Fig. 5-25B shows a comparison between mechanical filter bandwidth and typical LC-tuned circuit bandwidth. At each end are electromechanical transducers that consist of a bias magnet, coil, and drive(n) wire. These transducers act much like earphones or loudspeakers, but at IF rather than AF frequencies. The narrow bandpass is achieved with resonant metal disks coupled together with wire arms. These disks are cut to resonate at the center IF frequency.

The goal in constructing this project was to make a high-gain, 455- or 500-kHz IF amplifier with good selectivity characteristics to permit me to build a "super" AM radio receiver (to do some broadcast band DXing). This goal required an 8-kHz bandwidth, 455-kHz mechanical filter. Figure 5-26A shows the completed IF amplifier strip, and Figs. 5-26B (input side) and 5-26C (output side) show the actual internal circuit. The mechanical filter is the silver cylindrical object mounted to an L-bracket in Fig. 5-26B.

The input amplifier shown in Fig. 5-26B is shown as a circuit diagram in Fig. 5-26D. The amplifier is a grounded-gate junction field effect transistor (JFET). The popular MPF-102 is used (same as NTE-312). Input signal is coupled through an altered 455-kHz IF transformer. The transformer is modified by crushing the resonating capacitor, which is accessible from the bottom outside of the shielded case of the transformer. The "secondary" of the transformer is used as the primary in this case. The drain terminal of the JFET is connected to $V+$ through a 1000-μH (1 mH) RF choke (RFC1) and a resistor-capacitor decoupling network. The output of the amplifier is coupled to the input of the mechanical filter (FL$_1$) through a 0.1-μF dc-blocking capacitor (C_4). The 150-pF ceramic or mica trimmer capacitor (C_5) across the input of FL$_1$ is not necessary but is highly recommended. It will help adjust the flatness of the passband and peak the center frequency gain.

The output amplifier is shown in Fig. 5-26E. The gain is provided by a pair of MC-1350P gain-block IC devices. Interface to the output side of the mechanical filter is provided by another JFET (also an MPF-102), driven from a capacitive voltage divider (C_6/C_7).

Signal from the drain of Q_2 is passed to a relay circuit (K_1) that selects either a

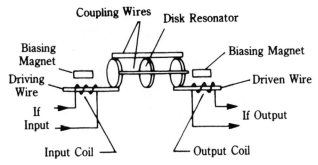

A. Components of a mechanical filter

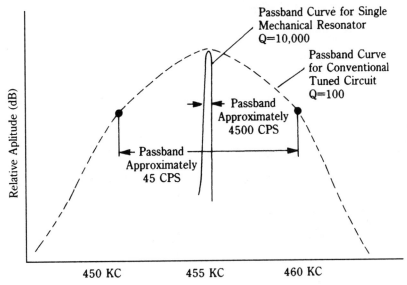

B. Typical passband curves of conventional tuned circuit and a single mechanical resonator

5-25 Mechanical filter: (A) construction and (B) passband.

U.S. Army Signal Corp repair manual for R-390 receiver.

shorted connection "straight through" or inserts a −20-dB attenuator in the signal path. This attenuator allows the receiver to operate with a high dynamic range in the presence of both large and weak signals (a necessity in an AM DX receiver!). You can either make a π-pad attenuator or use one of the ready-made varieties from Minicircuits Laboratories.

The attenuator relay (K_1) is a +5-V "TTL compatible" type in a DIP package. It is operated by applying a +5-V signal to a feedthrough capacitor (C_{21}) on the IF amplifier's metal case (Fig. 5-26A shows the several feedthrough capacitors used in this project). When the ATTN input is LOW (i.e., grounded, or <0.8 V), then the relay (K_1) is inoperative and the signal is passed through the shorted terminals. But

A

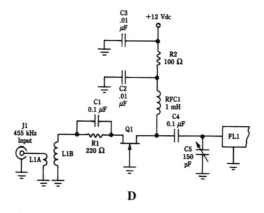

B   **C**

D

5-26 IF amplifier project; (A) completed project: (B) input end;
(C) output end; (D) input amplifier stage; and (E) output amplifier
circuit.

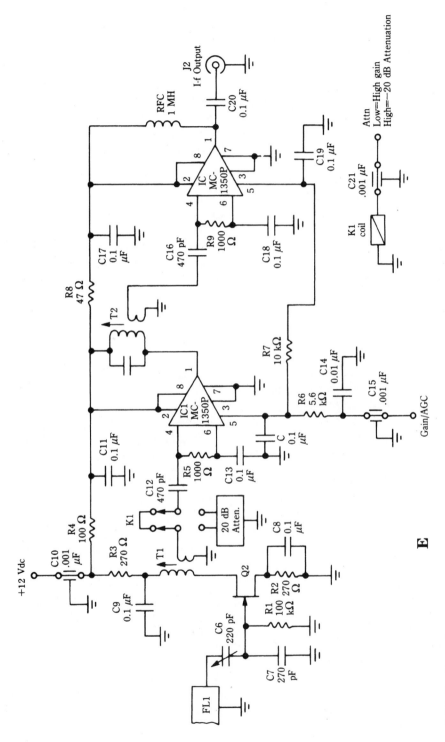

5-26 *Continued*

when the ATTN is high (i.e., 2.4 V < V < 5.2 V), then K_1 is energized and inserts the −20-dB attenuator in the path.

The IF amplifier of Figs. 5-25 and 5-26 provides about 80 dB of gain at 455 kHz.

Shortwave receiver project

This section shows a simple radio receiver kit that will provide startlingly good performance for a very low cost. It is not the equal of a $1000 communications receiver, but it provides reasonably good performance in the HF shortwave band for only a few dollars. Perhaps most important is the fact that it serves as a good building block to add on other circuits.

Construction projects have long been part of the radio hobby; today, it is easier than ever before. A number of companies make small kits that are readily built in an evening or over a weekend. But few of these "one-nighters" are so grandiose as to be called receivers. However, Ramsey Electronics [793 Canning Parkway, Victor, NY, 14564; (716)-924-4560] makes several kits for hobbyists in addition to their line of frequency counters (I use a Ramsey 600-MHz counter on my workbench). The cost of the receiver with a black plastic cabinet is in the $40 region, less if all you want is the "innards" (supply your own cabinet). Contact the company for current prices.

This receiver is battery-operated and covers a selected segment of approximately 2.5 MHz somewhere in the range of 4 to 10.5 MHz. The tuning limitation is based on the range of the tuning capacitor diode and the overall limits on the slug-tuned coil's inductance range. As it stands, the receiver will cover 40-m amateur band; the 8-MHz utilities, marine, and aviation band; and the 9-MHz international shortwave broadcast band.

The SR-1 shortwave receiver (Fig. 5-27) is based on the Signetics NE-602 RF converter chip. This chip provides the local oscillator and a double-balanced mixer in a single eight-pin mini-DIP package. In the case of the Ramsey SR-1, the input is untuned. Antenna signal is coupled to the broadband input transformer through a 5-kΩ "RF gain" control that actually functions as an attenuator [note: direct antenna circuit RF attenuation is an antique method that Ramsey resurrected].

The local oscillator circuit is inside the NE-602 chip but requires external tuning components. In the case of the SR-1, Ramsey elected to use a series-tuned variable-frequency oscillator. The inductor in the LC tuning network consists of an IF transformer with the internal capacitor disconnected. Only one winding of the transformer is used, but you might suspect that the other winding would make a dandy alternative output for the oscillator in case you wished to drive a digital frequency-counter display. The capacitor in the LC network is CR1, a varactor diode. That is, a voltage variable-capacitance diode. The varactor produces a capacitance that is proportional to the applied voltage. A potentiometer from the +9-Vdc battery supply is used to provide the tuning voltage.

The NE-602 device is designed to operate from potentials in the +4.5- to +6-Vdc range, so the 1-kΩ resistor in series with the $V+$ terminal (pin 8) is absolutely required. The dc power supply is a 9-Vdc "transistor radio" battery mounted directly to the printed circuit board (Fig. 5-28). If you want to make this radio receiver into an ac-operated model, or a mobile model, then it might be necessary to build an

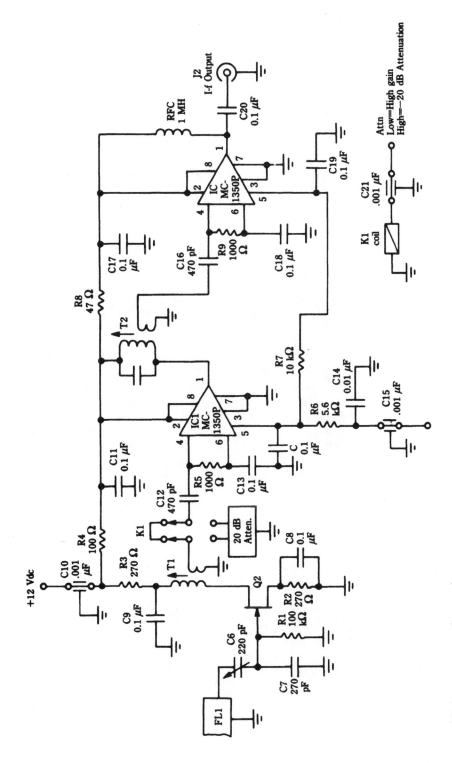

5-27 Shortwave receiver project.

external power supply. A type 7809 three-terminal voltage regulator IC will provide 9-Vdc regulated output from a +12-V (or higher) dc power supply.

The use of a voltage-regulated power supply, as described, will also help one problem with this design. The tuning is a function of applied voltage, so it will necessarily change when the battery voltage drops.

The receiver in Figs. 5-27 and 5-28 is a superheterodyne model, meaning that it converts the input RF frequency to a lower intermediate frequency (IF). The IF frequency is the difference between the RF frequency and the local oscillator (LO) frequency; that is: IF = RF − LO. In this project, the IF frequency is 260 kHz, so you can use components from Japanese AM car radios. However, there is a built-in limitation because of the 260-kHz IF frequency. The image frequency is too close to the RF frequency, so expect a terrible image response above about 10 MHz or so. Although the chip will convert up to 200 MHz, the low IF frequency is not amenable to higher-frequency operation.

5-28 Simple receiver project printed circuit.

In the model that I received, the audio output stage is a single 2N3904 transistor driving a set of earphones (not included). But the Ramsey people tell me that the design will soon be changed. The new circuit is the same as the old, except for replacement of the AF output amplifier with a high-gain LM-386 IC audio stage (Fig. 5-15B). This design provides higher gain and higher audio output power—a complaint that Ramsey tells me some builders made.

You can either build the receiver as is or use it as the basis for a project of your own. For example, you can replace the 260-kHz IF transformer and the tuning-circuit components to make the circuit work at higher frequencies. You could also alter the tuning circuit to change the range. Adding some capacitance in parallel with the varactor will reduce the minimum frequency below 4 MHz. Winding new input and LO coils will also change the range. You can add a beat-frequency oscillator at the detector diode to make CW/SSB reception possible.

Alternatively, you can replace the IF tuning with a low-pass filter (that cuts off everything above about 3 kHz) and use this circuit as a direct-conversion receiver. In direct-conversion circuits, the LO operates on the RF frequency so the difference is the audio modulation. The direct-conversion technique works well on CW and SSB, but for AM it must be exactly zero-beat with the received carrier or a beat note will be heard.

6
CHAPTER

Direct-conversion radio receivers

The direct conversion, or synchrodyne, receiver was invented in the late 1920s, but only with the advent of modern semiconductor technology has it come into its own as a real possibility for good-performance receivers. Although most designs are intended for novices, and lack certain features of high-grade superheterodyne receivers, the modern direct-conversion receiver (DCR) is capable of very decent performance. A case can be made for the assertion that the modern DCR is capable of performing as good as many midgrade ham and SWL communications receivers. Although that assertion might seem very bold, experience bears it out. Although no one, least of all me, would represent the DCR as capable of the best possible performance, modern DCR designs are no longer in the hobbyist curiosity category. In this chapter, you will find the basic theory of operation and some of the actual designs tried on the workbench.

Basic theory of operation

The DCR is similar to the superheterodyne in underlying concept: The receiver radio frequency (RF) signal is translated in frequency by nonlinear mixing with a local oscillator (LO) signal ("heterodyning"). Figure 6-1 shows the basic block diagram for the "front end" of both types of receiver. The mixer is a nonlinear element that combines the two signals, F_{RF} and F_{LO}. The output of the mixer contains a number of different frequencies that obey the relationship:

$$F_o = mF_{RF} \pm nF_{LO}, \qquad (6\text{-}1)$$

where

F_o is the output frequency
F_{RF} is the frequency of the received radio signal
F_{LO} is the frequency produced by the local oscillator
All frequencies are in the same units
m and n are integers $(0, 1, 2, 3 \ldots)$.

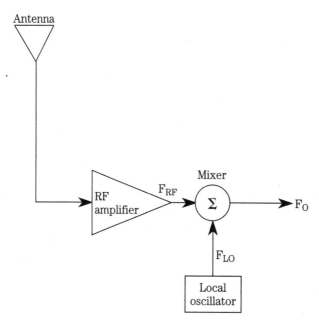

6-1 Basic mixer-local oscillator circuit.

All frequencies other than F_{RF} and F_{LO} are product frequencies. In general, we are only interested in the cases where m and n are either 0 or 1, so the output frequency spectrum of interest is limited to F_{RF} and F_{LO} plus the product frequencies $F_{RF} + F_{LO}$, and $F_{RF} - F_{LO}$. The latter two are called sum and difference intermediate frequencies (IF). Other products are certainly present, but for these purposes, they are regarded as negligible. They are not regarded as negligible in serious receiver design, however.

In a superheterodyne radio receiver, a tuned bandpass filter will select either the sum IF or the difference IF, while rejecting the other IF, the LO and RF signals. Most of the gain (which helps determine sensitivity) and the selectivity of the receiver are accomplished at the IF frequency. In older receivers, it was almost universally true that the difference IF frequency was selected (455 and 460 kHz being very common), but in modern communications receivers, either or both might be selected. For example, it is common to use a 9-MHz IF amplifier on high-frequency (HF) band shortwave receivers. On bands below 9 MHz, the sum IF is selected; on bands above 9 MHz, the difference IF is selected. A popular combination on amateur radio receivers uses a 9 MHz IF combined with a 5- to 5.5-MHz variable-frequency oscillator. To receive the 75- to 80-m band (3.5 to 4.0 MHz), the sum IF is used. The same combination of LO and IF frequencies will also receive the 20-m (14.0 to 14.4 MHz) band if the difference IF (i.e., $14.0 - 5 = 9$ MHz) is used.

In a DCR, on the other hand, only the difference IF frequency is used (see Fig. 6-2). Because the DCR LO operates at the same frequency as the RF carrier, or on a nearby frequency in the case of CW and SSB reception, the difference frequency represents the audio modulation of the radio signal. Amplitude-modulated (AM) sig-

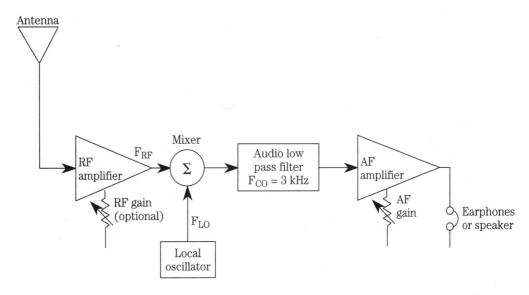

6-2 Partial block diagram for a direct-conversion receiver.

nals are accommodated by zero-beating the LO to the radio signal, making $F_{LO} = F_{RF}$(carrier). Thus, only the recovered upper and lower sidebands will pass through the system, and they are at the audio frequency.

For CW signals (Morse code on/off telegraphy) and single-sideband (SSB) signals, it is necessary to offset the LO frequency slightly to recover the signal. For the CW case, you must select a comfortable tone (which is an individual preference). In my own case, I am most comfortable using an 800-Hz (0.8 kHz) note when copying CW, so it will offset the LO from the RF by 800 Hz. For example, when copying a CW signal at, for example, 3650 kHz, the LO will be tuned to either 3649.2 or 3650.8 kHz. In either case, the beat note heard in the output is 800 Hz. Single-sideband reception requires an offset on the order of 1.8 to 2.8 kHz for proper reception.

As was true in the superheterodyne receiver, the majority of the gain and selectivity in the DCR is provided by the stages after the first mixer. Although the superheterodyne uses the IF amplifier chain for this purpose, followed by second detection and audio amplification, the DCR must use only the audio amplifier chain. Thus, it becomes necessary to provide some very high-gain audio amplifiers and audio bandpass filtering in the DCR design.

One implication of DCR operation is the lack of single-signal operation. Both CW and SSB signals will appear on both sides of the zero-beat point ($F_{RF} = F_{LO}$ exactly). Although this feature can be a problem, it has at least one charming attribute on SSB reception: the DCR will receive LSB signals on one side of zero-beat and USB signals on the other side of zero-beat. There have been attempts to provide single-signal reception of SSB signals on DCRs by using audio and VFO phasing circuits (in the manner of the phasing method of SSB generation). That approach greatly increases the complexity of the receiver, which might make other design approaches more reasonable than DCRs.

The most basic implementation of the DCR (Fig. 6-2) requires only a mixer stage, a local oscillator, and an audio amplifier. If the mixer has a high-enough output signal level, and high-impedance earphones are used to detect the audio, some designs can make do without the audio amplifier. These are, however, a rarity, and the one version that I tried did not work very well.

In some DCR designs, there will be optional RF input signal conditioning consisting of a low-pass filter, high-pass filter, or bandpass filter (as appropriate) to select the desired signal or reject undesired signals. Without some frequency selection at the front end, the mixer is wide open with respect to frequency and may be unable to prevent some unwanted signal, or spurious combinations of signals, from entering the receiver circuits. Some designs will include more than one style of filter. For example, a popular combination uses a single-staged tuned-resonant circuit at the input of the mixer to select the RF signal to be received and a high-pass filter with cut-off frequency F_{CO} of 2200 kHz that is designed to exclude AM broadcast band signals. The reason for such an arrangement is that the AM signal might be quite intense, usually being of local origin, and is therefore capable of overriding the minor selectivity provided by the tuned circuit.

The RF amplifier used in the front end is also optional and is used to provide extra gain and possibly some selectivity. The gain is needed to overcome losses or inherent insensitivity in the mixer design. Not all mixers require the RF amplifier, so it is frequently deleted in published designs. In general, RF amplifiers are used only in DCRs operating above 14 MHz. Below 14 MHz, signals tend to be relatively strong and human-made noise tends to be much stronger than inherent mixer noise.[1]

Problems associated with DCR designs

Over the years, there have been a number of articles in the popular technical press on DCR radio receiver designs, some of which are cited in this book. In addition, the major amateur radio communications handbooks typically discuss DCR designs. Several problems are cited as being common in the DCR receiver designs. While these problems are very real, careful design and construction can render them harmless. The list of observed problems includes hum, microphonics, poor dynamic range, low output power levels (which makes for uncomfortable listening), and unwanted detection of AM broadcast band signals ("AM breakthrough").[2]

Hum

Hum is caused by the alternating current (ac) power lines and may be radiated into the DCR through the antenna circuit, radiated into the wiring of the set, caused by ground loops, or communicated to the DCR circuits as ripple in the dc power supply. In the first two instances, the hum will have a frequency of 50 Hz (Europe) or 60 Hz (North America), which are the ac power-line frequencies; in the latter two cases, it will be 100 Hz (Europe) or 120 Hz (North America) if full-wave rectified dc

1 Ibid. (*ARRL Handbook,* pp. 6–8).

2 Rick Campbell, KK7B, "High Performance Direct-Conversion Receivers," *QST,* August 1992, pp. 19–28.

power supplies are used. All hum problems are aggravated by the high-gain audio amplifiers used in DCRs; for this reason, bandpass filtering that attenuates signals below 300 Hz[3] is highly recommended.

Hum signals received through the antenna circuits are best handled by use of high-pass RF filtering that does not admit energy in the 50/60 Hz region. The front end of many DCRs is wide open with respect to frequency, so improper choices can leave the receiver sensitive to hum. Some mixer circuits, such as the double balanced mixer (DBM) diode designs, are inherently insensitive to hum because of the nature of the inductors (and their cores) used in the associated inductors. If the hum is radiated into the circuit from the environment, then shielding the circuit against electrical fields will usually solve the problem.

Finally there is the case where hum is received via the dc power supply and is ripple-related (probably the most common cause). "Ripple" is the residual rectified ac (or "pulsating dc") riding on the output voltage of a dc power supply after all filtering and voltage regulation is done. The worst hum occurs when the dc power-supply ripple modulates the LO signal. In that case, hum becomes worse as frequency increases.[4] A combination of good voltage regulation (which tends to limit ripple as if it were a very large filter capacitor) and proper grounding techniques will solve the problem.

Figure 6-3 shows one grounding sin that must be avoided at all costs. In the properly grounded receiver, all of the grounds will be connected to point "B" ("single-point" or "star grounded"). In actual practice, however, the nature of printed circuit or perforated wire board point-to-point wiring is such that the "star" ground concept is not achievable in its fullest sense, so there might be one or two additional points of grounding. Care must be taken to prevent ground loops in such cases.

The great sin in Fig. 6-3 is grounding the dc power supply to the antenna input ground point while the amplifier is grounded elsewhere. Even with very low resistance ground tracks, a considerable signal is created when the dc current required by the DCR flows from "A" to "B" and then into the receiver. For example, if the receiver draws 50 mA, and the track has a dc resistance of 0.05 Ω (neither number is unreasonable), a voltage drop of 2.5 mV will exist. Although this voltage seems small, it is not so small when followed by the 80 to 120 dB of gain typically found in a radio receiver. Under these conditions, even the smallest ripple waveform riding on the dc power-supply voltage can cause massive hum in the receiver's output!

A solution to the hum problem is shown in Fig. 6-4. Single-point grounding is used in order to reduce ground loops. In addition, a toroidal decoupling choke is placed in the dc power supply leads.[5] This choke consists of two bifilar windings of 20 turns each over a toroidal core that has a permeability (μ) between 600 and 1500. The wire used to make the bifilar windings must be sufficiently large to accommodate the current requirements of the DCR circuitry.

---

3 300 Hz is considered the low end -3 dB point for speech waveforms; the upper -3 dB point is usually specified as 3000 Hz.

4 Ibid. (*ARRL Handbook*).

5 *ARRL Handbook* (*op. cit.*).

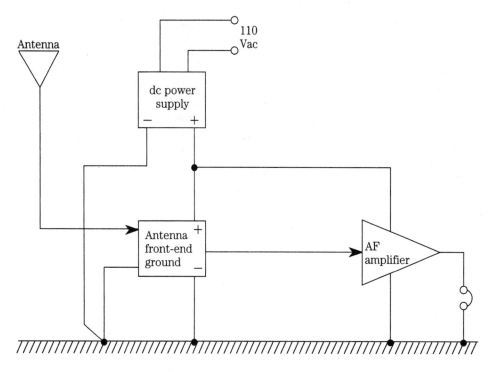

6-3 Improper wiring of stages in direct-conversion receiver.

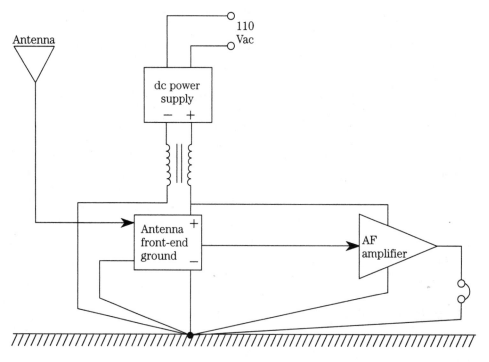

6-4 Proper "star" wiring of direct-conversion receiver stages.

Of course, using batteries for the dc power supply will solve the problem of ripple-induced hum. Even so, good grounding technique is a fundamental requirement for building a good receiver.

Microphonics

Microphonics are a highly sensitive, damped, "echolike" ringing sound in the output when the receiver is vibrated or jarred and continues until the vibration dies out. It is caused by modulation of the signal by the vibration. Microphonics are generated because of the 80- to 120-dB gain associated with the typical DCR. Any modulation of the signal, therefore, produces change in the output signal. Two principal sources of microphonics are movement of the frequency-setting components (mostly the inductor) in the LO and changes in coupling between circuit elements (including wires). Ordinarily, the latter would not occur; DCRs usually have a tremendous amount of gain, so such possibilities exist.

Dynamic range problems

Dynamic range is basically the difference between the maximum and minimum signals that the receiver can accommodate and is usually expressed in decibels (dB). A description of dynamic range found in Hayward and DeMaw describe it as:[6]

$$Dynamic\ Range\ = \frac{2(P_i - M_{DS})}{3}, \tag{6-2}$$

where

P_i is the signal level associated with the third-order intercept point of the receiver

M_{DS} is the minimum discernible signal.

Dynamic range is important in all bands, but there are cases where it is most important. An example is when receiving any frequency when the receiver is in close proximity to an AM broadcast station, a currently transmitting ham operator, or other active transmitter site, regardless of the frequency being received. This problem becomes especially acute when the front-end of the receiver is wideband; therefore, it provides little or no discrimination against the unwanted frequencies. However, also recognize that front-end tuning does not necessarily completely eliminate unwanted out-of-band signals that are very strong—especially if only one tuned ciruit is provided.

Although normally associated with very large signals, another case where dynamic range improves signals is on very crowded bands, a large number of weak, moderate, and strong signals are mixed together. One source asserts that the dynamic range of signals received on a 40-m (7 to 7.3 MHz) half-wavelength dipole is greater than the dynamic range of an audio compact disk.[7] Intermodulation products and other spurious signals on these bands are often stronger than legitimate signals. Although none of the myriad signals present would tax the receiver's capabilities by

---

6 *Solid-State Design for the Radio Amateur,* American Radio Relay League (Newington, CT, USA, 1977).

7 Ibid. (Campbell). Given the mixture of powerful international broadcasters and ham operators (both kilowatt and "QRP" stations), that claim is probably close to the mark.

itself, the net result of all of them being simultaneously present is poor performance. Being able to discriminate weak signals from strong signals, especially when using a wideband front end, is a benefit of having a high dynamic-range receiver. When the dynamic range of the receiver is insufficient for the task at hand, the output will be distorted and possibly very weak if desensitization occurs.

Several factors contribute to the dynamic range of the DCR, but the primary focus should be on the characteristics of the mixer circuit. Campbell provides three criteria for achieving a good dynamic range in the DCR.[8]

Any RF device (such as a mixer) should be terminated in its output impedance for maximum signal transfer, minimum reflections, and proper operation. Indeed, many circuits will not perform as advertised, unless they are properly terminated. Resistors in the signal path are to be avoided, especially prior to the high-gain audio section because resistors generate noise of their own. Even a "perfect" resistor will generate thermal noise (4KTBR), and practical resistors add some additional types of noise to the mix.

Restriction of the system bandpass prior to audio amplification is done because it reduces the unwanted noise components of the signal prior to processing in the high gain of the audio stages. The signal-to-noise ratio (SNR) is thereby improved. This same trick is used by some authorities to justify using an "antenna-to-mixer" approach on expensive shortwave superheterodyne receivers. In those designs, the main bandpass filtering is at the output of the mixer, and the mixer input is driven directly from the bandpass filters in the front-end of the receiver (no RF amplifier). This philosophy was used in the Squires-Sanders HF vacuum-tube receivers made in the United States some years ago. Although it was a novel approach at the time, it has become one of several possible standard approaches today.

Many designers of DCRs use a filter between the output of the mixer and the input of the preamplifiers. If the filter is matched to the impedance of the mixer and the input of the preamplifier, and if it contains no resistors in series with the signal path, then it largely meets Campbell's criteria. The diplexer filter (of which, more later) largely meets these criteria.

AM breakthrough

Stations in the AM broadcast band (540 to 1700 kHz), and in many regions of the world VLF broadcast band also, tend to be both relatively high-powered and very local to the receiver (thus very strong). As a result, many users of radio receivers for other bands are afflicted with the phenomena of *AM breakthrough*. This term refers to any of several phenomena, but the end result is AM interference to reception. Sometimes the problem is caused by simple overload of the front end of the receiver, causing desensitization. In other cases, there will be intermodulation problems, and in still others severe distortion of the desired signals. In a great many cases the AM BCB signal will be demodulated and interfere with the audio from the desired station. AM breakthrough is often accompanied by howling and screeching sounds from the

---

8 Ibid. (Campbell): (1) proper termination of the mixer over a wide band; (2) avoidance of resistors in the signal path; and (3) restriction of the system bandpass prior to the audio preamplifier (which also helps hum rejection).

receiver, although at other times, the effects are quite subtle. A frequently seen form of this problem is a strong background "din" caused by demodulation of the AM BCB signal in nonlinear elements in the circuit, particularly PN junctions.[9] The effects are usually more severe in simple DCR designs because the front end might be wideband.

Several different approaches are used to overcome AM breakthrough. First, the DCR should be well-shielded so that there is no "antenna effect" from circuit wiring. Even so, coaxial cable should be used from the DCR printed circuit board to the antenna connector. In addition to shielding, which by itself does not affect signals riding in on the antenna, there should be at least some filtering of the input signal in a LC network. The network can be either a bandpass filter that passes an entire segment of the spectrum but restricts the rest or is tuned to a single frequency.

Another approach, which can be used in conjunction with either or both of these, is to place a high-pass filter that rejects the AM broadcast band in series with the signal path. Figure 6-5 shows two such filters that can be made with easily available components. Both of these filters are designed to be installed right at the antenna connector, either inside or outside of the receiver cabinet.

The AM rejection filter of Fig. 6-5A can be built with ordinary disk ceramic, silver mica, or Panasonic V-series mylar capacitors. Good performance is achievable with 5% tolerance units, although it is also a good practice to match them using a digital capacitance meter or bridge. If silver mica units are selected, then it is possible to select 1000-pF (0.001 μF) capacitors, using two in parallel for the 0.002-μF capacitors (C_1 and C_3). The inductors of this circuit are each 3.3 μH and can be either shielded inductors with regular cores or unshielded with toroidal cores. One combination that I've used is the Amidon Associates type T-50-2 (RED) toroidal core wound with 29 turns of no. 26 enameled wire.

The filter in Fig. 6-5B is a little more complex, but it offers as much as 40 dB of AM suppression in the HF bands. It is a high-pass filter with a cutoff near 2200 kHz. Notice that the coils have different inductance values and are each in series with a capacitor. These coils are wound on Amidon T-37-2 toroidal forms.

Both of these filter circuits should be built inside of a well-shielded enclosure. Most well-made aluminum sheet metal or die-cast project boxes will do nicely. However, be wary of the sort of sheet metal box that has no overlapping edges at the mating surfaces of the two halves of the box. A well-made sheet metal box will have at least 5 mm of overlapping flange built onto either the top or bottom half of the box. These boxes are a bit more "RF tight" when joined together. Even so, additional screws might be necessary.

When laying out the filter on a perfboard or printed circuit board, be sure to use good construction practices. That is, lay out the components in a line from one end to the other, without excessive space between them and without doubling back so that the output is close to the input.

Low audio output

One of the frequent complaints about DCRs is that the audio level is too low for comfortable listening. The mixer output level is very low, so a DCR typically requires

---

9 *QST,* August 1980, pp. 14–19.

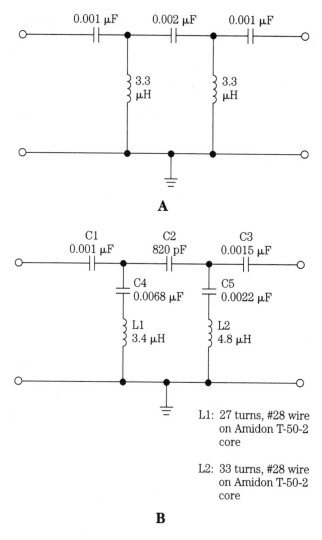

6-5 Two filters for eliminating AM broadcast band interference.

a large amount of gain in the audio amplifier chain to produce even minimal levels of power to earphones or loudspeakers. Additional gain and a reasonable power amplifier can be provided, but only at the risk of exacerbating all of those problems that come along with high gain in the first place. Attention to good layout, design, and grounding practices can greatly ease some of these problems without undue burden.

The use of proper filtering of the signal prior to the audio preamplifier will greatly enhance the enjoyment of listening to a DCR and reduce some of the problems associated with low power levels. The goal is to structure the bandpass of the receiver to that which is needed to pass the information content of the modulation but not the noise outside of that passband.

Figure 6-6 shows the block diagram for a simple, but reasonable, DCR that is based on the principles developed in this chapter. Later in this chapter is a closer, more detailed, look at some specific circuits to fill the blocks, but for now the block diagram description will suffice. Notice that this description is functional, rather than stage-by-stage, so in most designs some of the blocks will be combined into a single stage.

The "front end" of the DCR of Fig. 6-6 consists of an AM-suppression high-pass filter, followed by either a tuned filter or bandpass filter at the frequency of choice. An optional PIN diode attenuator circuit provides a degree of manual gain control for RF signals. The local oscillator is a variable-frequency oscillator (VFO) that tunes the entire band of interest. It must be relatively well designed and be free of drift and noise problems. The LO/VFO signal is mixed with the RF path signals from the chain of filters at the mixer circuit to produce a downconverted signal that contains the modulation that you are attempting to recover.

The function following the mixer provides bandpass filtering to limit the effects of out-of-band signals, noise, and other artifacts while also providing the mixer with appropriate impedance termination at all frequencies (in practice, the mixer will be well terminated above and below the audio "base band" of interest, typically 300 to

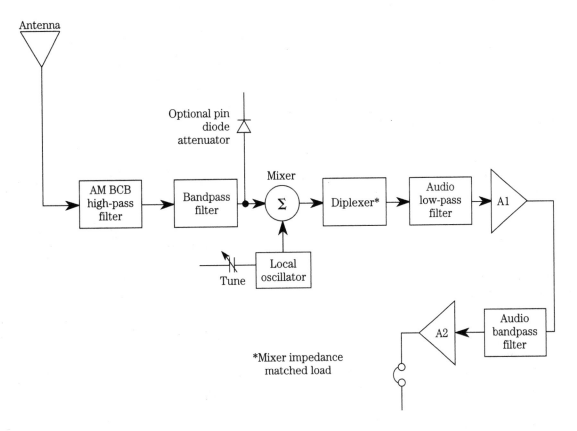

6-6 Complete block diagram for a direct-conversion receiver.

3000 Hz, but is not well terminated in the base band). Amplifiers A_1 and A_2 provide the bulk of the necessary audio frequency gain, and a filter between them keeps the bandwidth correct for good noise suppression. In practice, amplifiers A_1 and A_2, and the audio bandpass filter, will be all part of one stage, although such is not a requirement. Hayward and others advocate the use of a postfilter at the output of the high-gain stages. This filter should have a bandwidth matched to the system bandwidth, although one source recommends "slightly wider" without being too specific. The purpose of the postfilter is to reduce the noise generated in the high-gain audio amplifiers (which can be considerable). Following the high-gain stages is an audio power amplifier that boosts the signal sufficiently for either headphones or a loudspeaker. An audio volume control will be provided at this point, or perhaps earlier in the chain.

Mixer circuits in direct-conversion receivers

The principal element in any direct-conversion receiver (DCR) is the mixer. The mixer is a nonlinear circuit element that exhibits changes of impedance of cyclical excursions of the input signals. When mixing is linear, one signal will ride on the other (see Fig. 6-7A) as an algebraic sum (i.e., the two waveforms are additive), but the product (i.e., multiplicative) frequencies are not generated. In nonlinear mixing, the classic amplitude-modulated waveform (Fig. 6-7B) is produced when the two frequencies are widely separated. A mixer that produces product frequencies can be used either in DCRs or superheterodyne receivers. In superhet terminology, it is common to call the frequency translation mixer that produces the IF signal a *first detector* and the mixer that recovers the audio modulation either a *product detector* or a *second detector,* even though exactly the same sort of circuit can be used by either case.

Any number of mixer circuits are used in radio receivers, and most of them are candidates for use in direct-conversion receivers. In nearly all cases, the output circuit of the mixer will be a low-pass filter that passes audio frequencies but not RF frequencies.

Two issues dominate mixer selection for DCR service: sensitivity and dynamic range. The former determines how small a signal can be detected, and the latter determines the ratio between the minimum detectable signal and the maximum allowable signal. Some passive mixers produce so much loss, so much noise, and require so much signal strength to operate that they are simply not suited to DCR design (unless adequate preamplification is provided). Such detectors can sometimes be put to good use in superheterodyne receivers because they are preceded by the gain of the front end and the IF amplifier chains, which can be considerable.

There are two issues to resolve when selecting a mixer element for a DCR. First, there is always the possibility of radiation of the local oscillator (LO) signal back through the antenna. In order to prevent this problem, it is necessary to keep the mixer unilateralized (i.e., the signal flowing in only one direction through the circuit). Some mixers are inherently good in this respect, and others are a bit problematical. In cases where LO radiation might occur an RF amplifier should be used ahead of the mixer, regardless of whether it is needed for purposes of improving sensitivity or selectivity.

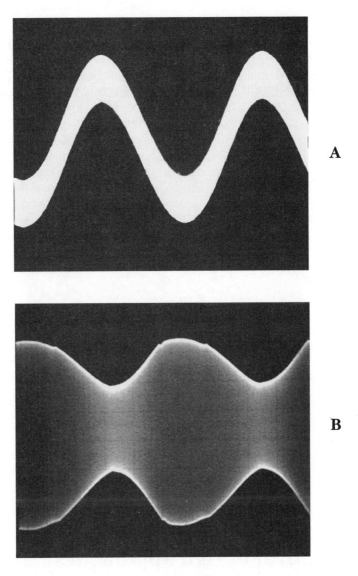

6-7 (A) Linearly mixed signals and (B) nonlinearly mixed signals result in the AM modulation waveform when two frequencies are wide apart.

The second problem that must be resolved is transmission of the RF and LO signals to the output of the mixer. You normally only want the first-order sum and difference frequencies. Many forms of the simple mixer circuit are particularly bad in this respect, and others are considerably better. Theoretically, any mixer can be used for the front end of the DCR; however, the simple half-wave rectifier diode envelope detectors are not at all recommended. Representative mixer circuits are covered in Chapter 10.

Considerations for good DCR designs

It probably does not surprise many readers that there are some principles of good design that result in superior DCR performance. Some of these principles were discussed by Campbell[10] and others.[11] Even relatively simple DCR designs, including those based on the Signetics NE-602 integrated circuit double-balanced modulator[12] and the popular LM-386 audio amplifier, have proven to be very sensitive and free of hum and microphonics, even though that combination is not without critics. Dillon's design, which was tested in the laboratories of the American Radio Relay League (ARRL), proved remarkably free of the problems often associated with simple DCR designs.[13]

One method for terminating the mixer is to place a resistor-capacitor (RC) network across the "IF OUT" terminals of the mixer and ground (see Fig. 6-8). The SBL-1 mixer is designed for 50-Ω input and output impedances, so the device is terminated in its characteristic impedance at RF frequencies by the 51-Ω resistor (R_1). Because capacitor C_1 has a value that produces a high reactance at audio frequencies (AF) and a low reactance at RF, the mixer is terminated for any residual LO and RF signal (which are absorbed by R_1), but AF is transmitted to the low-pass filter.

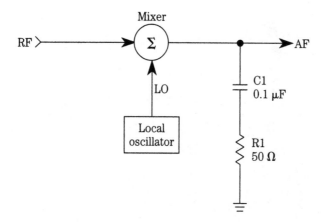

6-8 Mixer termination for high frequencies (above audio).

Some practical design approaches

The literature on DCRs has several different popular approaches, and each has its own place. Some of the simpler designs are based on the combination of a Signetics NE-602 device and a LM-386 IC audio section. Others are based on different

10 Rick Campbell, KK7B, "High Performance Direct-Conversion Receivers," *QST,* August 1992, pp. 19–28.

11 *QST,* August 1980, pp. 14–19; Paul G. Daulton, K5WMS, "The Explorer: HF Receiver for 40 and 80 Meters," *73 Amateur Radio Today,* August 1992, pp. 30–34; John Dillon, WA3RNC, "The Neophyte Receiver," *QST,* February 1988, pp. 14–18.

12 *Radio-Electronics,* April 1990, pp. 49–52; Joseph J. Carr, "NE-602 Primer," Elektor Electronics USA, Jan. 1992, pp. 20–25.

13 Ibid. (Dillon).

IC devices, such as the Signetics TDA7000 or some other product. These chips were designed for the cellular telephone and "cordless" telephone markets as receiver front ends. Still others are based on commercial or homebrew double-balanced modulators. This section examines several of these approaches.

The NE-602 type of DCR is relatively easy to build and provides reasonable performance for only a little effort. The NE-602 chip is relatively easy to obtain; for the most part, it is well-behaved in circuits (i.e., it does what it is supposed to do). It has about 20 dB of conversion gain, so it can help overcome some circuit losses, and it reduces slightly the amount of gain required of the audio amplifier that follows. The NE-602 can provide very good sensitivity; on the order of 0.3 μV is relatively easy to obtain, but it lacks something in dynamic range. Although the specifications of the device allow it to accept signals up to −15 dBm, at least one source recommended a maximum signal level of −25 dBm.[14] At higher input signal levels, the NE-602 tends to fall apart.[15] The newer NE-612 is basically the same chip but has improved dynamic range.

Figure 6-9 shows the basic circuit of the simplest form of NE-602 DCR. The input and LO circuits can be any of several configurations, although the one shown here is probably the most common.[16] The output signal is taken from either pin 4 or pin 5 and is fed to an audio amplifier. This circuit configuration will work, but it is not recommended. There will be a fairly large noise and image signal component and no filtering.

The Dillon design shown in Fig. 6-10 uses the push-pull outputs of the NE-602 (i.e., both pins 4 and 5) and is superior to the single-ended variety. According to Dillon, the balanced output approach improves the performance, especially in regard to AM BCB breakthrough rejection. Also helping the breakthrough problem is the use of a 0.047-μF capacitor across the output terminals of the NE-602.[17]

Daulton takes exception to the use of the NE-602 as the DCR front end and prefers instead to use the Signetics TDA-7000 chip. Although functionally similar to the NE-602, the TDA-7000 is more complex and is said to deliver superior performance with respect to dynamic range and signal-overload characteristics. Figure 6-11 shows a DCR front-end circuit that is based on the TDA-7000 after Daulton's design. This circuit uses the same balanced front end as other designs and, like the typical NE-602 design, uses the internal oscillator for the variable-frequency oscillator (VFO). The circuit following this front end should be of the sort typically found in the NE-602 designs. This particular variant uses the internal operational amplifiers of the TDA-7000 to provide active bandpass filtering.

A simplified variant on the Lawellen design[18] is shown in Fig. 6-12 (the original design included a QRP transmitter as well as the DCR). The front end of this variant consists of an RF transformer with a tuned secondary winding (L_{1A}). This secondary is tuned to resonance by capacitor C_1 and is tapped at the 50-Ω point in order to match the input impedance of the double-balanced mixer (DBM).

14 Ibid. (Covington).

15 Ibid. (Daulton).

16 Ibid. (Carr).

17 Telephone conversation between the author and John Dillon, 27 August 1992. See also Dillon's article (*op. cit.*).

18 Ibid. (Lawellen).

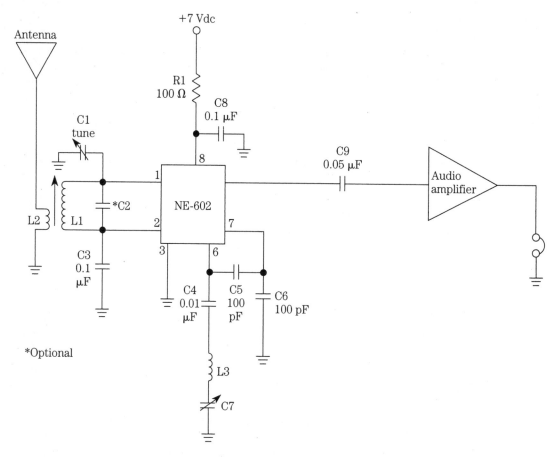

6-9 Partial schematic of NE-602 direct conversion receiver.

The particular DBM selected here is a Mini-Circuits SBL-1-1, although in the original article, Lawellen used a homebrew DBM made from diodes and toroidal transformers. The RF signal is input to pin 1 of the DBM (which can accommodate signal levels up to +1 dBm), and the local oscillator signal ("VFO IN") is applied to pin 8 through capacitor C_2. The VFO/LO signal must be on the order of +7 dBm.

The design of Fig. 6-12 uses two methods for matching the output of the mixer circuit. First, there is an RC network (R_1/C_3) that matches high frequencies to 51 Ω (the capacitor limits operation to high frequencies). The second method used here is to use a grounded-base input amplifier (Q_1) to the audio chain. Such an amplifier applies input signal across the emitter-base path and takes output signal from the collector-base path (the base being grounded for audio ac signals through capacitor C_5). This preamplifier is equipped with an active decoupler circuit, consisting of transistor Q_2 and its associated circuitry. The input side of the grounded-base audio amplifier consists of a LC low-pass filter (C_4/RFC$_1$) that passes audio frequencies, but not the residual VFO and RF signals from the DBM. In the original design, Lawellen followed the grounded-base amplifier with a direct-coupled operational amplifier active low-pass filter.

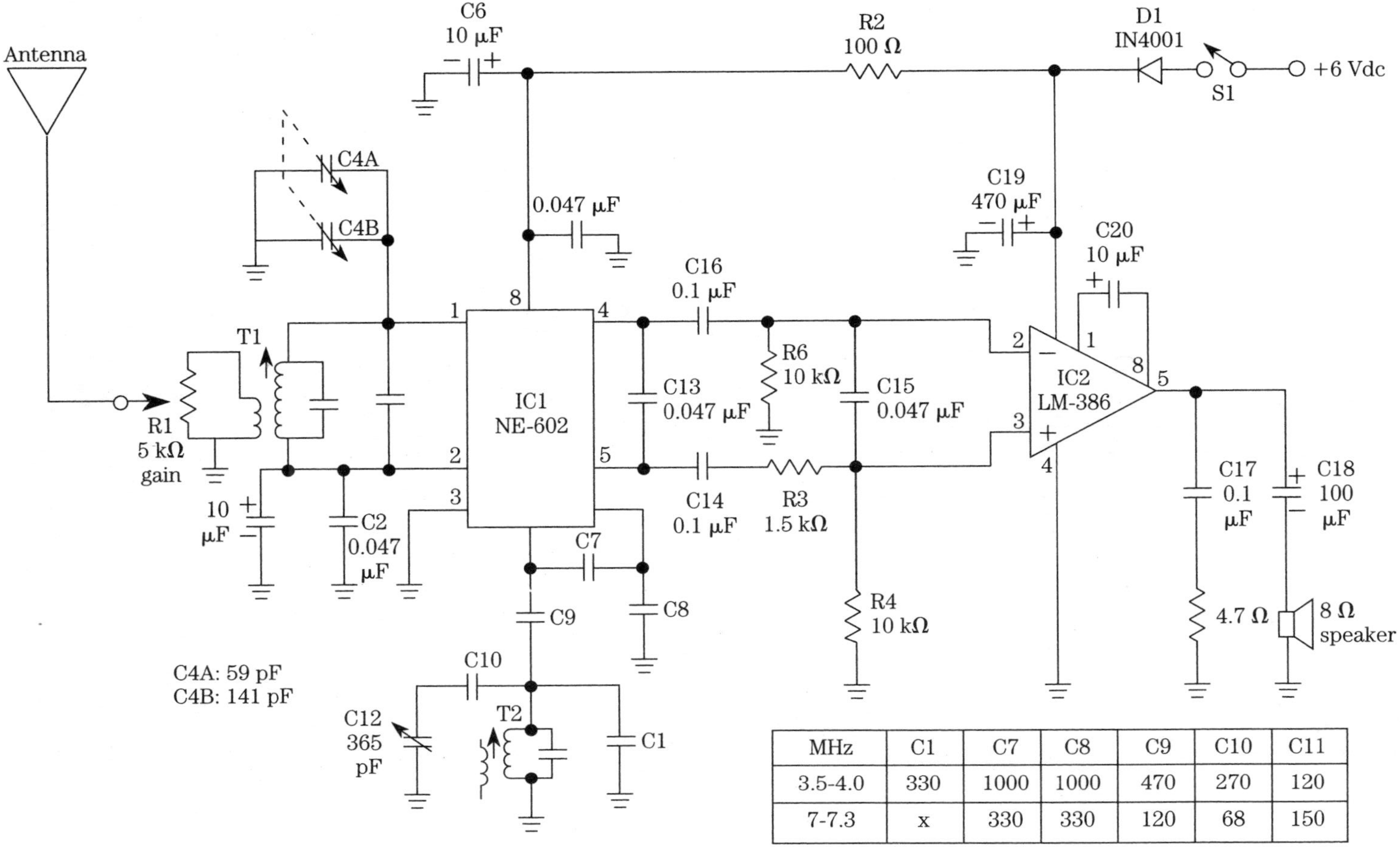

MHz	C1	C7	C8	C9	C10	C11
3.5-4.0	330	1000	1000	470	270	120
7-7.3	x	330	330	120	68	150

6-10 Complete schematic of direct-conversion receiver using NE-602.

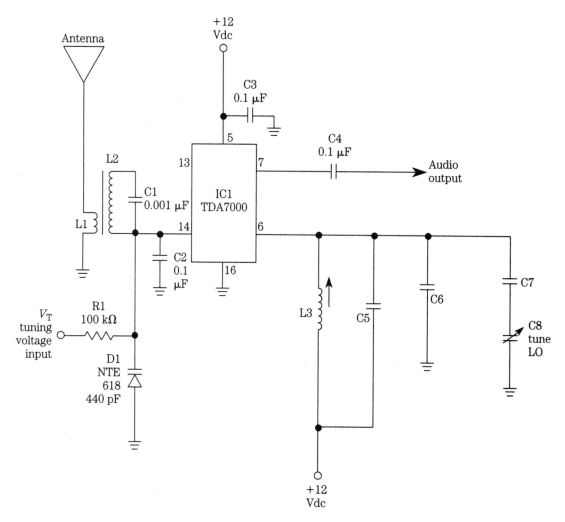

6-11 Direct-conversion receiver using TDA-7000 chip.

The Campbell design[19] extends the concepts from Lawellen. Figure 6-13 shows the block diagram for a portion of Campbell's direct conversion receiver. The front end consists of a double-balanced mixer followed by a matched diplexer filter that provides a 50-Ω input impedance from dc to 300 Hz and from 3000 Hz to some upper high-frequency beyond the range of interest. The diplexer also passes the standard communications audio bandwidth (300 to 3000 Hz) to the matched grounded base amplifier. Finally, there is an LC audio bandpass filter prior to sending the signal on to the high-gain audio amplifier stages.

Figure 6-14A shows the passive diplexer used by Campbell. It consists of several inductor, resistor, and capacitor elements that form both low-pass and high-pass filter sections. The values of the inductors (L_1, L_2, and L_3) are selected with their dc resistance in mind, so it is important to use the originally specified components or

19 Ibid. (Campbell).

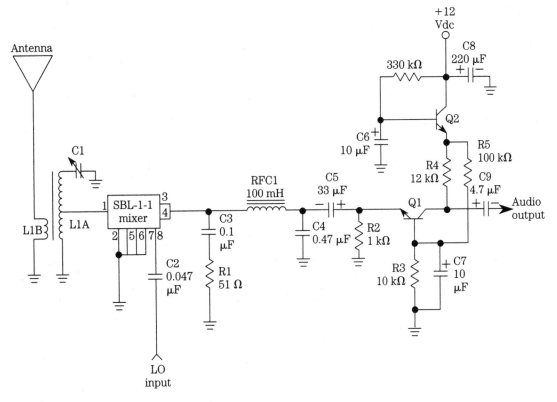

6-12 Direct-conversion mixer and first audio using the SBL-1-1 DBM.

their exact equivalents in replicating the project. Campbell used Toko Type 10RB inductors: L_1 is Toko 181LY-392J, L_2 is Toko 181LY-273J, and L_3 is Toko 181LY-273J. These coils are available from Digi-Key (P.O. Box 677, Thief River Falls, MN 56701-0677, USA; Voice 1-800-344-4539; FAX 218-681-3380).

Campbell's article supplied me with another example of the "digital myth," i.e., the concept that the digital implementation of a function is always superior to the analog version. He points out that the dynamic range of an LC filter is set by the inductors. The low-end is the thermal noise currents created by the circuit resistance (4KTBR), and the upper end is set by the saturation current of the inductor cores. For the parts selected by Campbell, he claims this range to be 180 dB. By contrast, an expensive 24-bit audio A/D converter provides only 144 dB of dynamic range.

The matched 50-Ω audio preamplifier is shown in Fig. 6-14B and is an improved version of the Lawellen circuit. According to Campbell, this circuit provides about 40 dB of gain and offers a noise figure of about 5 dB. The range of input signals that it will accommodate ranges from about 10 nV to 10 mV without undue distortion. These specifications make the amplifier a good match to the DBM. Like the Lawellen circuit, the Campbell circuit uses a grounded-base input amplifier (Q_1) and an active decoupler (Q_2). But Campbell also adds an emitter-follower/buffer amplifier (Q_3).

A set of three passive audio filters, which can be switched into or out of the circuit, is shown in Fig. 6-14C. These filters are designed for termination in an impedance of 500 Ω. Three different bandpasses are offered: 1, 3, and 4 kHz. The 4-kHz

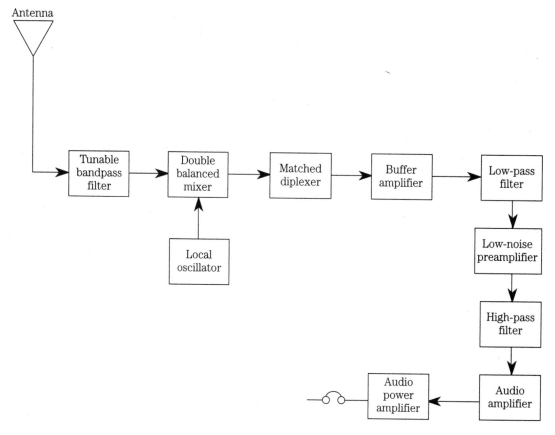

6-13 Block diagram for a complex direct-conversion receiver.

filter is a fifth-order Butterworth design, while the 3-kHz filter is a seventh-order Elliptical design after Niewiadomski.[20] The 1000-Hz design is scaled from the 3000-Hz design. Campbell claims that these filters offered a shape factor of 2.1:1, with an essentially flat passband ". . . with rounded corners, no ripple and no ringing."

Campbell implied the use of switching, as shown in Fig. 6-14C, but did not actually show the circuitry. As shown here, the switching involves use of a pair of ganged SP3P rotary switches. PIN diode switches can be used for this purpose.

A complex DCR was designed by Breed and reported in the amateur radio literature as a direct-conversion single-sideband receiver.[21] The single-sideband (SSB) mode is properly called *single-sideband suppressed-carrier amplitude modulation,* for it is a variant of AM that reduces the RF carrier and one of the two AM sidebands to negligible levels. This mode is used in HF transmissions because it reduces the bandwidth required by half and removes the carier that produces heterodyne squeals on the shortwave bands.

20 *Ham Radio,* Sept. 1985, pp. 17–30; cited in Campbell (*op. cit.*).
21 *QST,* Jan 1988, pp. 16–23.

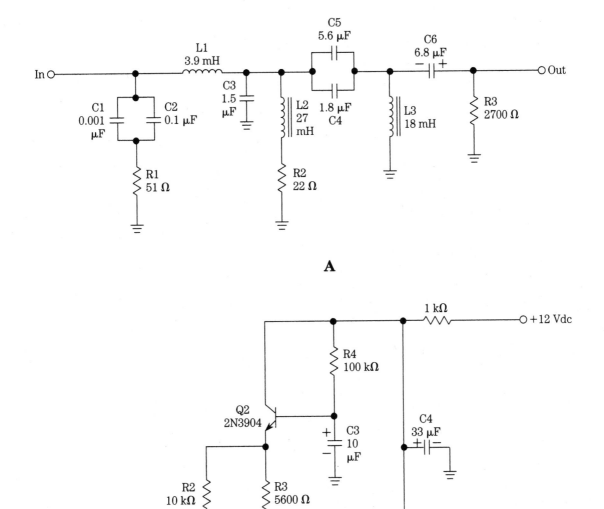

6-14 (A) Diplexer circuit for DBM; (B) audio postamplifier; and (C) three audio bandpass filters.

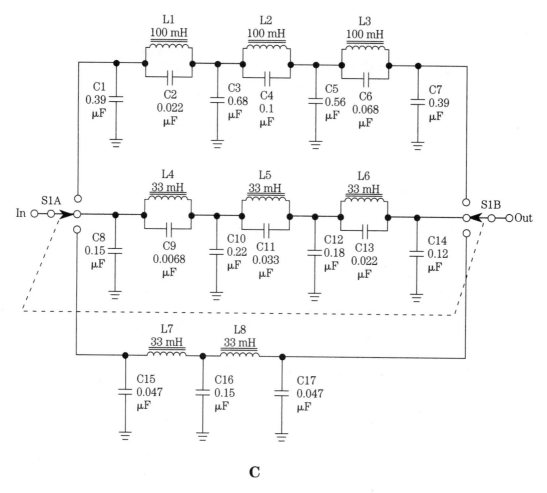

C

6-14 *Continued.*

There are two methods for generating SSB. The most common today uses a double-balanced modulator to combine a fixed carrier and the audio signal to produce a double-sideband suppressed-carrier output signal; the unneeded sideband is then removed by filtering. The older and more-complex variant uses a phasing method of SSB generation. Breed uses the inverse process to demodulate SSB signals in a clever, but complex, receiver design (Fig. 6-15). This circuit splits the incoming RF signal into two components, then feeds them both to separate mixers. These mixers are driven 90 degrees out of phase by a VFO that produces −45° and +45° outputs. The respective outputs of the mixers are amplified, then fed to bilateral 90° audio phase-shift networks, where they are recombined. The output of the phase-shift network is filtered in a low-pass filter and bandpass filter to provide the recovered modulation.

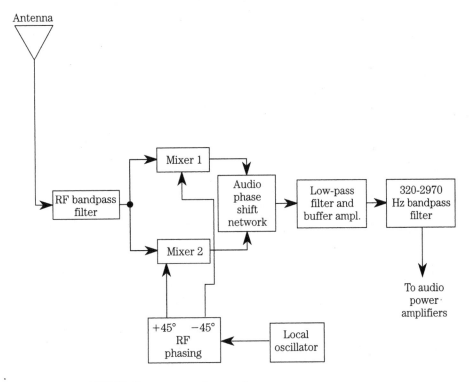

6-15 Phased CW/SSB direct-conversion receiver.

Audio circuits

The audio chain in the direct-conversion receiver tends to be very high gain in order to compensate for the low output levels usually found on the mixer circuits. The principal job of the audio amplifier is to increase the signal level by an amount that will create comfortable listening level while also tailoring the bandpass characteristics of the overall receiver to limit noise and other artifacts. Although any number of discrete and integrated circuit (IC) circuits are suitable, most designers today tend to use the IC versions that uses a discrete circuit with IC preamplifiers. Most other designers today use all-IC circuits. Figure 12-16 shows a simple LM-386 design while the published literature shows many other designs as well.[22]

The LM-386 design of Fig. 6-16 is the single-ended configuration for the LM-386 low-power audio stage. This IC device contains both preamplifiers and power amplifiers for a nominal output power of 250 mW.[23] The LM-386 series of audio power ICs are easy to use, but because of the high gain needed, it will oscillate if layout is not correct or if grounding is not proper. There are two basic circuit configurations for

---

22 *Radio-Electronics,* April 1990, pp. 53–57. See also the *National Semiconductor Linear Data Books* or the data books of other major IC manufacturers for applications information and device data.

23 Variants of the LM-386 include the LM-386-3, which produces 500 mW, and the LM-386-4 which produces 700 mW.

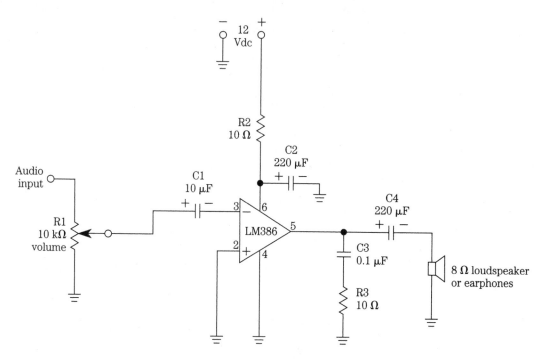

6-16 LM-386 audio power amplifier stage.

the LM-386. The differential version was shown in Fig. 6-10 (Dillon's design); Fig. 6-16 shows the more common single-ended design. The gain of the circuit can be either 46 dB ($\times$200) when capacitor C_2 is used or 26 dB ($\times$20) when C_2 is deleted (leave pins 1 and 8 open-circuited).

Local oscillator circuits for direct-conversion receivers

The local oscillator (LO) for a continuously tunable receiver of any description is basically a variable-frequency oscillator (VFO). Although higher-grade receivers today typically use frequency-synthesis techniques for generating the LO signal, the standard inductor-capacitor (LC) controlled VFO still has appeal for less-complex receivers. The VFO used for the LO in receivers is pretty much the same as the VFO in transmitters, so transmitter VFOs are frequently used. In some cases, however, a receiver LO has at least one specification that is more rigid than the transmitter equivalent: many receivers have a requirement for low FM phase noise. In the main, however, amateur radio applications of direct-conversion receivers typically use the transmitter VFO for the receiver as well.

Several different VFO designs are used for receiver LOs: Armstrong, Hartley, Colpitts/Clapp, and an amplitude-limiting design. The first three of these circuits are recognized according to the nature of their respective feedback networks, and the other is recognized by the special connection of a transformer. Notice that the Colpitts and Clapp are basically the same circuit, except that the Colpitts uses a parallel-tuned LC frequency-setting network and a Clapp oscillator uses a series-tuned LC network.

A common test chassis for projects

Part of the decision to build and test several direct-conversion receivers involved having a common chassis for all three designs. Although not very elegant, being made of scrap aluminum chassis and bottom plates from the "junk box," it was at least low cost and effective. Figure 6-17 shows the receiver test bed front panel. It is fitted with a Jackson Brothers calibrated dial with a 10:1 fast/slow vernier drive. The 6.35-mm (0.25 in) shaft coupling on the vernier drive is used to turn either a variable air-dielectric capacitor or potentiometer, depending on whether the DCR being built is a mechanically tuned or a voltage-tuned version. For most of the experiments, the voltage-tuned variety was used. Two additional controls are also provided, and both are potentiometers. The pot to the right of the tuning knob is a volume control, and that to the left is the RF tuning control (for voltage-tuned frontend circuits).

6-17 Photo of direct-conversion receiver.

Three circuits are provided for the test bed: two dc power supplies and an LM-386 audio power amplifier. One dc power supply (Fig. 6-18) is used to provide +12 Vdc (regulated) to the circuits of the DCRs used on the test bed. It uses a 7812 three-terminal IC voltage regulator and works from a +15-Vdc (or higher) source. In this project, raw power was provided by a series of three 6-V lantern dry-cell batteries connected in series. The second dc power supply (Fig. 6-19) consists of a 78L12 low-power, three-terminal voltage regulator and a potentiometer. The potentiometer is for main tuning and is ganged to the main dial of the chassis. In normal use, this potentiometer is used to tune the local oscillator (LO) of the DCR.

The audio amplifier is the LM-386 low-power "audio system on a chip" device. The LM-386 uses a minimum of external components and includes both the audio preamplifiers and the power amplifiers to produce between 250 mW and 700 mW of

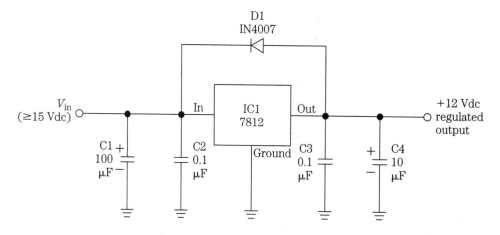

6-18 Dc power-supply circuit.

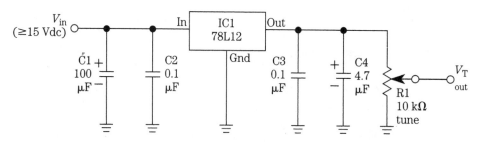

6-19 Voltage tuning power supply.

audio power, depending on the particular device specified. The circuit for the audio section is shown in Fig. 6-20, and its location on the test bed chassis is shown in Fig. 6-21. Notice that the audio section has its own +12-V regulator. This is an optional feature, but it does keep load variations in the audio amplifier from coupling to the rest of the circuitry. The audio section is the small circuit board on the left side of the chassis, right by the audio volume control.

Figure 6-21 shows the rear view of the DCR test bed chassis. As mentioned, the audio amplifier section is shown on the left. The gray metal box contains either the variable air-dielectric capacitor or the potentiometer and power supply (Fig. 6-19), either of which can be used to tune the local oscillator of the DCRs being tested. The +12-Vdc power supply is located on the small printed circuit board to the rear of the tuning box. Space is provided to the right of the tuning box for the circuitry of the DCR being tested. This board is changed from one project to the other.

Wiring board construction used two different methods. The audio amplifier, the tuning dc power supply, and +12-Vdc regulated power supply were built on RadioShack "universal" printed circuit boards. The DCRs, on the other hand, were wired using Vector perfboard with a hole grid on 0.100″ centers or an equivalent product.

Three different antennas were used for testing the DCRs in this chapter: a 5BTV Cushcraft 0.5-wavelength ham band vertical, an outdoor 30-m (100-ft) random-

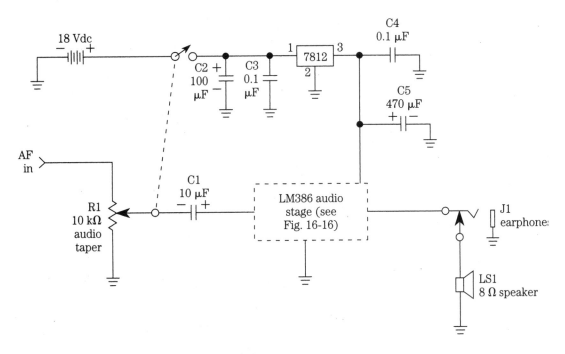

6-20 Dc power supply for LM-386 power amplifier.

6-21 Rear view of the DCR test bed chassis.

length end-fed wire, and a 6-m (20-ft) wire strung across the ceiling of my basement workshop. Interestingly enough, on the HF bands, there was not a large difference between the two outdoor antennas' performance and only slightly more difference between the outdoor antennas and the indoor antenna. On the VLF band, however, the random-length wire was clearly superior to the other two antennas.

References and notes

Gary A. Breed, "A New Breed of Receiver," *QST*, Jan. 1988, pp. 16–23.

The ARRL Handbook for the Radio Amateur—Sixty-Fifth Edition, American Radio Relay League (Newington, CT, USA, 1988).

Ibid. (*ARRL Handbook* pp. 6–8).

Rick Campbell, KK7B, "High Performance Direct-Conversion Receivers," *QST*, August 1992, pp. 19–28.

Ibid. (*ARRL Handbook*).

After W. Hayward in *ARRL Handbook* (*op. cit.*).

Wes Hayward and Doug DeMaw, *Solid-State Design for the Radio Amateur,* American Radio Relay League (Newington, CT, USA, 1977).

Ibid. (Campbell).

Ibid. (Campbell).

Roy W. Lawellen, W7EL, "An Optimized QRP Transceiver," *QST*, August 1980, pp. 14–19.

Rick Campbell, KK7B, "High Performance Direct-Conversion Receivers," *QST*, August 1992, pp. 19–28.

Roy W. Lawellen, W7EL, "An Optimized QRP Transceiver," *QST*, August 1980, pp. 14–19; Paul G. Daulton, K5WMS, "The Explorer: HF Receiver for 40 and 80 Meters," *73 Amateur Radio Today*, August 1992, pp. 30–34; John Dillon, WA3RNC, "The Neophyte Receiver," *QST*, February 1988, pp. 14–18.

Michael A. Covington, "Single-Chip Frequency Converter," *Radio-Electronics,* April 1990, pp. 49–52; Joseph J. Carr, "NE-602 Primer," *Elektor Electronics USA*, Jan. 1992, pp. 20–25.

Ibid. (Dillon).

Ibid. (Covington).

Ibid. (Daulton).

Ibid. (Carr).

Telephone conversation between the author and John Dillon, 27 August 1992. See also Dillon's article (*op. cit.*).

Ibid. (Lawellen).

Ibid. (Campbell).

S. Niewiadomski, "Passive Audio Filter Design," *Ham Radio,* Sept 1985, pp. 17–30; cited in Campbell (*op. cit.*).

This claim seems excessive. I've ordered the correct parts and will be investigating these filter designs within the next few months.

Gary A. Breed, "A New Breed of Receiver," *QST*, Jan 1988, pp. 16–23.

Ray Marston, "Audio Amplifier ICs," *Radio-Electronics,* April 1990, pp. 53–57. See also the *National Semiconductor Linear Data Books* or the data books of other major IC manufacturers for applications information and device data.

7
CHAPTER

RF amplifier and preselector circuits

Low-priced shortwave receivers often suffer from performance problems that are a direct result of the trade-offs that the manufacturers make to produce a low-cost model. In addition, older receivers often suffer the same problems, as do many homebrew radio receiver designs. Chief problems are sensitivity, selectivity, and image response.

Sensitivity is a measure of the receiver's ability to pick-up weak signals. Part of the cause of poor sensitivity is low-gain in the front end of the radio receiver, although the IF amplifier contributes most of the gain.

Selectivity is a measure of the ability of the receiver to (a) separate two closely spaced signals and (b) reject unwanted signals that are not on or near the desired frequency being tuned. The selectivity provided by a preselector is minimal for very closely spaced signals (that is the job of the IF selectivity in a receiver), but it is used for reducing the effects (e.g., input overloading) of large local signals . . . so fits the second half of the definition.

Image response affects only superheterodyne receivers (which most are) and is an inappropriate response to a signal at twice the receiver IF frequency that the receiver is tuned to. A superhet receiver converts the signal frequency (RF) to an intermediate frequency (IF) by mixing it with a local oscillator (LO) signal generated inside the receiver. The IF can be either the sum or difference between the LO and RF (i.e., LO + RF or LO − RF), but in most older receivers and nearly all low-cost receivers it is the difference (LO − RF). The problem is that there are always two frequencies that meet "difference" criteria: LO − RF and an image frequency (F_i) that is LO + IF. Thus, both F_i − LO and LO − RF are equal to the IF frequency. If the image frequency gets through the radio's front-end tuning to the mixer, it will appear in the output as a valid signal.

A cure for all of these is a little circuit called an *active preselector*. A preselector can be either active or passive. In both cases, however, the preselector includes an inductor/capacitor (LC) resonant circuit that is tuned to the frequency that the receiver is tuned to. The preselector is connected between the antenna and the receiver antenna input connector (Fig. 7-1A). Therefore, it adds a little more selectivity to the front end of the radio to help discriminate against unwanted signals. The inset to Fig. 7-1A shows the normal switching used in preselectors to allow it to be cut in or out of the circuit.

The difference between the active and passive designs is that the active design contains an RF amplifier stage, but the passive design does not. Thus, the active preselector also deals with the sensitivity problem of the receiver. This chapter looks at one passive and several active RF preselector circuits that you can build and adapt to your own needs.

The passive preselector circuit is shown in Fig. 7-1B. This circuit contains only inductors and capacitors, with each tank circuit shielded from each other. The individual tuned circuits are coupled using the common reactance method (i.e., capacitors C_2 and C_3 carry RF from one circuit to another). The input coil and the output

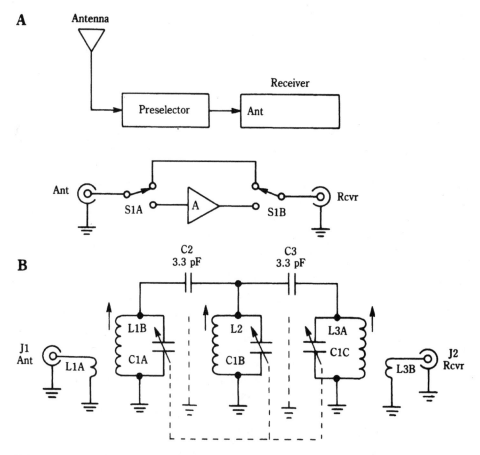

7-1 (A) Preselector is connected as an RF amplifier ahead of the receiver and (B) passive preselector circuit.

coil contain link coupling to transfer the RF signal in and out of the circuit respectively.

The tuning capacitors used in the passive preselector should be ganged to a common shaft. A proper unit would have three identical capacitors on the same shaft, with trimmer capacitors on each section (which allows the sections to be aligned to track each other). The trimmer capacitors are adjusted to track the high end of the band, and the slug-tuned inductors are adjusted for low-end tracking.

The active preselector circuits are based on either of two devices: the MPF-102 junction field-effect transistor (JFET) and the 40673 metal-oxide semiconductor field-effect transistor (MOSFET). Both of these devices are easily available from both mail-order sources and from local distributor replacement lines (for example, the MPF-102 is the NTE-312 and the 40673 is the NTE-222). These transistors were selected because they are both easily obtained and are well-behaved into the VHF region.

Preselectors should be built inside shielded metal boxes in order to prevent RF leakage around the device. Select boxes that are either die-cast or are made of sheet metal and have an overlapping lip. Do not use a low-cost tab-fit sheet-metal box.

JFET preselector circuits

Figure 7-2 shows the most basic form of JFET preselector. This circuit will work into the low-VHF region. This circuit is in the common source configuration, so the input signal is applied to the gate and the output signal is taken from the drain. The

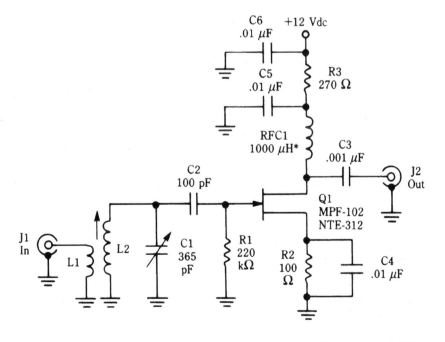

*100 μH for low-band VHF

7-2 JFET active preselector circuit.

source bias is supplied by the voltage drop across resistor R_2, and the drain load is supplied by a series combination of a resistor (R_3) and a radio frequency choke (RFC1). The RFC should be 1000 μH (1 mH) at the AM broadcast band and HF (shortwave), and 100 μH in the low-VHF region (>30 MHz). At VLF frequencies (below the AM broadcast band), use 2.5 mH for RFC1 and increase all 0.01-μF capacitor values to 0.1 μF. All capacitors are either disk ceramic or one of the newer "poly" capacitors (if rated for VHF service—not all are!).

The input circuit is tuned to the RF frequency, but the output circuit is untuned. The reason for the lack of output tuning is that tuning both the input and the output permits the JFET to oscillate at the RF frequency—and you don't want that. Other possible causes of oscillation include layout and a self-resonance frequency of the RFC that is too near the RF frequency (select another choke).

The input circuit consists of an RF transformer that has a tuned secondary (L_2/C_1). The variable capacitor (C_1) is the tuning control. Although the value shown is the standard 365-pF "AM broadcast variable," any form of variable can be used if the inductor is tailored to it. These components are related by:

$$F = \frac{1}{2\pi\sqrt{LC}}, \qquad (7\text{-}1)$$

where

F is the frequency in hertz
L is the inductance in henrys
C is the capacitance in farads.

Be sure to convert inductances from microhenrys to henrys and picofarads to farads. Allow approximately 10 pF to account for stray capacitances, although keep in mind that this number is a guess that might have to be adjusted (it is a function of your layout, among other things). We can also solve Eq. (7-1) for either L or C:

$$L = \frac{1}{39.5\,F^2 C} \qquad (7\text{-}2)$$

and

$$C = \frac{1}{39.5\,F^2 L}. \qquad (7\text{-}3)$$

Space does not warrant making a sample calculation, you can check the results for yourself. In a sample calculation, I wanted to know how much inductance is required to resonate 100 pF (90 pF capacitor plus 10 pF stray) to 10 MHz WWV. The solution, when all numbers are converted to hertz and farads, results in 0.00000253 H, or 2.53 μH. Keep in mind that the calculated numbers are close, but are nonetheless approximate—and the circuit might need to be tweaked on the bench.

The inductor (L1/L2) can be either a variable inductor (as shown) from a distributor, such as Digi-Key, or "homebrewed" on a toroidal core. Most people will want to use the T-50-6 (RED) or T-68-6 (RED) toroids (Amidon Associates) for shortwave applications. The number of turns required for the toroid is calculated from $N = 100 \times [L_{\mu H}\backslash A_L]^{1/2}$, where $L_{\mu H}$ is in microhenrys and A_L is 49 for T-50-RED and 57 for T-68-RED. Example: a 2.53-μH coil needed for L_2 (Fig. 7-2) wound on a T-50-RED core requires 23 turns. Use nos. 26 or 28 enameled wire for the winding. Make L_1 approximately 4 to 7 turns over the same form as L_2.

Figure 7-3 shows two methods for tuning both the input and output circuits of the JFET transistor. In both cases, the JFET is wired in the common gate configuration, so the signal is applied to the source and the output is taken from the drain. The dotted line indicates that the output and input tuning capacitors are ganged to the same shaft.

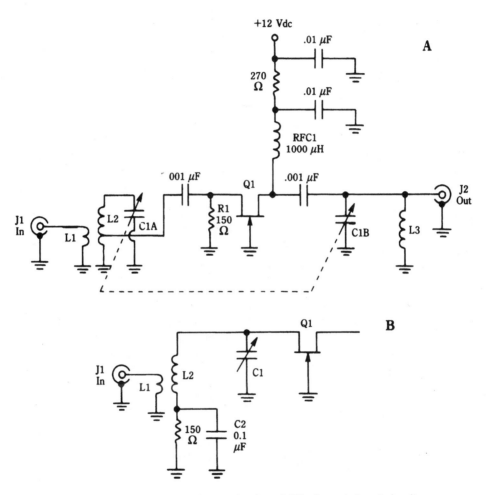

7-3 (A) Grounded-gate JFET preselector circuit and (B) alternate input circuit.

The source circuit of the JFET is low-impedance, so some means must be provided to match the circuit to the tuned circuit. In Fig. 7-3A, a tapped inductor is used for L_2 (tapped at 1/3 the coil winding); in Fig. 7-3B, a slightly different configuration is used.

The circuit in Fig. 7-4 is a VHF preamplifier that uses two JFET devices connected in cascode (i.e., input device Q_1 is in common source and is direct coupled to

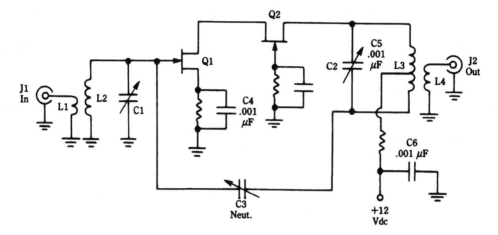

7-4 Cascode active preselector circuit.

the common gate output device Q_2). In order to prevent self-oscillation of the circuit, a neutralization capacitor (NEUT) is provided. This capacitor is adjusted to keep the circuit from oscillating at any frequency within the band of operation. In general, this circuit is tuned to a single channel by the action of L_2/C_1 and L_3/C_2.

MOSFET preselector circuits

The 40673 MOSFET used in the following preselector circuit (Fig. 7-5) is inexpensive and easily available. It is a dual-gate MOSFET. The signal is applied to gate G_1, and gate G_2 is either biased to a fixed positive voltage or connected to a variable dc voltage that serves as a gain control signal. The dc network is similar to that of the previous (JFET) circuits, with the exception that a resistor voltage divider (R_3/R_4) is needed to bias gate G_2.

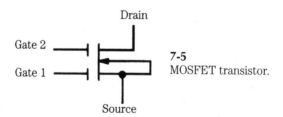

7-5
MOSFET transistor.

There are three tuned circuits for this preselector project, so it will produce a large amount of selectivity improvement and image rejection. The gain of the device will also provide additional sensitivity. All three tuning capacitors (C_{1A}, C_{16}, and C_{1C}) are ganged to the same shaft (Fig. 7-6) for "single-knob tuning." The trimmer capacitors (C_2, C_3, and C_4) are used to adjust the tracking of the three tuned circuits (i.e., ensure that they are all tuned to the same frequency at any given setting of C_{1A}–C_{1C}).

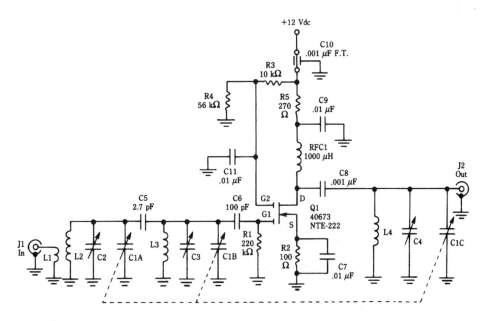

7-6 MOSFET active preselector circuit.

The inductors are of the same sort as described. It is permissible to put L_1/L_2 and L_3 in close proximity to each other, but these should be separated from L_4 in order to prevent unwanted oscillation because of feedback arising from coil coupling.

The circuit in Fig. 7-7 is a little different. In addition to using only input tuning (which lessens the potential for oscillation), it also uses voltage tuning. The hard-to-find variable capacitors are replaced with varactor diodes, also called *voltage variable-capacitance diodes*. These PN junction diodes exhibit a capacitance that is a function of the applied reverse bias potential, V_T. Although the original circuit was built and tested for the AM broadcast band (540 to 1610 kHz), it can be changed to any band by correctly selecting the inductor values. The designated varactor (NTE-618) offers a capacitance range of 440 pF down to 15 pF over the voltage range of 0 to +18 Vdc.

The inductors might be either "store-bought" types or wound over toroidal cores. I used a toroid for L_1/L_2 (forming a fixed inductance for L_2) and "store-bought" adjustable inductors for L_3 and L_4. There is no reason, however, why these same inductors cannot be used for all three uses. Unfortunately, not all values are available in the form that has a low-impedance primary winding to permit antenna coupling.

Figure 7-7 shows the connection points for the dc power (+9 to +12 Vdc). These points are 0.001-μF ceramic feedthrough capacitors. These are a little hard to find locally, but can be bought from Newark Electronics or other distributors.

In both of the MOSFET circuits, the fixed bias network used to place gate G_2 at a positive dc potential can be replaced with a variable voltage circuit, such as Fig. 7-8. The potentiometer in Fig. 7-8 can be used as an RF gain control to reduce

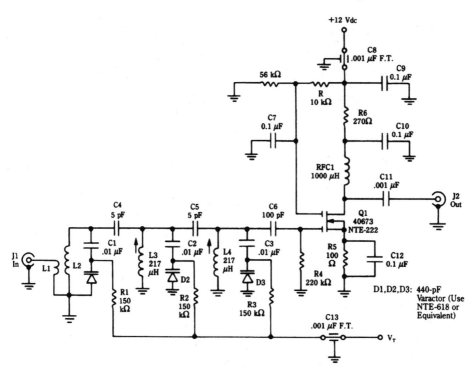

7-7 Varactor-tuned MOSFET active preselector circuit.

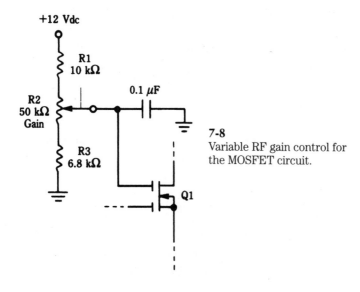

7-8
Variable RF gain control for
the MOSFET circuit.

gain on strong signals and increase it on weak signals. This feature allows the active preselector to be custom set to prevent overloading from strong signals.

Noise and preselectors

The weakest signal that you can detect is determined mainly by the noise level in the receiver. Some noise arrives from outside sources, and other noise is generated inside the receiver. At the VHF/UHF range, the internal noise is predominant, so it is common to use a low-noise preamplifier ahead of the receiver. This preamplifier will reduce the noise figure for the entire receiver. If you select a commercial ready-built or kit VHF preamplifier, such as the units sold by Hamtronics, Inc. (65 Moul Rd., Hilton, NY 14468-9535; Phone: 716-392-9430), be sure to specify the low-noise variety for the first amplifier in the system.

The low-noise amplifier (LNA) should be mounted on the antenna if it is wideband and at the receiver if it is tunable (notice: the term *preselector* only applied to tuned versions, while *preamplifier* could denote either tuned or wideband models). Of course, if your receiver is used only for one frequency, then it can also be mounted at the antenna. The reason for mounting the antenna right at the antenna is to build up the signal and improve the signal-to-noise ratio (SNR) prior to feeding the signal into the transmission line, where losses cause it to weaken somewhat.

Broadband RF preamplifier for VLF, LF, and AM BCB

In many situations, a broadband (as opposed to tuned) RF amplifier is needed. Typical applications include boosting the output of RF signal generators (which tend to be normally quite low level), antenna preamplification, loop antenna amplifier, and in the front ends of receivers. A number of different circuits published, including some by me, but one failing that I've noted on most of them is that they often lack response at the low end of the frequency range. Many designs offer -3 dB frequency response limits of 3 to 30 MHz, or 1 to 30 MHz, but rarely are the VLF, LF, or even the entire AM broadcast band (540 to 1700 kHz) covered.

The original need for this amplifier was that I needed an amplifier to boost AM BCB DX signals. Many otherwise fine communications or entertainment-grade "general coverage" receivers operate from 100 kHz to 30 MHz or so, and that range initially sounds really good to the VLF through AM BCB DXer. But when examined closer, it turns out that the receiver lacks sensitivity on the bands below either 2 or 3 MHz, so it fails somewhat in the lower end of the spectrum. Although most listening on the AM BCB is to powerful local stations (where receivers with no RF amplifier and a loopstick antenna will work nicely), those who are interested in DXing are not well served. In addition to the receiver, I wanted to boost my signal generator 50-Ω output to make it easier to develop some AM and VLF projects that I am working on and to provide a preamplifier for a square-loop antenna that tunes the AM BCB.

Several requirements were developed for the RF amplifier. First, it had to retain the 50-Ω input and output impedances that are standard in RF systems. Second, it had to have a high dynamic range and third-order intercept point in order to cope with the bone-crunching signal levels on the AM BCB. One of the problems of the AM BCB is that those sought after DX stations tend to be buried under multikilowatt local stations on adjacent channels. That's why high dynamic range, high-intercept point loop antennas tend to be required in these applications. I also wanted the amplifier to cover at least two octaves (4:1 frequency ratio), but it achieved a decade (10:1) response (250 to 2500 kHz).

Furthermore, the amplifier circuit had to be easily modifiable to cover other frequency ranges up to 30 MHz. This last requirement would make the amplifier useful to a large number of readers as well as extend its usefulness to me.

Consider a number of issues when designing an RF amplifier for the front end of a receiver. The dynamic range and intercept point requirements were mentioned previously. Another issue is the amount of distortion products (related to third-order intercept point) that are generated in the amplifier. It does no good to have a high capability on the preamplifier only to overload the receiver with a lot of extraneous RF energy it can't handle—energy that was generated by the preamplifier, not from the stations being received. These considerations point to the use of a push–pull RF amplifier design.

Push–pull RF amplifiers

The basic concept of a push–pull amplifier is demonstrated in Fig. 7-9. This type of circuit consists of two identical amplifiers that each process half the input sine-wave signal. In the circuit shown, this job is accomplished by using a center-tapped transformer at the input to split the signal and another at the output to recombine the signals from the two transistors. The transformer splits the signal because its center tap is grounded; thus, it serves as the common for the signals applied to the two transistors. Because of normal transformer action, the signal polarity at end "A" will be opposite that at end "B" when the center tap ("CT") is grounded. Thus, the two amplifiers are driven 180° out of phase with each other; one will be turning on while the other is turning off and vice versa.

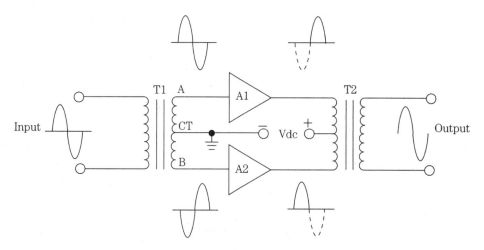

7-9 Push–pull broadband amplifier block diagram.

The push–pull amplifier circuit is balanced, and as a result it has a very interesting property: even-order harmonics are canceled in the output, so the amplifier output signal will be cleaner than for a single-ended amplifier using the same active amplifier devices.

There are two general categories of push–pull RF amplifiers: *tuned amplifiers* and *wideband amplifiers*. The tuned amplifier will have the inductance of the input and output transformers resonated to some specific frequency. In some circuits, the nontapped winding might be tuned, but in others, a configuration such as Fig. 7-10 might be used. In this circuit, both halves of the tapped side of the transformer are individually tuned to the desired resonant frequency. Where variable tuning is desired, a split-stator capacitor might be used to supply both capacitances.

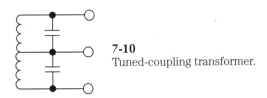

7-10
Tuned-coupling transformer.

The broadband category of circuit is shown in Fig. 7-11A. In this type of circuit, a special transformer is usually needed. The transformer must be a broadband RF transformer, which means that it must be wound on a suitable core so that the windings are bifilar or trifilar. The particular transformer in Fig. 7-11A has three windings, of which one is much smaller than the others. These must be trifilar wound for part of the way and bifilar wound the rest of the way. This means that all three windings are kept parallel until no more turns are required of the coupling link. Then, the remaining two windings are kept parallel until they are completed. Figure 7-11B shows an example for the case where the core of the transformer is a ferrite or powdered-iron toroid.

Actual circuit details

The actual RF circuit is shown in Fig. 7-12; it is derived from a similar circuit found in Doug DeMaw's excellent book *W1FB's QRP Notebook* (ARRL, 225 Main Street, Newington, CT 06111). The active amplifier devices are JFETs that are intended for service from dc to VHF. The device selected can be the ever-popular MPF-102 or its replacement equivalent from the SK, ECG, or NTE lines of devices. Also useful is the 2N4416 device. The particular device that I used was the NTE-451 JFET transistor. This device offers a transconductance of 4000 microsiemens, a drain current of 4 to 10 mA, and a power dissipation of 310 mW, with a noise figure of 4 dB maximum.

The JFET devices are connected to a pair of similar transformers, T_1 and T_2. The source bias resistor (R_1) for the JFETs, and its associated bypass capacitor (C_1), are connected to the center tap on the secondary winding of transformer T_1. Similarly, the +9-Vdc power-supply voltage is applied through a limiting resistor (R_2) to the center tap on the primary of transformer T_2.

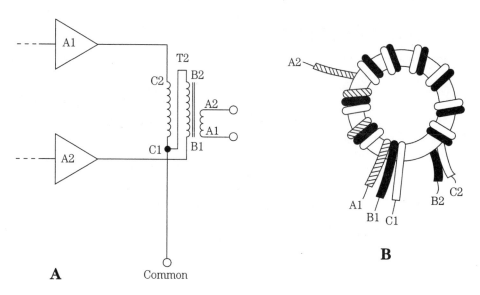

7-11 (A) Untuned (broadband) coupling transformer and (B) winding the transformer on a toroid.

Take special note of those two transformers. These transformers are known generally as *wideband transmission-line transformers* and can be wound on either toroid or binocular ferrite or powdered iron cores. For the project at hand, because of the low frequencies involved, I selected a type BN-43-202 binocular core. The type-43 material used in this core is a good selection for the frequency range involved. The core can be obtained from either Amidon Associates or Ocean State Electronics [P.O. Box 1458, 6 Industrial Drive, Westerly, RI 02891; Phones 401-596-3080 (voice), 401-596-3590 (fax) or 800-866-6626 (orders only)]. Three windings are on each transformer. In each case, the "B" and "C" windings are 12 turns of no. 30 AWG enameled wire wound in a bifilar manner. The coupling link in each is winding "A." The "A" winding on transformer T_1 consists of four turns of no. 36 AWG enameled wire, and on T_2 it consists of two turns of the same wire. The reason for the difference is that the number of turns in each is determined by the impedance-matching job it must do (T_1 has a 1:9 pri/sec ratio, and T_2 has a 36:1 pri/sec ratio). Neither the source nor drain impedances of this circuit are 50 Ω (the system impedance), so there must be an impedance-transformation function. If the two amplifiers in the circuit were of the sort that had 50-Ω input and output impedances, such as the Mini-Circuits MAR-1 through MAR-8 devices, then winding "A" in both transformers would be identical to windings "B" and "C." In that case, the impedance ratio of the transformers would be 1:1:1.

The detail for transformers T_1 and T_2 is shown in Fig. 7-13. I elected to build a header of printed circuit perforated board for this part; the board holes are on 0.100 in centers. The PC type of perfboard has a square or circular printed circuit soldering pad at each hole. A section of perfboard was cut with a matrix of 5 × 9 holes. Vector Electronics push terminals are inserted from the unprinted side then soldered into place. These terminals serve as anchors for the wires that will form the windings of the transformer. Two terminals are placed at one end of the header, and three are placed at the opposite end.

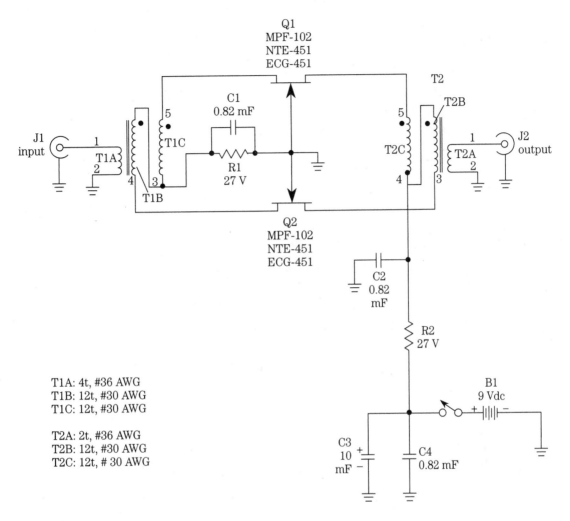

7-12 JFET push–pull broadband amplifier.

The coupling winding is connected to pins 1 and 2 of the header and is wound first on each transformer. Strip the insulation from a length of no. 36 AWG enameled wire for about ¼″ from one end. This can be done by scraping with a scalpel or X-acto knife or by burning with the tip of a soldering pencil. Ensure that the exposed end is tinned with solder then wrap it around terminal 1 of the header. Pass the wire through the first hole of the binocular core, across the barrier between the two holes, and then through the second hole. This U-shaped turn counts as one turn. To make transformer T_1 pass the wire through both sets of holes three more times (to make four turns). The wire should be back at the same end of the header as it started. Cut the wire to allow a short length to connect to pin 2. Clean the insulation off this free end, tin the exposed portion, then wrap it around pin 2 and solder. The primary of T1 is now completed.

The two secondary windings are wound together in the bifilar manner and consist of 12 turns each of no. 30 AWG enameled wire. The best approach seems

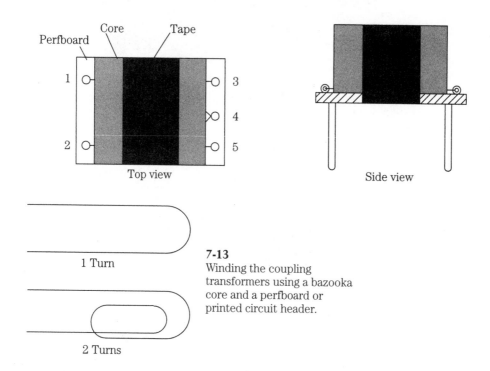

7-13
Winding the coupling
transformers using a bazooka
core and a perfboard or
printed circuit header.

to be twisting the two wires together. I use an electric drill to accomplish this job. Two pieces of wire, each 30″ long, are joined together and chucked up in an electric drill. The other ends of the wire are joined together and anchored in a bench vise or some other holding mechanism. I then back off and hold the drill in one hand, until the wire is nearly taut. Turning on the drill causes the two wires to twist together. Keep twisting them until you obtain a pitch of about 8 to 12 twists per inch.

It is *very important* to use a drill that has a variable speed control so that the drill chuck can be made to turn very slowly. It is also *very important* that you follow certain safety rules—especially with regard to your eyesight, when making twisted pairs of wire. Be absolutely sure to wear either safety glasses or goggles while doing this operation. If the wire breaks, and that is a common problem, then it will whip around as the drill chuck turns. Although no. 36 wire doesn't seem to be very substantial, at high speed it can severely injure an eye.

To start the secondary windings, scrap all of the insulation off both wires at one end of the twisted pair and tin the exposed ends with solder. Solder one of these wires to pin 3 of the header and the other to pin 4. Pass the wire through the hole of the core closest to pin 3, around the barrier, then through the second hole, returning to the same end of the header where you started. That constitutes one turn. Now do it 11 more times until all 12 turns are wound. When the 12 turns are completed, cut the twisted pair wires off to leave about $1/2″$ free. Scrape and tin the ends of these wires.

Connecting the free ends of the twisted wire is easy, but you will need an ohmmeter or continuity tester to see which wire goes where. Identify the end that is connected at its other end to pin 3 of the header and connect this wire to pin 4. The remaining wire should be the one that was connected at its other end to pin 4 earlier; this wire should be connected to pin 5 of the header.

Transformer T_2 is made in the identical manner as transformer T_1, but with only two turns on the coupling winding, rather than four. In this case, the coupling winding is the secondary, and the other two form two halves of the primary. Wind the two-turn secondary first, as was done with the four-turn primary on T_1.

The amplifier can be built on the same sort of perforated board as was used to make the headers for the transformers. Indeed, the headers and the board can be cut from the same stock. The size of the board will depend somewhat on the exact box you select to mount it in. For my purposes, the box was a Hammond $3 \times 5.5 \times 1.5''$ cabinet. Allowing room for the 9-Vdc battery at one end, and the input/output jacks and power switch at the other, left me with $2.5 \times 3.5''$ of available space in which to build the circuit (Fig. 7-14). For those who wish to experiment with a printed circuit board.

7-14 Completed preamplifier.

I built the circuit from the output end backward toward the input, so transformer T_2 was mounted first with pins 1 and 2 toward the end of the perfboard. Next, the two JFET devices were mounted, and then T_1 was soldered into place. After that, the two resistors and capacitors were added to the circuit. Connecting the elements together and providing push terminals for the input, output, dc power supply ground, and the +9 Vdc finished the board.

Because the input and output jacks are so close together, and because the dc power wire from the battery to the switch had to run the length of the box, I decided to use a shield partition to keep the input and output separated. This partition was made from 1-in brass stock. This material can be purchased at almost any hobby shop that caters to model builders. The RG-174/U coaxial cable between the input jack on the front panel and the input terminals on the perfboard run on the outside of the shield partition.

Variations on the theme

Three variations on the circuit extend the usefulness for many different readers. First, there are those who want to use the amplifier at the output of a loop antenna that is remote mounted. It isn't easy to go up to the roof or attic to turn on the amplifier any time you wish to use the loop antenna. Therefore, it is better to install the 9-Vdc power source at the receiver end and pass the dc power up the coaxial cable to the amplifier and antenna. This method is shown in Fig. 7-15. At the receiver end, RF is isolated from the dc power source by a 10-mH RF choke (RFC_2), and the dc is kept from affecting the receiver input (which could short it to ground!) by using a blocking capacitor (C_4). All of these components should be mounted inside of a shielded box. At the amplifier end, lift the grounded side of the T_2 secondary and connect it to RFC_2, which is then connected to the +9-Vdc terminal on the perfboard. A decoupling capacitor (C_3) keeps the "cold" end of the T_2 secondary at ground potential for RF, while keeping it isolated from ground for dc.

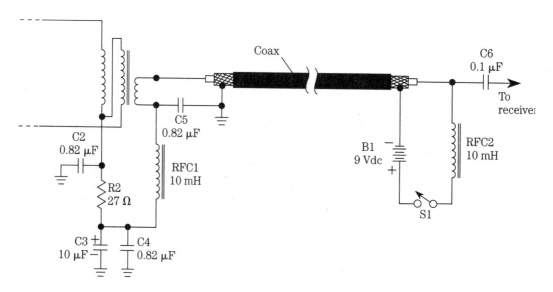

7-15 Powering a remote preamplifier.

A second variation is to build the amplifier for shortwave bands. This can be accomplished easily enough. First, reduce all capacitors to 0.1 μF. Second, build the transformers (T_1 and T_2) on a toroid core rather than the binocular core. In the original design (*op. cit.*), a type TF-37-43 ferrite core was used with the same 12:12:2 and 12:12:4 turns scheme as used.

Alternatively, select a powdered-iron core, such as T-50-2 (RED) or T-50-6 (YEL). I suspect that about 20 turns will be needed for the large windings, 4 turns for the "A" winding on T_2, and 7 turns for the "A" winding on T_1. You can experiment with various cores and turns counts to optimize for the specific section of the shortwave spectrum that you wish to cover.

The third variation is to make the amplifier operate on a much lower frequency (e.g., well down into the VLF region). The principal changes needed are in the cores used for transformers T_1 and T_2, the number of turns of wire needed, and the capacitors needed. The type 43 core will work down to 10 kHz or so but requires a lot more turns to work efficiently in that region. The type 73 material, which is found in the BN-73-202 core, will provide an A_L value of 8500 as opposed to 2890 for the BN-43-202 device used in this chapter. Doubling the number of turns in each winding is a good starting point for amplifiers below 200 kHz. The type 73 core works down to 1 kHz, so with a reasonable number of turns should work in the 20- to 100-Hz range as well.

Broadband RF amplifier (50-Ω input and output)

This project is a highly useful RF amplifier that can be used in a variety of ways. It can be used as a preamplifier for receivers operating in the 3- to 30-MHz shortwave band. It can also be used as a postamplifier following filters, mixers, and other devices that have an attenuation factor. It is common, for example, to find that mixers and crystal filters have a signal loss of 5 to 8 dB (this is called *insertion loss*). An amplifier following these devices will overcome that loss. The amplifier can also be used to boost the output level of signal generator and oscillator circuits. In this service, it can be used either alone, in its own shielded container, or as part of another circuit containing an oscillator circuit.

The circuit is shown in Fig. 7-16A. This circuit was originated by Hayward and used extensively by Doug DeMaw in various projects. The transistor (Q_1) is a 2N5179 broadband RF transistor. It can be replaced by the NTE-316 or ECG-316 devices, if the original is not available to you. The NTE and ECG devices are intended for service and maintenance replacement applications, so they are often sold by the local electronic parts distributors.

This amplifier has two important characteristics: the degenerative feedback in the emitter circuit and the feedback from collector to base. Degenerative, or negative, feedback is used in amplifiers to reduce distortion (i.e., make it more linear) and to stabilize the amplifier. One of the negative feedback mechanisms of this amplifier is seen in the emitter. The emitter resistance consists of two resistors, R_5 is 10 Ω and R_6 is 100 Ω. In most amplifier circuits, the emitter resistor is bypassed by a capacitor to set the emitter of the transistor at ground potential for RF signals, while keeping it at the dc level set by the resistance. In normal situations, the reactance of the capacitor should be not more than one-tenth the resistance of the emitter resistor. The 10-Ω portion of the total resistance is left unbypassed, forming a small amount of negative feedback.

The collector to base feedback is accomplished by two means. First, a resistor-capacitor network ($R_1/R_3/C_2$) is used; second, a 1:1 broadband RF transformer (T_1) is used. This transformer can be homemade. Wind 15 bifilar turns of no. 26 enameled wire on a toroidal core, such as the T-50-2 (RED) or T-50-6 (YEL); smaller cores can also be used.

The circuit can be built on perforated wireboard that has a grid of holes on 0.100-in centers. Alternatively, you can use the printed circuit board pattern shown in Fig. 7-16B. In this version of the project, the PC board is designed for use with a

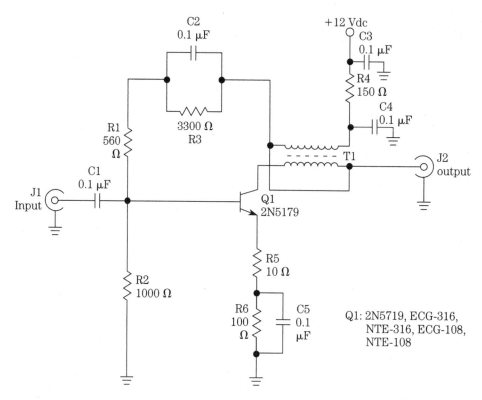

7-16 Feedback NPN transistor preamplifier.

Mini-Circuits 1:1 broadband RF transformer. Alternatively, use a homebrew transformer made on a small toroidal core. Use the size 37 core, with no. 36 enameled wire. As in the previous case, make the two windings bifilar.

Broadband or tuned RF/IF amplifier using the MC-1350P

The MC-1350P is a variant of the MC-1590 device, but unlike the 1590, it is available in the popular and easy-to-use eight-pin mini-DIP package. It has gain sufficient to make a 30-dB amplifier, although it is a bit finicky and tends to oscillate if the circuit is not built correctly. Layout, in other words, can be a very critical factor because of the gain.

If you cannot find the MC-1350P, use the NTE-746 or ECG-746. These devices are MC-1350Ps, but are sold in the service and maintenance replacement lines and are usually available locally.

Figure 7-17A shows the basic circuit for the MC-1350P amplifier. The signal is applied to the −IN input, pin 4, and the +IN input is decoupled to ground with a 0.1-μF capacitor. All capacitors in this circuit, except C_6 and C_7, should be disk ceramic or one of the newer dielectrics that are competent at RF frequencies to 30 MHz. A capacitor in series with the input terminal, C_1, is used to prevent dc riding on the signal from affecting the internal circuitry of the MC-1350P.

The output circuitry is connected to pin 1 of the MC-1350P. Because this circuit is broadband, the output impedance load is an RF choke (L_1). For most HF application, L_1 can be a 1-mH choke, although for the lower end of the shortwave region and the AM broadcast band, use a 2.5-mH choke. The same circuit can be used for 455-kHz IF amplifier service if the coil (L_1) is made 10 mH.

Pin 5 of the MC-1350P device is used for gain control. This terminal needs to see a voltage of +5 to +9 V, with the maximum gain at the +5-V end of the range (this is opposite what is seen in other chips). The gain-control pin is bypassed for RF signals.

The dc power supply is connected to pins 8 and 2 simultaneously. These pins are decoupled to ground for RF by capacitor C_4. The ground for both signals and dc power are at pins 3 and 7. The $V+$ is isolated somewhat by a 100-Ω resistor (R_3) in series with the dc power-supply line. The $V+$ line is decoupled on either side of this resistor by electrolytic capacitors. C_6 should be a 4.7- to 10-μF tantalum capacitor, and C_7 is a 68-μF (or greater) tantalum or aluminum electrolytic capacitor.

A partial circuit with an alternate output circuit is shown in Fig. 7-17B. This circuit is tuned, rather than broadband, so it could be used for IF amplification or RF amplification at specific frequencies. Capacitor C_8 is connected in parallel with the inductance of L_1, tuning L_1 to a specific frequency. In order to keep the circuit from oscillating the resonant tank circuit is "de-Q-ed" by connecting a 2.2-kΩ resistor in parallel with the tank circuit. Although considered optional in Fig. 7-17A, it is not optional in this circuit if you want to prevent oscillation.

Figure 7-17C shows a printed circuit board pattern that can be used for building the circuit of Fig. 7-17A. The spacing of the holes for the inductor are designed to accommodate the Toko line of fixed inductors from Digi-Key (use sizes 7 or 10).

The MC-1350P device has a disgusting tendency to oscillate at higher gains. One perfboard version of the Fig. 7-17A circuit that I built would not produce more than 16 dB without breaking into oscillation. One tactic to prevent the oscillation is to use a shield between the input and output of the MC-1350P. Extra holes are on the printed circuit pattern to anchor a shield. The shield should be made of copper or brass stock sheet metal, such as the type that can be bought at hobby shops. Cut a small notch along one edge of a piece of 1″ stock. The notch should be just large enough to fit over the MC-1350P without shorting out. The location of the shield is shown by the dotted line in Fig. 7-17A. Notice that it is bent a little bit in order to fit from the two ground pins on the MC-1350P (i.e., pins 2 and 7).

VLF preamplifier

The VLF bands run from 5 or 10 to 500 kHz or just about everything below the AM broadcast band. The frequencies above about 300 kHz can be accommodated by circuitry not unlike 455-kHz IF amplifiers. But as frequency decreases, it becomes more of a problem to build a good preamplifier. This project is a preamplifier designed for use from 5 to 100 kHz. This band contains a lot of Navy communications, as well as the most accurate time and frequency station, operated by the National Institute of Science and Technology (NIST), WWVB. The operating frequency of WWVB is a very accurate 60 kHz and is used as a frequency standard in many situations. The signal of WWVB is also used to update electronic clocks.

The circuit of Fig. 7-18A is similar to Fig. 7-17A, except that a large-value RF choke (L_1) is used across the $-$IN and $+$IN terminals. Both of these chokes are 120-mH 10 Toko units. If oscillation occurs, then select a different value (82 or 100 mH) for one of these chokes. The problem is that the chokes have a capacitance between

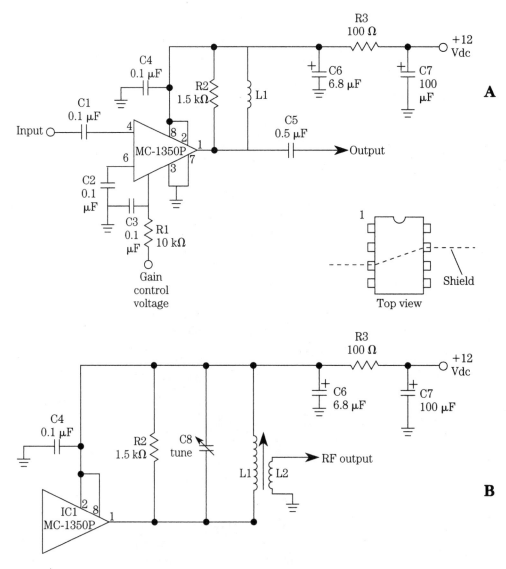

7-17 (A) MC-1350P preamplifier; (B) alternate output circuit; and (C) printed circuit board pattern.

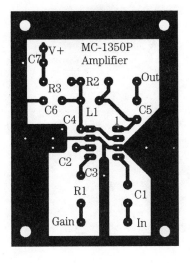

C

7-17
Continued.

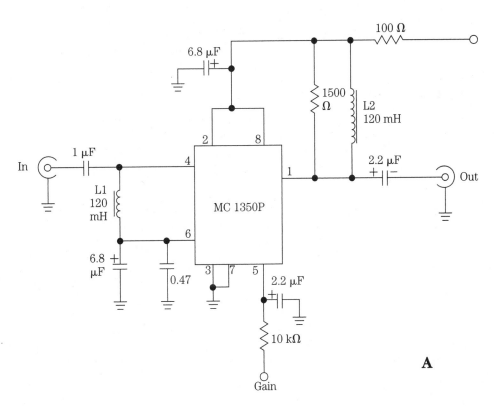

A

7-18 (A) Circuit for a MC-1350P preamplifier (alternate version) and (B) printed circuit pattern.

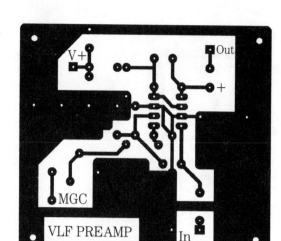

7-18
Continued.

B

windings and that capacitance can resonate the choke to a frequency (this is the "self-resonance" factor). By using different values of chokes in the input and output circuits, their self-resonant frequencies are moved away from each other, reducing the chance of oscillation.

Conclusion

A preselector can improve the performance of your receiver, no matter whether you listen to VLF, AM broadcast band, shortwave, or VHF/UHF bands. The circuits presented in this chapter allow you to "roll your own" and be successful at it.

8
CHAPTER

Building IF amplifiers

Most of the gain and selectivity of a superheterodyne radio receiver are in the intermediate frequency (IF) amplifier. The IF amplifier is, therefore, a high-gain, narrow-bandwidth amplifier. Typically, IF power gains run in the 60- to 120-dB range, depending on the receiver design. It usually has far narrower bandwidth than the RF amplifier. For example, 2.8 KHz for an SSB receiver and 500 Hz for a CW receiver.

The purpose of the IF amplifier is to provide gain and selectivity to the receiver. The selectivity portion of the equation is provided by any of several different types of filter circuit. Figure 8-1 shows several different filters used in IF amplifiers. The classical circuit is shown in Fig. 8-1A. This transformer is shown with taps, but it may also exist without the taps. The taps provide a low-impedance connection while retaining the overall advantages of high-impedance tuned circuits. Note that the capacitors are usually inside the transformer shielded can. A slightly different version is shown in Fig. 8-1B. This transformer differs from the one previous in that the secondary winding is capacitor coupled to its load and that capacitor may or may not be resonating. Still a third version is had by making the secondary winding an untuned low-impedance loop.

A somewhat different approach is shown in Fig. 8-1C. This transformer is a series-resonant tapped circuit on the input end and parallel-tuned and tapped on the output end. It requires two shield cans to implement this approach.

Finally, we have the representation shown in Fig. 8-1D. This symbol represents any of several mechanical or crystal filters. Such filters usually give a much narrower bandwidth than the LC filters discussed previously.

Amplifier circuits

A simple IF amplifier is shown in Fig. 8-2. A simple AM band radio may have one such stage, while FM receivers, shortwave receivers, and other types of communications receiver may have two to four stages such as Fig. 8-2. This IF amplifier is based on the type of LC filter circuits discussed in Fig. 8-1A. Transformer T_1 has a low-impedance tap on its secondary connected to the base of a transistor (Q_1). Similarly with T_2, but in this case the primary winding is tapped for the collector of the transistor.

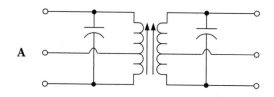

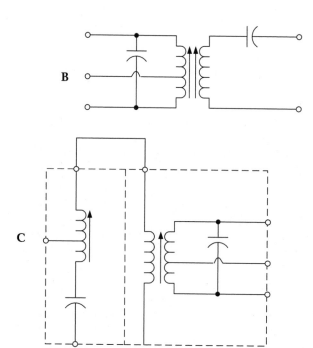

8-1
Various IF filters.

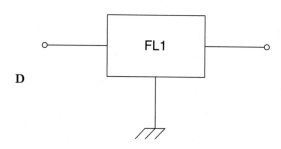

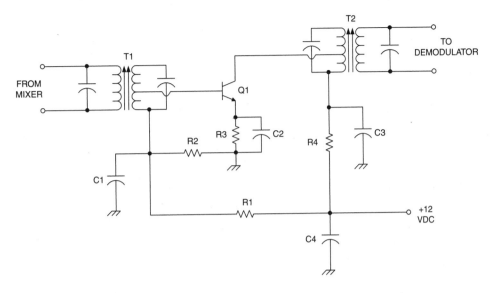

8-2 Simple transistor IF amplifier.

The bias for the transistor is provided by resistors R_1 and R_2, with capacitor C_1 being used to place the cold end of the T_1 secondary at ground potential for ac signals. Resistor R_3 is used to provide a bit of stability to the circuit. Capacitor C_2 is used to keep the emitter of the transistor at ground potential for ac signals, while keeping it at a potential of $I_E R_3$ for dc. The reactance of capacitor C_2 should be $<R_{3/10}$. Resistor R_4 forms part of the collector load for transistor Q_1. Capacitors C_3 and C_4 are used to bypass and decouple the circuit.

Figure 8-3 shows a gain-controlled version of Fig. 8-2. This particular circuit is designed for a low frequency (240 to 500 KHz), although with certain component value changes it could be used for higher frequencies as well. Another difference between this circuit and the circuit of Fig. 8-2 is that this circuit has double-tapped transformers for T_1 and T_2.

The major change is the provision for automatic gain control. A capacitor is used to sample the signal for some sort of AGC circuit. Furthermore, the circuit has an additional resistor (R_3) to the dc control voltage of the AGC circuit.

Cascode pair amplifier

A cascode pair amplifier is shown in Fig. 8-4. This amplifier uses two transistors (both JFETs) in an arrangement that puts Q_1 in the common source configuration and Q_2 in the common gate configuration. The two transistors are direct-coupled. Input and output tuning is accomplished by a pair of LC filters ($L_2 C_1$ and $L_3 C_2$). To keep this circuit from oscillating at the IF frequency a neutralization capacitor (C_3) is provided. This capacitor is connected from the output LC filter on Q_2 to the input of Q_1.

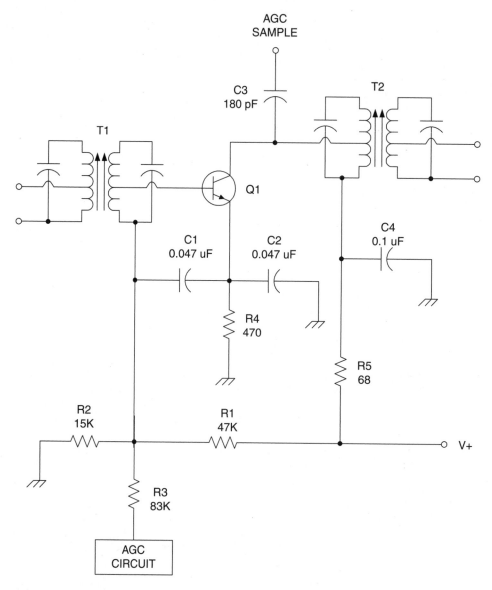

8-3 Simple IF amplifier with AGC.

"Universal" IF amplifier

The IF amplifier in Fig. 8-5 is based on the popular MC-1350P integrated circuit. This chip is easily available through any of the major mail order parts houses and many small ones. It is basically a variation on the LM-1490 and LM-1590 type of circuit but is a little easier to apply.

If you have difficulty locating MC-1350P devices, the exact same chip is available in the service replacement lines such as ECG and NTE. These parts lines are sold at

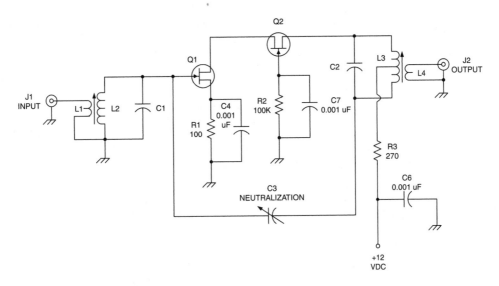

8-4 Cascode JFET IF amplifier.

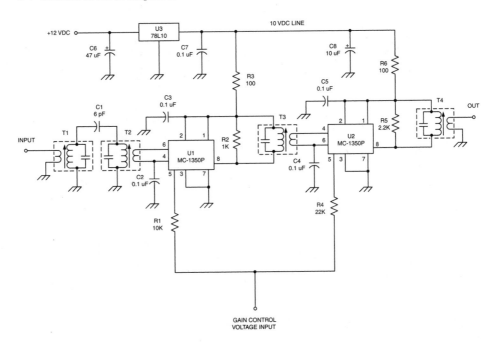

8-5 IF amplifier with crystal filter.

local electronics parts distributors and are intended for the service repair shop trade. I used actual MC-1350P chips in one version and NTE-746 (same as ECG-746) chips in the other without any difference in performance. The NTE and ECG chips are actually purchased from the sources of the original devices and then renumbered.

The circuit is shown in Fig. 8-5. Two MC-1350P devices in cascade are used. Each device has a differential input (pins 4 and 6). These pins are connected to the link windings on IF transformers (e.g., T_2 at device U_1 and T_3 at U_2). In both cases, one of the input pins are grounded for AC (i.e., RF and IF) signals through a bypass capacitor (C_2 and C_4).

In the past I've had difficulties applying the MC-1350P devices when two were used in cascade. The problem is that these are high-gain chips and any coupling at all will cause oscillation. I've built several really good MC-1350P oscillators—the problem is that I was building IF *amplifiers*. The problem is basically solved by two tactics that I'd ignored in the past. This time I reversed the connections to the input terminals on the two devices. Note that pin 4 is bypassed to ground on U_1, while on U_2 it is pin 6. The other tactic is to use different value resistors at pin 5.

Pin 5 on the MC-1350P device is the gain control pin. It is used to provide either manual gain control (MGC) or automatic gain control (AGC). The voltage applied to this pin should be between $+3$ and $+9$ V, with the highest gain being at $+3$ V and nearly zero gain at $+9$ V.

The outputs of the MC-1350P are connected to the primaries of T_3 and T_4. Each output circuit has a resistor (R_2 and R_5) across the transformer winding. The transformers used are standard "transistor radio" IF transformers provided that the impedance matching requirements are met.

The DC power is applied to the MC-1350P devices through pins 1 and 2, which are connected together. Bypass capacitors C_3 and C_5 are used to decouple the DC power lines and thereby prevent oscillation. All of the bypass capacitors (C_2, C_3, C_4, and C_5) should be mounted as close to the bodies of U_1 and U_2 as possible. They can be disk ceramic devices, or some of the newer dielectric capacitors, provided of course that they are rated for operation at the frequency you select. Most capacitors will work to 10.7 MHz, but if you go to 50 MHz or so, some capacitor types might show too much reactance (disk ceramic devices work fine at those frequencies, however).

The dc power supply should be regulated at some voltage between $+9$ and $+15$ Vdc. More gain can be obtained at $+15$ Vdc, but I used $+10$ Vdc with good results. In each power line there is a 100-Ω resistor (R_3 and R_6), which help provide some isolation between the two devices. Feedback via the power line is one source of oscillation in high-frequency circuits.

The dc power line is decoupled by two capacitors. C_7 is a 0.1-μF capacitor and is used to decouple high frequencies that either get in through the regulator or try to couple from chip to chip via the dc power line (which is why they are called "decoupling" capacitors). The other capacitor (C_8) is a 10- to 100-μF device used to smooth out any variations in the dc power or to decouple low frequencies that the 0.1 μF doesn't take out effectively.

The RF/IF input circuit deserves some comment. I elected to use a double-tuned arrangement. This type of circuit is of a category that are coupled via a mutual reactance. Various versions of this type of circuit are known, but I elected to use the version that uses a capacitive reactance (C_1) at the "hot" end of the LC tank circuits. Coupling in and out of the network is provided by the transformer coupling links.

The power and gain control connections are bought through the aluminum box wall through 1000-pF feedthrough capacitors. Two kinds are available, one solder-in and the other screw thread-mounted. For aluminum boxes the screw thread is needed because it is difficult to solder to aluminum. Both types are available in either 1000- or 2000-pF values, either of which can be used in this application. If you elect to use some other form of connector, then add disk ceramic capacitors (0.001 µF) to the connector, right across the pins as close as possible to the connector.

There are several ways to make this circuit work at other frequencies. If you want to use a standard IF frequency up to 45 MHz or so, then select one with the configuration shown in Fig. 8-5 (these are standard).

If you want to make the circuit operate in the HF band on a frequency other than 10.7 MHz, then it's possible to use the 10.7-MHz transformer. If the desired frequency is less than 10.7 MHz, then add a small value fixed or trimmer capacitor in parallel with the tuned winding. It will add to the built-in capacitance, reducing the resonant frequency. I don't know how low you can go, but I've had good results at the 40-m amateur band (7–7.3 MHz) using additional capacitance across a 10.7-MHz IF transformer.

On frequencies higher than 10.7 MHz you must take some more drastic action. Take one of the transformers and turn it over so that you can see the pins. In the middle of the bottom header, between the two rows of pins, there will be an indentation containing the tuning capacitor. It is a small tubular ceramic capacitor (you may need a magnifying glass to see it well if your eyes are like mine). If it is color-coded you can obtain the value using your knowledge of the standard color code. Take a small screwdriver and crush the capacitor. Clean out all of the debris to prevent shorts at a later time. You now have an untuned transformer with an inductance right around 2 µH. Using this information, you can calculate the required capacitance using this formula:

$$C = \frac{2.53 \times 10^{10}}{f^2 L_{\mu H}} \, (\mu F) \tag{8-1}$$

where
  C is the capacitance in picofarads
  f is the desired frequency in kilohertz (kHz)
  $L_{\mu H}$ is the inductance in microhenrys.

Equation (8-1) is based on the standard L-C resonance equation, solved for C, and with all constants and conversion factors rolled into the numerator. If you know the capacitance that must be used, and need to calculate the inductance, then swap the L and C terms in Eq. (8-1).

If the original capacitor was marked as to value or color-coded, then you can calculate the approximate capacitance needed by taking the ratio of the old frequency to the new frequency and then square it. The square of the frequency ratio is the capacitance ratio, so multiply the old capacitance by the square of the frequency ratio to find the new value. For example, suppose a 110-pF capacitor is used for 10.7 MHz, and you want to make a 20.5-MHz coil. The ratio is $(10.7/20.5)^2 = 0.272$ MHz. The new capacitance will be about 0.272×110 pF = 30 pF. For other frequencies, you might consider using homebrew toroid inductors.

A variation on the theme is to make the circuit wideband. This can be done for a wide portion of the HF spectrum by removing the capacitors from the transformers and not replacing them with some other capacitor. In that case, IF filtering is done at the input (between the IF amplifier and the mixer circuit).

Coupling to other filters

Crystal and mechanical filters require certain coupling methods. Figure 8-6 shows a method for coupling to a crystal filter connected between two bipolar transistors. Each stage of the amplifier is a common emitter bipolar transistor amplifier, biased by R_1/R_2 and R_5/R_6. The connection to the filter circuit is direct because the filter is not sensitive to dc (such cannot be said of mechanical filters).

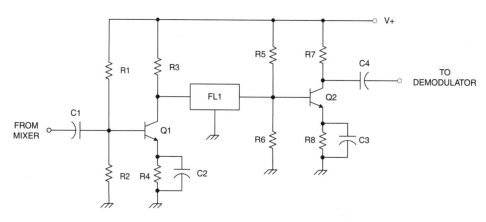

8-6 Universal IF amplifier.

Figure 8-7 shows a different approach that accommodates mechanical filters as well as crystal filters. The particular circuit shown is for very high frequency IF amplifiers (e.g., 50 MHz), but with changes to the values of the components, this IF amplifier could be used from VLF through VHF regions. The resonant frequency of this circuit is set by L_1 and C_1 and the filter circuit. The amplifier is a MOSFET transistor device connected in the common source configuration. Gate G_1 is used for the signal and G_2 is used for DC bias and gain control (see "AGC").

The filter is connected to the filter through a capacitor to block the dc at the drain of the MOSFET device (similar capacitors would be used in bipolar circuits as well). The output of the filter may or may not be capacitor coupled depending upon the design of the circuits to follow this one.

IC IF amplifiers

The universal IF amplifier presented earlier is an example of an integrated circuit (IC) IF amplifier. In this section we will take a look at several additional IC IF amplifier circuits.

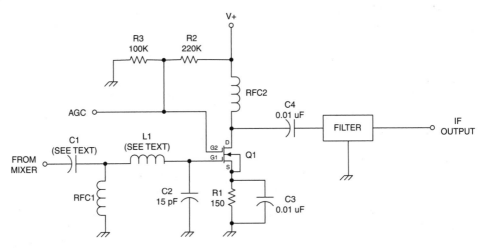

8-7 MOSFET IF amplifier with AGC.

MC-1590 circuit

Figure 8-8 shows an amplifier based on the 1490 or 1590 chips. This particular circuit works well in the VHF region (30 to 80 MHz). Input signal is coupled to the IC through capacitor C_1. Tuning is accomplished by C_2 and L_1, which forms a parallel

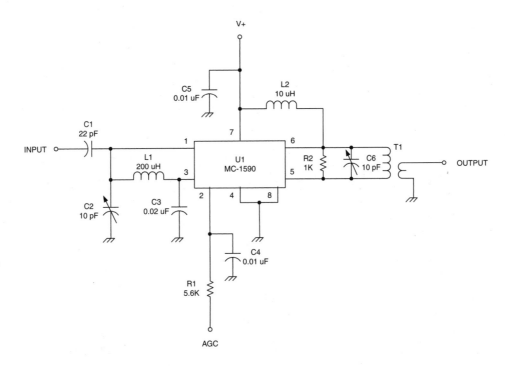

8-8 MC-1590 IF amplifier.

resonant tank circuit. Capacitor C_3 sets the unused differential input of the 1590 chip to ground potential, while retaining its dc level.

Output tuning in Fig. 8-8 reflects the fact that the 1590 chip is differential output as well as differential input. The LC-tuned circuit, consisting of the primary of T_1 and capacitor C_6, is parallel-resonant and is connected between pins 5 and 6. A resistor across the tank circuit reduces its loaded Q, which has the effect of broadening the response of the circuit.

V+ power is applied to the chip both through the V+ terminal and pin 6 through the coil L_2. Pin 2 is used as an AGC gain-control terminal.

SL560C circuits

The SL560C is basically a gain block that can be used at RF and IF frequencies. Figure 8-9 shows a circuit based on the SL560C. The input of the SL560C is differential, but this is a single-ended circuit. That requires the unused input to be bypassed to ground through capacitor C_3. Because this is a wideband circuit, there is no tuning associated with the input or output circuitry. The input circuitry consists of a 0.02-μF coupling capacitor and an RF choke (RFC$_1$).

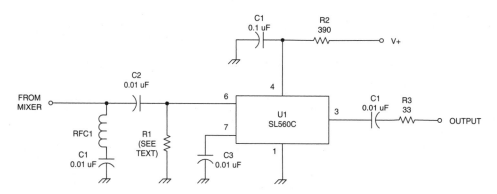

8-9 SL-560C IF amplifier.

A tuned circuit version of the circuit is shown in Fig. 8-10. This circuit replaces the input circuitry with a tuned circuit (T_1) and places a transformer (T_2) in the output circuit. Also different is that the V+ circuit in this case uses a zener diode to regulate the dc voltage.

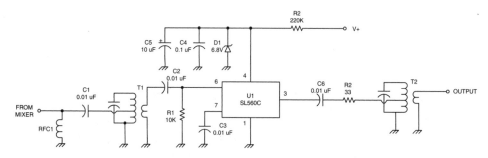

8-10 SL560C IF amplifier (tuned circuit version).

IF processing ICs

There are several forms of IC used for IF processing. The CA3189E (Fig. 8-11) is one of several that are used in broadcast and communications receivers. The input circuitry consists of a filter, although LC-tuned circuits could be used as well. In this version, the filter circuit is coupled via a pair of capacitors (C_1 and C_2) to the CA3189E. The input impedance is set by resistor R_1 and should reflect the needs of the filter rather than the IC (filters don't produce the same response when mismatched).

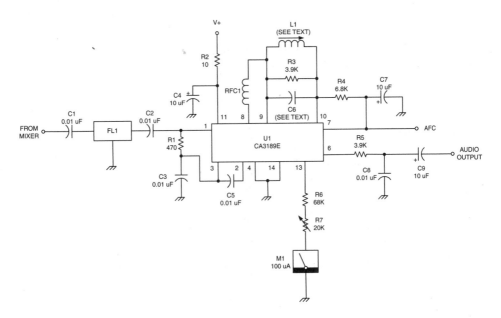

8-11 CA3189E FM IF system.

The CA3189E is an IF gain block and demodulator circuit, all in one IC. Coil L_1 and C_6 are used for the quadrature detector and their value depends on the frequency used. There are three outputs used on the CA3189E. The audio output is derived from the demodulator, as is the automatic frequency control (AFC) output. There is also a signal-strength output that can be used to drive an S-meter (M_1) or left blank.

Successive detection logarithmic amplifiers

Where signal level information is required, or where instantaneous outputs are required over a wide range of input signal levels, a logarithmic amplifier might be used. Linear amplifiers have a gain limit of about 100 dB, but only have head room of 3 to 6 dB in some circumstances. One solution to the problem is the logarithmic amplifier. Radar receivers frequently use log amps in the IF amplifier stages.

The successive detection method is used because it is difficult to produce high-gain logarithmic amplifiers. The successive detection method uses several log amps and then detects all outputs and sums these outputs together. Each amplifier has an output voltage equal to:

$$V_o = k \log V_{in}, \tag{8-2}$$

where

V_o is the output voltage
V_{in} is the input voltage
k is a constant.

The circuit shown in Fig. 8-12A is such an amplifier, while Fig. 8-12B shows a hypothetical output-vs-input characteristic for the amplifier. Because the circuit uses four or more stages of nonlinear amplification, signals are amplitude-compressed in this amplifier. Because of this, the amplifier is sometimes called a *compression amplifier.*

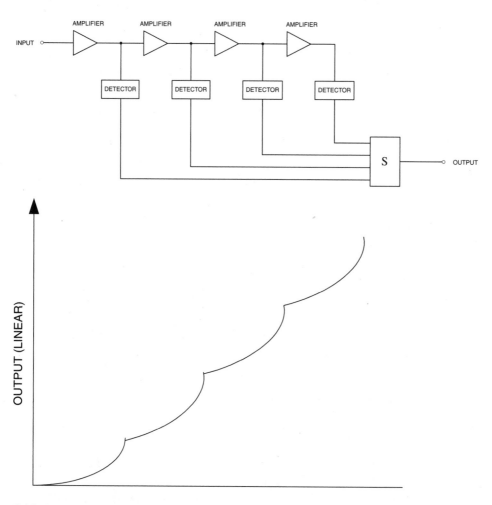

8-12 Output of compression amplifiers.

Filter switching in IF amplifiers

Switching filters is necessary to accommodate different modes of transmission. AM requires 4 or 6 KHz on shortwave and 8 KHz on the AM broadcast band (BCB). Single-sideband requires 2.8 KHz, RTTY/RATT requires 1.8 KHz, and CW requires either 270 or 500 Hz. Similarly, FM might require 150 KHz for the FM BCB and as little as 5 KHz on landmobile communications equipment. These various modes of transmission require different filters, and those filters have to be switched in and out of the circuit.

The switching could be done directly, but that requires either a coaxial cable between the switch and the filter or placing the switch at the site of the filters. A better solution is found in Fig. 8-13: diode switching. A diode has the unique ability to pass a small ac signal on top of the dc bias. In Fig. 8-13 switch S_1 is used to apply the proper polarity voltage to the diodes in the circuit. In one sense of S_1, the positive voltage is applied to D_4 through the primary winding of T_2, through RFC$_4$ and RFC$_2$, to D_3 and then through the secondary winding of T_1 to ground. This same current flow reverse biases diodes D_1 and D_2. The response is to turn on filter FL$_2$. Similarly when S_1 is turned to the other position. In that case D$_1$ and D$_2$ are forward biased, selecting FL1, and D3-D4 are reverse biased.

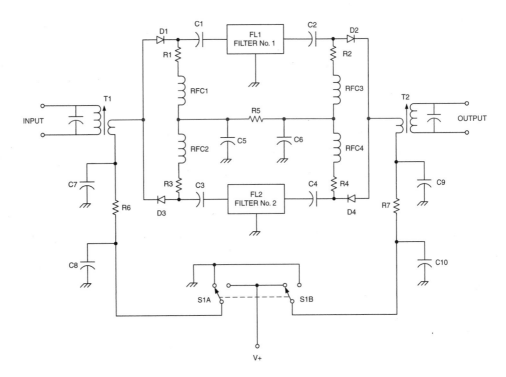

8-13 Dual filter scheme.

References

The ARRL Handbook for Radio Amateurs (1998; CD-ROM version), Chapter 16.

Bottom, V. E. (1982). *Introduction to Quartz Crystal Unit Design.* New York: Van Nostrand–Reinhold.

Carver, B. (1993). "High Performance Crystal Filter Design." *Communications Quarterly,* Winter.

Colin (1958). "Narrow Bandpass Filters Using Identical Crystals Design by the Image Parameter Method" (in French). *Cables and Transmission,* Vol. 21, April 1967.

Cohn, S. (1957). "Direct Coupled Resonator Filters." *Proceedings of IRE,* February.

Cohn, S. (1959). "Dissipation Loss in Multiple Coupled Resonators." *Proceedings of IRE,* August.

Dishal, ***. (1965). "Modern Network Theory Design of Single Sideband Crystal Ladder Filters." *Proceedings of IEEE,* Vol. 53, September.

Gottfried, H. L. (1958). "An Inexpensive Crystal-Filter I.F. Amplifier." *QST,* February.

Hamish (1962). "An Introduction to Crystal Filters." *RSGB Bulletin,* January and February.

Hardcastle, J. A. (1978). "Some Experiments with High-Frequency Ladder Crystal Filters." *QST,* December; and *Radio Communications* (RSGB) January, February, and September (1977).

Hardcastle, J. A. (1980). "Ladder Crystal Filter Design." *QST,* November, pp. 22–23.

Hayward, W. (1982). "A Unified Approach to the Design of Crystal Ladder Filters." *QST,* May, pp. 21–27.

Hayward, W. (1987). "Designing and Building Simple Crystal Filters." *QST,* July, pp. 24–29.

Makhinson, J. (1995). "Designing and Building High-Performance Crystal Ladder Filters," *QEX,* No. 155, January 1995.

Pochet,*** . (1976). "Crystal Ladder Filters," Technical Topics, *Radio Communications,* September; and *Wireless World* July (1977).

"Refinements in Crystal Ladder Filter Design." *QEX,* No. 160, June 1995 (CD-ROM version).

Sabin, W. E. (1996). "The Mechanical Filter in HF Receiver Design." *QEX,* March.

Sykes. "A New Approach to the Design of High-Frequency Crystal Filters." *Bell System Monograph 3180.* n.d.

Van Roberts, W. B. (1953). "Magnetostriction Devices and Mechanical Filters for Radio Frequencies Part I: Magnetostriction Resonators." *QST,* June.

Van Roberts, W. B. (1953). "Magnetostriction Devices and Mechanical Filters for Radio Frequencies Part II: Filter Applications." *QST,* July.

Van Roberts, W. B. (1953). "Magnetostriction Devices and Mechanical Filters for Radio Frequencies Part III: Mechanical Filters." *QST,* August.

Vester, B. H. (1959). "Surplus-Crystal High-Frequency Filters: Selectivity at Low Cost." *QST,* January, pp. 24–27.

Zverev, A. L. (1967). *Handbook of Filter Synthesis.* New York: Wiley.

Interpreting radio receiver specifications

Radio receivers are a key element in radio communications and broadcasting systems. This chapter presents some of the more important receiver specification parameters. It will help you understand receiver spec sheets and lab test results.

A radio receiver must perform two basic functions:

- It must respond to, detect, and demodulate desired signals
- It must not respond to, detect, or be adversely affected by undesired signals.

Both functions are necessary, and weakness in either makes a receiver a poor bargain. The receiver's performance specifications tell us how well the manufacturer claims that their product does these two functions.

A hypothetical radio receiver

Figure 9-1 shows the block diagram of a simple communications receiver. We will use this hypothetical receiver as the basic generic framework for evaluating receiver performance. The design in Fig. 9-1 is called a *superheterodyne receiver* and is representative of a large class of radio receivers; it covers the vast majority of receivers on the market. Other designs, such as the tuned radio frequency (TRF) and direct-conversion receivers (DCR), are simply not in widespread commercial use today.

Heterodyning

The main attribute of the superheterodyne receiver is that it converts the radio signal's RF frequency to a standard frequency for further processing. Although today the new frequency, called the *intermediate frequency* (IF), might be either higher or lower than the RF frequencies, early superheterodyne receivers were always down-converted to a lower IF frequency (IF < RF). The reason was purely practical; in those days, higher frequencies were more difficult to process than lower

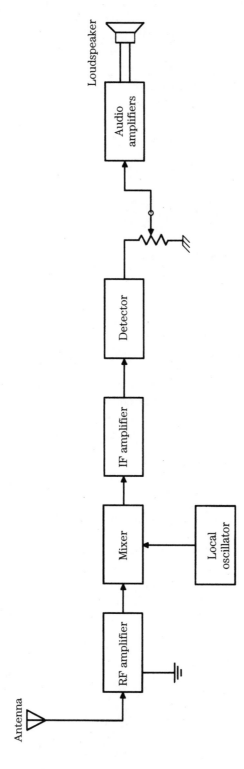

9-1 Block diagram of a single-conversion superheterodyne receiver.

frequencies. Even today, because variable tuned circuits still tend to offer different performance over the band being tuned, converting to a single IF frequency and obtaining most of the gain and selectivity functions at the IF allows more uniform overall performance over the entire range being tuned.

A superheterodyne receiver works by frequency converting ("heterodyning"— the "super" part is 1920s vintage advertising hype) the RF signal. This occurs by nonlinearly mixing the incoming RF signal with a local oscillator (LO) signal. When this process is done, disregarding noise, the output spectrum will contain a large variety of signals according to:

$$F_O = mF_{RF} \pm nF_{LO}, \tag{9-1}$$

where

F_{RF} is the frequency of the RF signal
F_{LO} is the frequency of the local oscillator
m and n are either zero or integers $(0, 1, 2, 3 \ldots n)$.

Equation (9-1) means that there will be a large number of signals at the output of the mixer, although for the most part the only ones that are of immediate concern to understanding basic superheterodyne operation are those for which m and n are either 0 or 1. Thus, for our present purpose, the output of the mixer will be the fundamentals (F_{RF} and F_{LO}) and the second-order products ($F_{LO} - F_{RF}$ and $F_{LO} + F_{RF}$), as seen in Fig. 9-2. Some mixers, notably those described as double-balanced mixers (DBM), suppress F_{RF} and F_{LO} in the mixer output, so only the second-order sum and difference frequencies exist with any appreciable amplitude. This case is simplistic and is used only for this discussion. Later on, we will look at what happens when third-order ($2F_1 \pm F_2$ and $2F_2 \pm F_1$) and fifth-order ($3F_1 \pm 2F_2$ and $3F_2 \pm 2F_1$) become large.

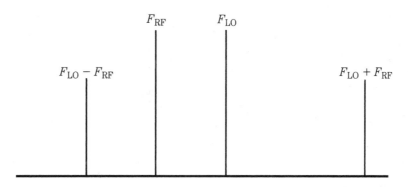

9-2 Combining the LO and RF frequencies in a mixer product second-order products of LO − RF and LO + RF.

Notice that the local oscillator frequency can be either higher than the RF frequency (high-side injection) or lower than the RF frequency (low-side injection). There is ordinarily no practical reason to prefer one over the other, except that it will make a difference whether the main tuning dial reads high-to-low or low-to-high.

The candidates for IF are the sum (LO + RF) and difference (LO − RF) second-order products found at the output of the mixer. A high-Q tuned circuit following the mixer will select which of the two are used. Consider an example. Suppose a superhet radio has an IF frequency of 455 kHz and the tuning range is 540 to 1700 kHz. Because the IF is lower than any frequency within the tuning range, the difference frequency will be selected for the IF. The local oscillator is set to be high-side injection, so it will tune from 540 + 455 = 995 kHz to 1700 + 455 = 2155 kHz.

Front-end circuits

The principal task of the "front-end" section of the receiver in Fig. 9-1 is to perform the frequency conversion. But in many radio receivers, there are additional functions. In some cases (but not all), an RF amplifier will be used ahead of the mixer. Typically, these amplifiers have a gain of 3 to 10 dB, with 5 to 6 dB being very common. The tuning for the RF amplifier is sometimes a broad bandpass fixed frequency filter that admits an entire band. In other cases, it is a narrow band, but variable frequency, tuned circuit.

Intermediate frequency (IF) amplifier

The IF amplifier is responsible for providing most of the gain in the receiver as well as the narrowest bandpass filtering. It is a high-gain, usually multistaged, single-frequency tuned radio-frequency amplifier. For example, one receiver block diagram lists 120 dB of gain from antenna terminals to audio output, of which 85 dB are provided in the 8.83-MHz IF amplifier chain. In the example of Fig. 9-1, the receiver is a single-conversion design, so there is only one IF amplifier section.

Detector

The detector demodulates the RF signal and recovers whatever audio (or other information) is to be heard by the listener. In a straight AM receiver, the detector will be an ordinary half-wave rectifier and ripple filter and is called an *envelope detector.* In other detectors, notably double-sideband suppressed carrier (DSBSC), single-sideband suppressed carrier (SSBSC or SSB), or continuous wave (CW or Morse telegraphy), a second local oscillator, usually called a *beat frequency oscillator* (BFO), operating near the IF frequency, is heterodyned with the IF signal. The resultant difference signal is the recovered audio. That type of detector is called a *product detector.* Many AM receivers today have a sophisticated synchronous detector rather than the simple envelope detector.

Audio amplifiers

The audio amplifiers are used to finish the signal processing. They also boost the output of the detector to a usable level to drive a loudspeaker or set of earphones. The audio amplifiers are sometimes used to provide additional filtering. It is quite common to find narrowband filters to restrict audio bandwidth or notch filters to eliminate interfering signals that make it through the IF amplifiers intact.

Three basic areas of receiver performance must be considered. Although interrelated, they are sufficiently different to merit individual consideration: noise, static, and dynamic. We will look at all of these areas, but first let's look at the units of measure that we will use in this series.

Units of measure

Input signal voltage

Input signal levels, when specified as a voltage, are typically stated in either microvolts (μV) or nanovolts (nV); the volt is simply too large a unit for practical use on radio receivers. Signal input voltage (or sometimes power level) is often used as part of the sensitivity specification or as a test condition for measuring certain other performance parameters.

Two forms of signal voltage that are used for input voltage specification: source voltage (V_{EMF}) and potential difference (V_{PD}), as illustrated in Fig. 9-3 (after Dyer, 1993). The source voltage (V_{EMF}) is the open terminal (no load) voltage of the signal generator or source while the potential difference (V_{PD}) is the voltage that appears across the receiver antenna terminals with the load connected (the load is the receiver antenna input impedance, R_{in}). When $R_s = R_{in}$, the preferred "matched impedances" case in radio receiver systems, the value of V_{PD} is one-half V_{EMF}. This can be seen in Fig. 9-3 by noting that R_s and R_{in} form a voltage divider network driven by V_{EMF} and with V_{PD} as the output.

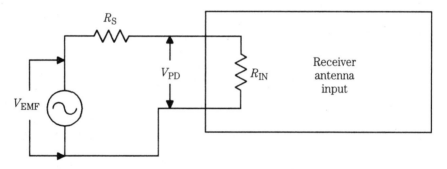

9-3 Equivalent circuit of a receiver input network.

dBm

This unit refers to decibels relative to 1 mW dissipated in a 50-Ω resistive impedance (defined as the 0-dBm reference level) and is calculated from 10 log ($P_W/0.001$) or 10 log (P_{MW}). In the noise voltage case, 0.028 μV in 50 Ω, the power is $V^2/50$, or 5.6×10^{-10} W, which is 5.6×10^{-7} mW. In dBm notation, this value is 10 log (5.6×10^{-7}), or -62.5 dBm.

dBmV

This unit is used in television receiver systems in which the system impedance is 75 Ω rather than the 50 Ω normally used in other RF systems. It refers to the signal voltage, measured in decibels, with respect to a signal level of 1 millivolt (1 mV) across a 75-Ω resistance (0 dBmv). In many TV specs, 1 mV is the full quieting signal that produces no "snow" (i.e., noise) in the displayed picture.

dBμV

This unit refers to a signal voltage, measured in decibels, relative to 1 μV developed across a 50-Ω resistive impedance (0 dBμV). For the case of our noise signal voltage, the level is 0.028 μV, which is the same as -31.1 dBμV. The voltage used for this measurement is usually the V_{EMF}, so to find V_{PD}, divide it by two after converting dBμV to μV. To convert dBμV to dBm, merely subtract 113; i.e., 100 dBμV = -13 dBm.

It requires only a little algebra to convert signal levels from one unit of measure to another. This job is sometimes necessary when a receiver manufacturer mixes methods in the same specifications sheet. In the case of dBm and dBμV, 0 dBμV is 1 μV V_{EMF}, or a V_{PD} of 0.5 μV, applied across 50 Ω, so the power dissipated is 5×10^{-15} watts, or -113 dBm. To convert from dBμV to dBm add 107 dB to the dBm figure.

Noise

A radio receiver must detect signals in the presence of noise. The signal-to-noise ratio (SNR) is the key here because a signal must be above the noise level before it can be successfully detected and used.

Noise comes in a number of different guises, but for sake of this discussion, you can divide them into two classes[1]: sources external to the receiver and sources internal to the receiver. You can do little about the external noise sources, for they consist of natural and human-made electromagnetic signals that fall within the passband of the receiver. Figure 9-4 shows an approximation of the external noise situation from the middle of the AM broadcast band to the low end of the VHF region. You must select a receiver that can cope with external noise sources—especially if the noise sources are strong.

Some natural external noise sources are extraterrestrial. For example, if you aim a beam antenna at the eastern horizon prior to sunrise, a distinct rise of noise level occurs as the Sun slips above the horizon—especially in the VHF region. The reverse occurs in the west at sunset, but is less dramatic, probably because atmospheric ionization decays much slower than it is generated. During World War II, it is reported that British radar operators noted an increase in received noise level any time the Milky Way was above the horizon, decreasing the range at which they could detect in-bound bombers. There is also some well-known, easily observed noise from the planet Jupiter in the 18- to 30-MHz band (Carr, 1994 and 1995).

The receiver's internal noise sources are affected by the design of the receiver. Ideal receivers produce no noise of their own, so the output signal from the ideal receiver would contain only the noise that was present at the input along with the radio signal. But real receiver circuits produce a certain level of internal noise of their own. Even a simple fixed-value resistor is noisy. Figure 9-5A shows the equivalent circuit for an ideal, noise-free resistor. Figure 9-5B shows a practical real-world re-

---

1 It is said that people are divided into two classes: those who divide everything into two classes and those who don't.

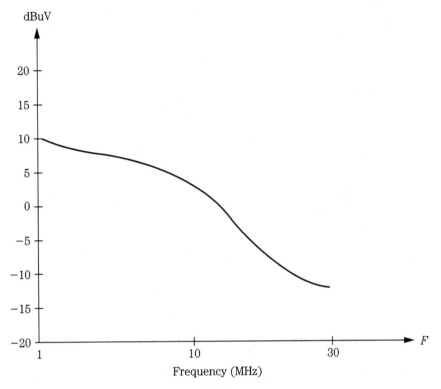

dBuV

Frequency (MHz)

9-4 External noise from about 1 to 140 MHz.

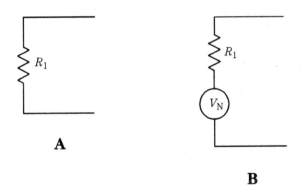

A

B

9-5 (A) Ideal resistor and (B) resistor with noise thermal source.

sistor. The noise in the real-world resistor is represented in Fig. 9-5B by a noise voltage source, V_n, in series with the ideal, noise-free resistance, R_i. At any temperature above Absolute Zero (0°K or about $-273°C$) electrons in any material are in constant random motion. Because of the inherent randomness of that motion, however, there is no detectable current in any one direction. In other words, electron drift in any single direction is canceled over even short time periods by equal drift in the opposite direction. Electron motions are therefore statistically decorrelated. There is, however, a continuous series of random current pulses generated in the material, and those pulses are seen by the outside world as noise signals.

If a shielded 50-Ω resistor is connected across the antenna input terminals of a radio receiver (Fig. 9-6), the noise level at the receiver output will increase by a predictable amount over the short-circuit noise level. Noise signals of this type are called by several names: *thermal agitation noise, thermal noise,* or *Johnson noise.* This type of noise is also called *white noise* because it has a very broadband (nearly gaussian) spectral density. The thermal noise spectrum is dominated by midfrequencies and is essentially flat. The term *white noise* is a metaphor developed from white light, which is composed of all visible color frequencies. The expression for such noise is:

$$V_n = \sqrt{4\,K\,T\,B\,R},\qquad(9\text{-}2)$$

where

V_n is the noise potential in volts (V)
K is Boltzmann's constant (1.38×10^{-23} J/°K)
T is the temperature in degrees Kelvin (°K), normally set to 290 or 300°K by convention.
R is the resistance in ohms (Ω)
B is the bandwidth in hertz (Hz).

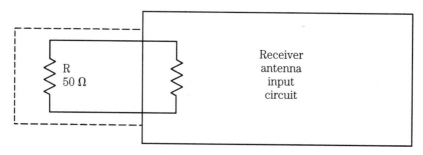

9-6 Noise test set-up for demonstration purposes only.

Example

Find the noise voltage in a circuit containing only a 50-Ω resistor if the bandwidth is 1000 Hz.

$$V_n = \sqrt{4\,K\,T\,B\,R}$$

$$V_n = \sqrt{(4)(1.38 \times 10^{-23}\text{J/K})\,(290\text{ K})\,(1000\text{ Hz})\,(50\ \Omega)}$$

$$V_n = \sqrt{8 \times 10^{-16}} = 2.8 \times 10^{-8}\text{ volts} = 0.028\ \mu\text{V}.$$

Table 9-1 and Fig. 9-7 show noise values for a 50-Ω resistor at various bandwidths out to 5 and 10 kHz, respectively.

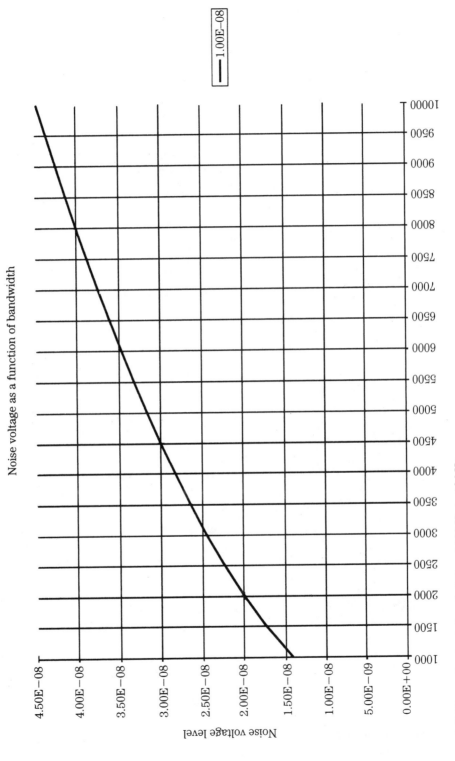

Noise voltage as a function of bandwidth

9-7 Thermal noise in bandwidths of 100 Hz to 10 kHz.

**Table 9-1. Noise levels
tabulated for three bandwidths
commonly found in receivers**

BW (Hz)	Noise
500	1.00E–08
1000	1.41E–08
1500	1.73E–08
2000	2.00E–08
2500	2.24E–08
3000	2.45E–08
3500	2.65E–08
4000	2.83E–08
4500	3.00E–08
5000	3.16E–08
5500	3.32E–08
6000	3.46E–08
6500	3.61E–08
7000	3.74E–08
7500	3.87E–08
8000	4.00E–08
8500	4.12E–08
9000	4.24E–08
9500	4.36E–08
10000	4.47E–08

Because different bandwidths are used for different reception modes, it is common practice to delete the bandwidth factor in Eq. (9-2) and write it in the form:

$$V_n = \sqrt{4\,K\,T\,R}\; V/\sqrt{\text{Hz}}. \tag{9-3}$$

With Eq. (9-3), you can find the noise voltage for any particular bandwidth by taking its square root and multiplying it by the equation. This equation is essentially the solution of the previous equation normalized for a 1-Hz bandwidth.

Signal-to-noise ratio (SNR or S_n)

Receivers can be evaluated on the basis of signal-to-noise ratio (*S/N* or "SNR"), sometimes denoted S_n. The goal of the designer is to enhance the SNR as much as possible. Ultimately, the minimum signal level detectable at the output of an amplifier or radio receiver is that level which appears just above the noise floor level. Therefore, the lower the system noise floor, the smaller the minimum allowable signal.

Noise factor, noise figure, and noise temperature

The noise performance of a receiver or amplifier can be defined in three different, but related, ways: noise factor (F_n), noise figure (N_F), and equivalent noise temperature (T_e); these properties are definable as a simple ratio, decibel ratio, or Kelvin temperature, respectively.

Noise factor (F_n) For components such as resistors, the noise factor is the ratio of the noise produced by a real resistor to the simple thermal noise of an ideal resistor. The noise factor of a radio receiver (or any system) is the ratio of output noise power (P_{no}) to input noise power (P_{ni}):

$$F_n = \frac{P_{no}}{P_{ni}}\bigg|\, T = 290°K \qquad (9\text{-}4)$$

In order to make comparisons easier, the noise factor is usually measured at the standard temperature (T_o) of 290°K (standardized room temperature); although in some countries 299°K or 300°K are commonly used (the differences are negligible).

It is also possible to define noise factor F_n in terms of output and input *S/N* ratio:

$$F_n = \frac{S_{ni}}{S_{no}}, \qquad (9\text{-}5)$$

where

  S_{ni} is the input signal-to-noise ratio

  S_{no} is the output signal-to-noise ratio.

Noise figure (N_F) The noise figure is a frequently used measure of a receiver's "goodness," or its departure from "idealness." Thus, it is a figure of merit. The noise figure is the noise factor converted to decibel notation:

$$N_F = 10 \log (F_n), \qquad (9\text{-}6)$$

where

  N_F = the noise figure in decibels (dB)

  F_n = the noise factor

  log refers to the system of base-10 logarithms.

Noise temperature (T_e) The noise "temperature" is a means for specifying noise in terms of an equivalent temperature. Evaluating the noise equations shows that the noise power is directly proportional to temperature in degrees Kelvin and also that noise power collapses to zero at the temperature of Absolute Zero (0°K).

Notice that equivalent noise temperature T_e is not the physical temperature of the amplifier but rather a theoretical construct that is an equivalent temperature that produces that amount of noise power. The noise temperature is related to the noise factor by:

$$T_e = (F_n - 1)\, T_o \qquad (9\text{-}7)$$

and to noise figure by:

$$T_e = KT_o \log^{-1}\left(\frac{\text{N.F.}}{10}\right) - 1. \qquad (9\text{-}8)$$

Noise temperature is often specified for receivers and amplifiers in combination with, or in lieu of the noise figure.

Noise in cascade amplifiers

A noise signal is seen by any following amplifier as a valid input signal. Each stage in the cascade chain (Fig. 9-8) amplifies both signals and noise from previous

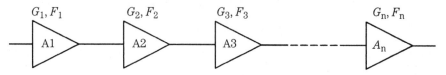

9-8 Cascade amplifier chain.

stages and also contributes some additional noise of its own. Thus, in a cascade amplifier, the final stage sees an input signal that consists of the original signal and noise amplified by each successive stage plus the noise contributed by earlier stages. The overall noise factor for a cascade amplifier can be calculated from Friis' noise equation:

$$F_n = F_1 + \frac{F_2 - 1}{G_1} + \frac{F_3 - 1}{G_1 G_2} + \cdots + \frac{F_n - 1}{G_1 G_2 \ldots G_{n-1}}, \tag{9-9}$$

where

F_n is the overall noise factor of N stages in cascade
F_1 is the noise factor of stage 1
F_2 is the noise factor of stage 2
F_n is the noise factor of the nth stage
G_1 is the gain of stage 1
G_2 is the gain of stage 2
G_{n-1} is the gain of stage $(n - 1)$.

As you can see from Friis' equation, the noise factor of the entire cascade chain is dominated by the noise contribution of the first stage or two. High-gain, low-noise amplifiers typically use a low-noise amplifier circuit for only the first stage or two in the cascade chain. Thus, you will find a LNA at the feedpoint of a satellite receiver's dish antenna, and possibly another one at the input of the receiver module itself, but other amplifiers in the chain are more modest.

The matter of signal-to-noise ratio (*S/N*) is sometimes treated in different ways that each attempt to crank some reality into the process. The signal-plus-noise-to-noise ratio (*S + N/N*) is found quite often. As the ratios get higher, the *S/N* and *S + N/N* converge (only about 0.5-dB difference at ratios as little as 10 dB). Still another variant is the SINAD (signal-plus-noise-plus-distortion-to-noise) ratio. The SINAD measurement takes into account most of the factors that can deteriorate reception.

Receiver noise floor

The noise floor of the receiver is a statement of the amount of noise produced by the receiver's internal circuitry and directly affects the sensitivity. The noise floor is typically expressed in dBm. The noise floor specification is evaluated as follows: the more negative the better. The best receivers have noise floor numbers of less than −130 dBm, and some very good receivers offer numbers of −115 dBm to −130 dBm.

The noise floor is directly dependent on the bandwidth used to make the measurement. Receiver advertisements usually specify the bandwidth, but be careful to note whether the bandwidth that produced the very good performance numbers is also the bandwidth that you'll need for the mode of transmission you want to receive.

If, for example, you are interested only in weak 6-kHz wide AM signals and the noise floor is specified for a 250-Hz CW filter then the noise floor might be too high for your use.

Static measures of performance

The two principal static levels of performance for radio receivers are sensitivity and selectivity. The sensitivity refers to the level of input signal required to produce a usable output signal (variously defined). The selectivity refers to the ability of the receiver to reject adjacent channel signals (again, variously defined). Look at both of these factors. Remember, however, that in modern, high-performance radio receivers the static measures of performance might also be the least relevant compared with the dynamic measures.

Sensitivity

Sensitivity is a measure of the receiver's ability to pick up ("detect") signals and is usually specified in microvolts (μV). A typical specification might be "0.5-μV sensitivity." The question to ask is "relative to what?" The sensitivity number in microvolts is meaningless unless the test conditions are specified. For most commercial receivers, the usual test condition is the sensitivity required to produce a 10-dB signal-plus-noise-to-noise ($S + N/N$) ratio in the mode of interest. For example, if only one sensitivity figure is given, one must find out what bandwidth is being used: 5 to 6 kHz for AM, 2.6 to 3 kHz for single sideband, 1.8 kHz for radioteletype, or 200 to 500 Hz for CW.

Bandwidth affects sensitivity measurements. Indeed, one place where "creative spec writing" takes place for commercial receivers is that the advertisements will cite the sensitivity for a narrow-bandwidth mode (e.g., CW), and the other specifications are cited for a more commonly used wider bandwidth mode (e.g., SSB). In one particularly egregious example, an advert claimed a sensitivity number that was applicable to the CW mode bandwidth only, yet the 270-Hz CW filter was an expensive option that had to be specially ordered separately!

The amount of sensitivity improvement is seen by some simple numbers. Recall that a claim of "$X \mu V$" sensitivity refers to some standard such as "$X \mu V$ to produce a 10-dB signal-to-noise ratio." Consider the case where the main mode for a high-frequency (HF) shortwave receiver is AM (for international broadcasting), and the sensitivity is 1.9 μV for 10 dB SNR, and the bandwidth is 5 kHz. If the bandwidth were reduced to 2.8 kHz for SSB then the sensitivity improves by the square root of the ratio, or $\sqrt{5/2.8}$. If the bandwidth is further reduced to 270 Hz (i.e., 0.27 kHz) for CW then the sensitivity for 10 dB SNR is $\sqrt{5/0.27}$. The 1.9-μV AM sensitivity therefore translates to 1.42 μV for SSB and 0.44 μV for CW. If only the CW version is given then the receiver might be made to look a whole lot better, even though the typical user might never use the CW mode (note differences in Fig. 9-9).

The sensitivity differences also explain why weak SSB signals can be heard under conditions when AM signals of similar strength have disappeared into the noise or why the CW mode has as much as 20-dB advantage over SSB, *ceterus paribus*.

In some receivers, the difference in mode (AM, SSB, RTTY, CW, etc.) can conceivably result in sensitivity differences that are more than the differences in the

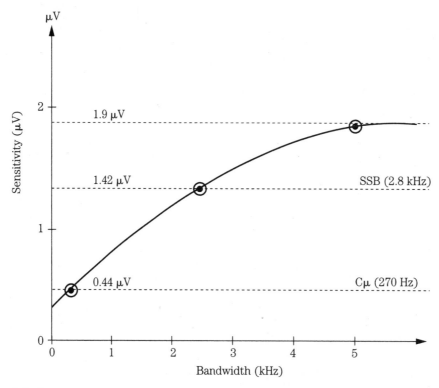

9-9 Comparison of signal levels required for 10-dB SNR for 270-Hz and 2.8 and 5-kHz bandwidths.

bandwidths associated with the various modes. The reason is that there is sometimes a "processing gain" associated with the type of detector circuit used to demodulate the signal at the output of the IF amplifier. A simple AM envelope detector is lossy because it consists of a simple diode (1N60, etc.) and an RC filter (a passive circuit). Other detectors (product detector for SSB and synchronous AM detectors) have their own signal gain, so they might produce better sensitivity numbers than the bandwidth suggests.

Another indication of sensitivity is minimum detectable signal (MDS) and is usually specified in dBm. This signal level is the signal power at the antenna input terminal of the receiver required to produce some standard $S + N/N$ ratio, such as 3 or 10 dB (Fig. 9-10). In radar receivers, the MDS is usually described in terms of a single pulse return and a specified $S+N/N$ ratio. Also, in radar receivers, the sensitivity can be improved by integrating multiple pulses. If N return pulses are integrated, then the sensitivity is improved by a factor of N if coherent detection is used, and $\sqrt{N}$ if noncoherent detection is used.

Modulated signals represent a special case. For those sensitivities, it is common to specify the conditions under which the measurement is made. For example, in AM receivers, the sensitivity to achieve 10-dB SNR is measured with the input signal modulated 30% by a 400- or 1000-Hz sinusoidal tone.

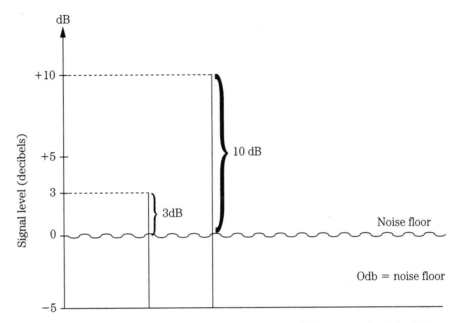

9-10 Minimum detectable signal (MDS) defined for two different standards (3-dB SNR and 10-dB SNR).

An alternate method is sometimes used for AM sensitivity measurements—especially in servicing consumer radio receivers (where SNR might be a little hard to measure with the equipment normally available to technicians who work on those radios). This is the "standard output conditions" method. Some manuals specify the audio signal power or audio signal voltage at some critical point, when the 30% modulated RF carrier is present. In one automobile radio receiver, the sensitivity was specified as "X μV to produce 400 mW across 8-Ω resistive load substituted for the loudspeaker when the signal generator is modulated 30% with a 400-Hz audio tone." The cryptic note on the schematic showed an output sine wave across the loudspeaker with the label "400 mW in 8 Ω (1.79 volts), @30% mod. 400 Hz, 1 μV RF." The sensitivity is sometimes measured essentially the same way, but the signal levels will specify the voltage level that will appear at the top of the volume control, or output of the detector/filter, when the standard signal is applied. Thus, there are two ways seen for specifying AM sensitivity: 10 dB SNR and standard output conditions.

There are also two ways to specify FM receiver sensitivity. The first is the 10-dB SNR method (i.e., the number of microvolts of signal at the input terminals required to produce a 10-dB SNR when the carrier is modulated by a standard amount). The measure of FM modulation is deviation expressed in kilohertz. Sometimes, the full deviation for that class of receiver is used and for others a value that is 25 to 35% of full deviation is specified.

The second way to measure FM sensitivity is the level of signal required to reduce the no-signal noise level by 20 dB. This is the "20-dB quieting sensitivity of the receiver." If you tune between signals on a FM receiver, you will hear a loud "hiss" signal. Some of that noise is externally generated, while some is internally generated. When a FM signal appears in the passband, that hiss is suppressed—even if the FM carrier is unmodulated. The quieting sensitivity of a FM receiver is a statement of the number of microvolts required to produce some standard quieting level, usually 20 dB.

Pulse receivers, such as radar and pulse communications units, often use the tangential sensitivity as the measure of performance, which is the amplitude of pulse signal required to raise the noise level by its own RMS amplitude (Fig. 9-11).

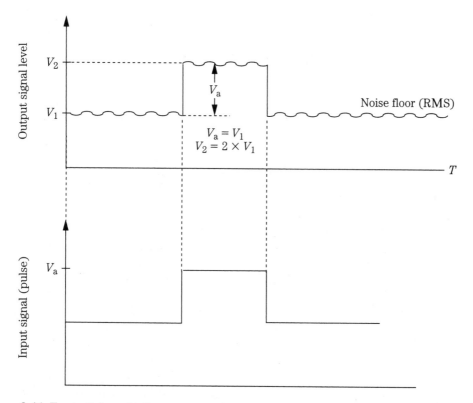

9-11 Tangential sensitivity measurement.

Selectivity

Although no receiver specification is unimportant, if one had to choose between sensitivity and selectivity, the proper choice most of the time would be to take selectivity.

Selectivity is the measure of a receiver's ability to reject adjacent channel interference. Or put another way, it's the ability to reject interference from signals on frequencies close to the desired signal frequency.

To understand selectivity requirements, you must first understand a little bit of the nature of real radio signals. An unmodulated radio carrier theoretically has an infinitesimal (near-zero) bandwidth (although all real unmodulated carriers have a very narrow, but nonzero, bandwidth because they are modulated by noise and other artifacts). As soon as the radio signal is modulated to carry information, however, the bandwidth spreads.

An on/off telegraphy (CW) signal spreads out either side of the carrier frequency an amount that depends on the sending speed and the shape of the keying waveform.

Figure 9-12 compares two radioteletype signals with 200 Hz (Fig. 9-12A) and 800 Hz (Fig. 9-12B) mark/space separation. Notice how the sidebands spread out from the main mark and space signals.

An AM signal spreads out an amount equal to twice the highest audio modulating frequencies. For example, a communications AM transmitter will have audio components from 300 to 3000 Hz, so the AM waveform will occupy a spectrum that is equal to the carrier frequency (F) plus/minus the audio bandwidth ($F \pm 3000$ Hz in the case cited).

An FM carrier spreads out according to the deviation. For example, a narrowband FM landmobile transmitter with 5-kHz deviation spreads out ±5 kHz and FM broadcast transmitters spread out ±75 kHz.

An implication of the radio signal bandwidth is that the receiver must have sufficient bandwidth to recover all of the signal. Otherwise, information might be lost

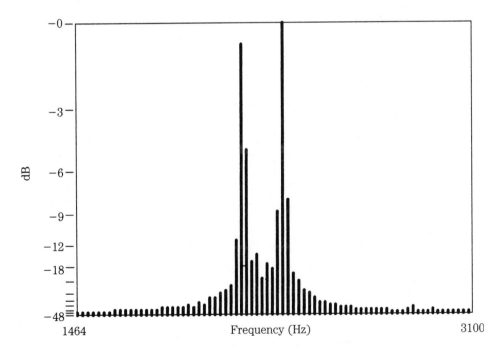

A

9-12 (A) Spectrum for 200-Hz RTTY and (B) spectrum for 800-Hz RTTY.

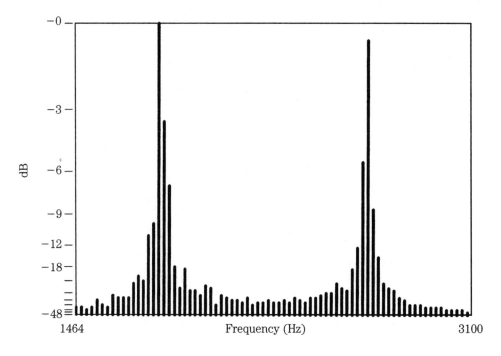

B

9-12 *Continued.*

and the output is distorted. On the other hand, allowing too much bandwidth increases the noise picked up by the receiver and thereby deteriorates the SNR. The goal of the selectivity system of the receiver is to match the bandwidth of the receiver to that of the signal. That is why receivers will use a bandwidth of 270 or 500 Hz for CW, 2 to 3 kHz for SSB, and 4 to 6 kHz for AM signals. They allow you to match the receiver bandwidth to the transmission type.

The selectivity of a receiver has a number of aspects that must be considered: front-end bandwidth, IF bandwidth, IF shape factor, and the ultimate (distant frequency) rejection.

Front-end bandwidth

The "front end" of a modern superheterodyne radio receiver is the circuitry between the antenna input terminal and the output of the first mixer stage. Front-end selectivity is important because it keeps out-of-band signals from afflicting the receiver. For example, AM broadcast band transmitters located nearby can easily overload a poorly designed shortwave receiver. Even if these signals are not heard by the operator (as they often are), they can desensitize a receiver or create harmonics and intermodulation products that show up as "birdies" or other types of interference on the receiver. Strong local signals can take a lot of the receiver's dynamic range and thereby make it harder to hear weak signals.

In some "crystal video" microwave receivers, that front end might be wide open without any selectivity at all, but in nearly all other receivers, some form of frequency selection will be present.

Two forms of frequency selection are typically found. A designer might choose to use only one of them in a design. Alternatively, both might be used in the design, but separately (operator selection). Or finally, both might be used together. These forms can be called the *resonant frequency filter* (Fig. 9-13A) and *bandpass filter* (Fig. 9-13B) approaches.

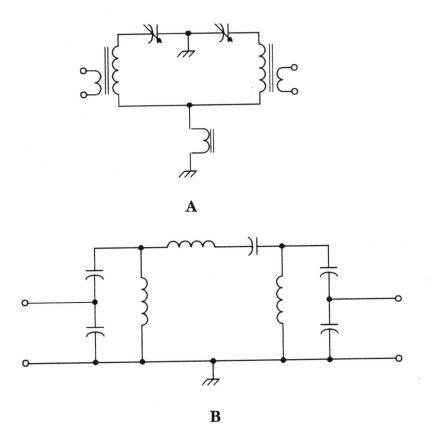

A

B

9-13 (A) Resonant filter and (B) bandpass filter.

The resonant frequency approach uses LC elements tuned to the desired frequency to select which RF signals reach the mixer. In some receivers, these LC elements are designed to track with the local oscillator that sets the operating frequency. That's why you see two-section variable capacitors for AM broadcast receivers with two different capacitance ranges for the two sections; one section tunes the LO and the other section tunes the tracking RF input. In other designs, a separate tuning knob ("preselector" or "antenna") is used.

The other approach uses a suboctave bandpass filter to admit only a portion of the RF spectrum into the front end. For example, a shortwave receiver that is designed to take the HF spectrum in 1-MHz pieces might have an array of RF input bandpass filters that are each 1 MHz wide (e.g., 9 to 10 MHz).

In addition to these reasons, front-end selectivity also helps improve a receiver's image rejection and 1st IF rejection capabilities.

Image rejection

An image in a superheterodyne receiver is a signal that appears at twice the IF distant from the desired RF signal and is located on the opposite side of the LO frequency from the desired RF signal. In Fig. 9-14, a superheterodyne operates with a 455 kHz (i.e., 0.455 MHz) IF and is turned to 24.0 MHz (F_{RF}). Because this receiver uses low-side LO injection, the LO frequency F_{LO} is 24.0–0.455, or 23.545 MHz. If a signal appears at twice the IF below the RF (i.e., 910 kHz below F_{RF}) and reaches the mixer then it too has a difference frequency of 455 kHz so it will pass right through the IF filtering as a valid signal. The image rejection specification tells how well this image frequency is suppressed. Normally, anything over about 70 dB is considered good.

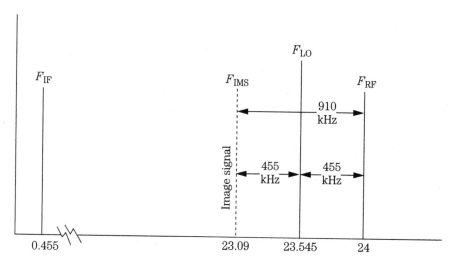

9-14 The image response problem.

Tactics to reduce image response vary with the design of the receiver. The best approach, at design time, is to select an IF frequency that is high enough that the image frequency will fall outside the passband of the receiver front end. Some modern HF receivers use an IF of 8.83, 9, or 10.7 MHz or something similar. For image rejection, these frequencies are considerably better than 455-kHz receivers in the higher HF bands. However, a common trend is to *double conversion.* In most such designs, the first IF frequency is considerably higher than the RF, being in the range 35 to 60 MHz (50 MHz is common). This high IF makes it possible to suppress the VHF images with a simple low-pass filter. If the 24.0-MHz signal (above) were first up converted to 50 MHz (74-MHz LO), for example, the image would be at 124 MHz. The second conversion brings the IF down to one of the frequencies mentioned, or even 455 kHz. The lower frequencies are preferable to 50 MHz for bandwidth selectivity reasons because good-quality crystal, ceramic, or mechanical filters in those ranges filters are easily available.

First IF rejection

The first IF rejection specification refers to how well a receiver rejects radio signals operating on the receiver's first IF frequency. For example, if your receiver has a first IF of 8.83 MHz, it must be able to reject radio signals operating on that frequency when the receiver is tuned to a different frequency. Although the shielding of the receiver is also an issue with respect to this performance, the front-end selectivity affects how well the receiver performs against first IF signals. If there is no front-end selectivity to discriminate against signals at the IF frequency, then they arrive at the input of the mixer unimpeded. Depending on the design of the mixer, they then pass directly through to the high gain IF amplifiers and be heard in the receiver output.

IF bandwidth

Most of the selectivity of the receiver is provided by the filtering in the IF amplifier section. The filtering might be LC filters (especially if the principal IF is a low frequency, such as 50 kHz), a ceramic resonator, a crystal filter, or a mechanical filter. Of these, the mechanical filter is usually regarded as best, with the crystal filter and ceramic filters coming in next.

The IF bandwidth is expressed in kilohertz and is measured from the points on the IF frequency response curve where gain drops off −3 dB from the mid-band value (Fig. 9-15). This is why you will sometimes see selectivity referred to in terms such as "6 kHz between −3 dB points."

The IF bandwidth must be matched to the bandwidth of the received signal for best performance. If a too-wide bandwidth is selected then the received signal will

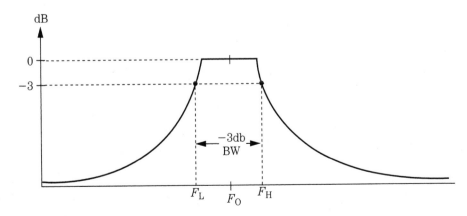

9-15 IF bandwidth selectivity.

be noisy and SNR deteriorates. If it is too narrow then you might experience diffi-culties recovering all of the information that was transmitted. For example, an AM broadcast band radio signal has audio components to 5 kHz, so the signal occupies up to 10 kHz of spectrum space (F $\pm$ 5 kHz). If a 2.8-kHz SSB IF filter is selected it will tend to sound "mushy" and distorted.

IF passband shape factor

The shape factor is a measure of the steepness of the receiver's IF passband and is taken by measuring the ratio of the bandwidth at -6 dB to the bandwidth at -60 dB. The general rule is that the closer these numbers are to each other, the bet-ter the receiver. Anything in the 1:1.5 to 1:1.9 region can be considered high quality, and anything worse than 1:3 is not worth looking at for "serious" receiver uses. If the numbers are between 1:1.9 and 1:3 then the receiver could be regarded as being middling but useful.

The importance of shape factor is that it modifies the bandwidth. The cited bandwidth (e.g., 2.8 kHz for SSB) does not take into account the effects of strong signals that are just beyond those limits. Such signals can easily "punch through" the IF selectivity if the IF passband "skirts" are not steep. After all, the steeper they are, the closer a strong signal can be without messing up the receiver's operation. Thus, selecting a receiver with a shape factor as close to the 1:1 ideal as possible (it's not, by the way) will result in a more usable radio.

Distant frequency ("ultimate") rejection

This specification tells something about the receiver's ability to reject very strong signals that are located well outside the receiver's IF passband. This number is stated in negative decibels, and the higher the number the better. An excellent receiver will have values in the -60- to -90-dB range, a middling receiver will see numbers in the -45- to -60-dB range, and a terrible receiver will be -44 or worse.

Stability

The stability specification measures how much the receiver frequency drifts as time elapses or temperature changes. The LO drift sets the overall stability of the re-ceiver. This specification is usually given in terms of short-term drift and long-term drift (e.g., from crystal aging). The short-term drift is important in daily operation, and the long-term drift ultimately affects general dial calibration.

If the receiver is VFO controlled, or uses partial frequency synthesis (which combines VFO with crystal oscillators), then the stability is dominated by the VFO stability. In fully synthesized receivers, the stability is governed by the master refer-ence crystal oscillator. If either an oven-controlled crystal oscillator (OCXO) or a temperature-compensated crystal oscillator (TCXO) is used for the master refer-ence then stability on the order of 1 part in 108/°C is achievable.

For most users, the short-term stability is what is most important—especially when tuning SSB, ECSS, or RTTY signals. A common spec for a good receiver will be 50 Hz/h after a 3-h warm-up or 100 Hz/h after a 15-min warm-up. The smaller the drift the better the receiver.

The foundation of good stability is at design time. The local oscillator, or VFO portion of a synthesizer, must be operated in a cool, temperature-stable location within the equipment and must have the correct type of components. Capacitor temperature coefficients are often selected in order to cancel out temperature-related drift in inductance values.

Postdesign time changes can also help, but these are less likely to be possible today than in the past. The chief cause of drift problems is heat. In the days of tube oscillators, the heat of the tubes produced lots of heat, which created drift. One popular receiver of the 1950s was an excellent performer, except for the drift of the VFO/LO. A popular modification was to insert metal panels backed with insulating material between the heat source (an IF amplifier chain) and the VFO case. Rather dramatic improvement resulted. In most modern solid-state receivers, however, space is not so easily found for such modifications.

One phenomenon seen on low-cost receivers and homebrew receivers of doubtful merit is mechanical frequency shifts. Although not seen on most modern receivers (even some very cheap designs), it was once a serious problem on less costly models. This problem is usually seen on VFO-controlled receivers in which vibration to the receiver cabinet imparts movement to either the inductor (L) or capacitor (C) element in a LC VFO. Mechanically stabilizing these components will work wonders.

AGC range and threshold

Modern communications receivers must be able to handle signals over the range of about 1,000,000:1. Tuning across a band occupied by signals of wildly varying strengths is hard on the ears and hard on the receiver's performance. As a result, most modern receivers have an automatic gain control (AGC) circuit that smooths out these changes. The AGC will reduce gain for strong signals and increase it for weak signals (AGC can be turned off on most HF communications receivers). The AGC range is the change of input signal (in dBμV) from some reference level (e.g., 1 μV$_{EMF}$) to the input level that produces a 2-dB change in output level. Ranges of 90 to 110 dB are common.

The AGC threshold is the signal level at which the AGC begins to operate. If it is set too low, the receiver gain will respond to noise and irritate the user. If set too high, then the user will experience irritating shifts of output level as the band is tuned. AGC thresholds of 0.7 to 2.5 μV are common on decent receivers, with the better receivers being in the 0.7- to 1-μV range.

Another AGC specification sometimes seen deals with the speed of the AGC. Although sometimes specified in milliseconds, it is also frequently specified in subjective terms like "fast" and "slow." This specification refers to how fast the AGC responds to changes in signal strength. If set too fast, then rapidly keyed signals (e.g., CW) or noise transients will cause unnerving large shifts in receiver gain. If set too slow, then the receiver might as well not have an AGC. Many receivers provide two or more selections in order to accommodate different types of signals.

Now, let's examine what some authorities believe are the most important specifications for receivers used on crowded bands or in high electromagnetic interference (EMI) environments: the dynamic measures of performance. These include intermodulation distortion, 1-dB compression point, third-order intercept point, and blocking.

Dynamic performance

The dynamic performance specifications of a radio receiver are those that deal with how the receiver performs in the presence of very strong signals—either cochannel or adjacent channel. Until about the 1960s, dynamic performance was somewhat less important than static performance for most users. However, today the role of dynamic performance is probably more critical than static performance because of crowded band conditions.

There are at least two reasons for this change in outlook (Dyer, 1993). First, in the 1960s, receiver designs evolved from tubes to solid state. The new solid-state amplifiers were somewhat easier to drive into nonlinearity than tube designs. Second, there has been a tremendous increase in radio-frequency signals on the air. There are far more transmitting stations than ever before, and there are far more sources of electromagnetic interference (EMI, the pollution of the air waves) than in prior decades. With the advent of new and expanded wireless services available to an ever widening market, the situation can only worsen. For this reason, it is now necessary to pay more attention to the dynamic performance of receivers than in the past.

Intermodulation products

Understanding the dynamic performance of the receiver requires knowledge of intermodulation products (IP) and how they affect receiver operation. Whenever two signals are mixed together in a nonlinear circuit, a number of products are created according to $mF_1 \pm nF_2$, where m and n are either integers or zero. Mixing can occur in either the mixer stage of a receiver front end or in the RF amplifier (or any outboard preamplifiers used ahead of the receiver) if the RF amplifier is overdriven by a strong signal.

It is also theoretically possible for corrosion on antenna connections, or even rusted antenna screw terminals to create IPs under certain circumstances. In some alleged cases, a rusty downspout on a house rain gutter caused reradiated mixed signals. However, all such cases that I've heard of have that distant third- or fourth-party quality ("I know a guy whose best friend's brother-in-law saw . . .") that suggests a profound unconfirmability. I know of no first-hand accounts verified by a technically competent person.

The spurious IP signals are shown graphically in Fig. 9-16. Given input signal frequencies of F_1 and F_2, the main IPs are:

Second order:	$F_1 \pm F_2$
Third order:	$2F_1 \pm F_2$
	$2F_2 \pm F_1$
Fifth order:	$3F_1 \pm 2F_2$
	$3F_2 \pm 2F_1$

of these, the third-order difference signals ($2F_1 - F_2$ and $2F_2 - F_1$) are the most serious because they typically fall close to the RF frequency.

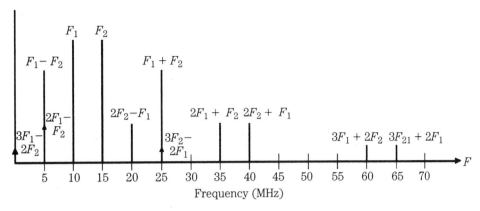

9-16 Intermodulation products of second-order, third-order, and fifth-order.

When an amplifier or receiver is overdriven, the second-order content of the output signal increases as the square of the input signal level, and the third-order responses increase as the cube of the input signal level.

Consider the case where two HF signals, $F_1 = 10$ MHz and $F_2 = 15$ MHz, are mixed together. The second-order IPs are 5 and 25 MHz; the third-order IPs are 5, 20, 35, and 40 MHz; and the fifth-order IPs are 0, 25, 60, and 65 MHz. If any of these are inside the passband of the receiver then they can cause problems. One such problem is the emergence of "phantom" signals at the IP frequencies. This effect is often seen when two strong signals (F_1 and F_2) exist and can affect the front end of the receiver, and one of the IPs falls close to a desired signal frequency, F_d. If the receiver were tuned to 5 MHz, for example, a spurious signal would be found from the $F_1 - F_2$ pair.

Another example is seen from strong in-band, adjacent-channel signals. Consider a case where the receiver is tuned to a station at 9610 kHz, and there are also very strong signals at 9600 and 9605 kHz. The near (in-band) IP products are:

Third order: 9595 kHz ($\Delta F = 15$ kHz)
9610 kHz ($\Delta F = 0$ kHz)   [On channel!]
Fifth order: 9590 kHz ($\Delta F = 20$ kHz)
9615 kHz ($\Delta F = 5$ kHz).

Notice that one third-order product is on the same frequency as the desired signal and could easily cause interference if the amplitude is sufficiently high. Other third- and fifth-order products might be within the range where interference could occur, especially on receivers with wide bandwidths.

The IP orders are theoretically infinite because there are no bounds on either m or n. However, in practical terms, because each successively higher order IP is reduced in amplitude compared with its next lower order mate, only the second-order, third-order, and fifth-order products usually assume any importance. Indeed, only the third-order is normally used in receiver specifications sheets.

–1-dB compression point

An amplifier produces an output signal that has a higher amplitude than the input signal. The transfer function of the amplifier (indeed, any circuit with output and input) is the ratio OUT/IN, so for the power amplification of a receiver RF amplifier, it is P_o/P_{in} (or, in terms of voltage, V_o/V_{in}). Any real amplifier will saturate, given a strong enough input signal (see Fig. 9-17). The dotted line represents the theoretical output level for all values of input signal (the slope of the line represents the gain of the amplifier). As the amplifier saturates (solid line), however, the actual gain begins to depart from the theoretical at some level of input signal (P_{in1}). The –1-dB compression point is that output level at which the actual gain departs from the theoretical gain by –1 dB.

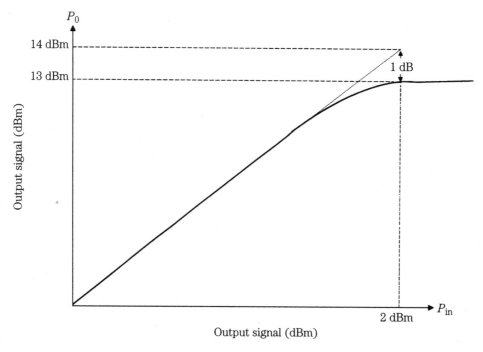

9-17 1-dB compression point phenomenon.

The –1-dB compression point is important when considering either the RF amplifier ahead of the mixer (if any) or any outboard preamplifiers that are used. The –1-dB compression point is the point at which intermodulation products begin to emerge as a serious problem. Also, harmonics are generated when an amplifier goes into compression. A sine wave is a "pure" signal because it has no harmonics (all other waveshapes have a fundamental plus harmonic frequencies). When a sine wave is distorted, however, harmonics arise. The effect of the compression phenomenon is to distort the signal by clipping the peaks, thus raising the harmonics and intermodulation distortion products.

Third-order intercept point

It can be claimed that the third-order intercept point (TIP) is the single most important specification of a receiver's dynamic performance because it predicts the performance as regards intermodulation, cross-modulation, and blocking desensitization.

Third-order (and higher) intermodulation products (IP) are normally very weak and do not exceed the receiver noise floor when the receiver is operating in the linear region. But as input signal levels increase, forcing the front end of the receiver toward the saturated nonlinear region, the IP emerges from the noise (Fig. 9-18) and begins to cause problems. When this happens, new spurious signals appear on the band and self-generated interference begins to arise.

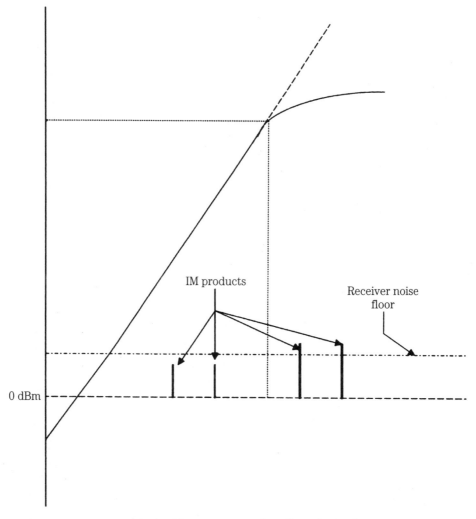

9-18 As signal input levels rise, IP spurs emerge from the noise level.

Figure 9-19 shows a plot of the output signal vs the fundamental input signal. Notice the output compression effect that was shown earlier in Fig. 9-17. The dotted gain line continuing above the saturation region shows the theoretical output that would be produced if the gain did not clip. It is the nature of third-order products in the output signal to emerge from the noise at a certain input level and increase as the cube of the input level. Thus, the slope of the third-order line increases 3 dB for every 1-dB increase in the response to the fundamental signal. Although the output response of the third-order line saturates similarly to that of the fundamental signal, the gain line can be continued to a point where it intersects the gain line of the fundamental signal. This point is the third-order intercept point (TOIP).

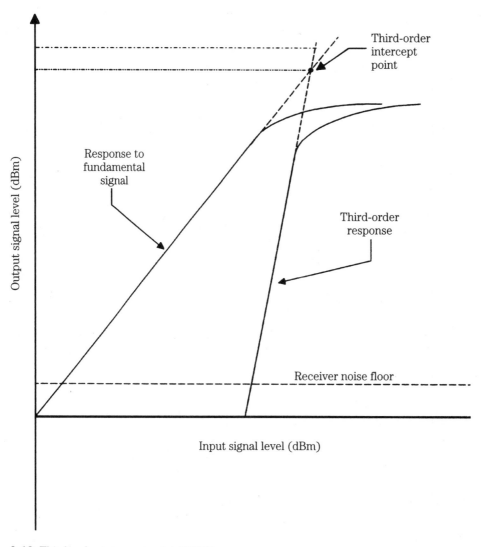

9-19 Third-order intercept point (TOIP).

Interestingly enough, one receiver feature that can help reduce IP levels back down under the noise is the use of a front-end attenuator (a.k.a. input attenuator). Even a few dB of input attenuation is often enough to drop the IPs back into the noise while afflicting the desired signals only a small amount.

Other effects that reduce the overload caused by a strong signal also help. I've seen situations in amateur radio operations where the apparent third-order performance of a receiver improved dramatically when a lower gain antenna was used. The late J. H. Thorne (K4NFU) demonstrated the effect to me on a VHF band using a spectrum analyzer/receiver. This instrument is a swept receiver that displays an output on an oscilloscope screen that is amplitude vs frequency, so a single signal shows as a spike. A strong, local ham 2-m band repeater came on the air every few seconds, and you could observe the second- and third-order IPs along with the fundamental repeater signal. Other strong signals were also on the air, but just outside the 2-m ham band. Inserting a 6-dB barrel attenuator in the input ("antenna") line eliminated the IP products, showing just the actual signals. Rotating a directional antenna away from the direction of the interfering signal will also accomplish this effect in many cases.

Preamplifiers are popular receiver accessories, but can often reduce, rather than enhance, performance. Two problems commonly occur (assuming that the preamp is a low-noise device; if not, then there are three). The best-known problem is that the preamp amplifies noise as much as signals, and although it makes the signal louder, it also makes the noise louder by the same amount. Because the signal-to-noise ratio is important, you cannot improve the situation. Indeed, if the preamp is itself noisy, it will deteriorate the SNR. The other problem is less well-known, but potentially more devastating. If the increased signal levels applied to the receiver drive the receiver nonlinear, then IPs begin to emerge. Poole (1995) reported that another 2-m operator informed him that Poole's transmission was being heard at several spots on the band. Poole discovered that the other chap was using two preamps in cascade (!) to achieve more gain and that disconnecting them caused "Poole's spurs" to evaporate. That was clearly a case of a preamplifier messing up, rather than improving, a receiver's performance.

When evaluating receivers, a TOIP of +5 to +20 dBm is excellent performance, up to +27 dBm is relatively easily achievable, and +35 dBm has been achieved with good design; anything greater than +50 dBm is close to miraculous (but attainable). Receivers are still regarded as good performers in the 0- to +5-dBm range and as middling performers in the −10- to 0-dBm range. Anything below −10 dBm is a not-too-wonderful machine to own. A general rule is to buy the best third-order intercept performance that you can afford—especially if strong signal sources are in your vicinity.

Dynamic range

The dynamic range of a radio receiver is the range from the minimum discernible signal to the maximum allowable signal (measured in decibels, dB). Although this simplistic definition is conceptually easy to understand, in the concrete it's a little more complex. Several definitions of dynamic range are used (Dyer, 1993).

One definition of dynamic range is that it is the input signal difference between the sensitivity figure (e.g., 0.5 μV for 10 dB S + N/N) and the level that drives the receiver far enough into saturation to create a certain amount of distortion in the output. This definition was common on consumer broadcast band receivers at one time (especially automobile radios, where dynamic range was somewhat more important because of mobility). A related definition takes the range as the distance in decibels from the sensitivity level and the −1-dB compression point. Still another definition, the blocking dynamic range, which is the range of signals from the sensitivity level to the blocking level (see below).

A problem with these definitions is that they represent single-signal cases so they do not address the receiver's dynamic characteristics. Dyer (1993) provides both a "loose" and a more formal definition that is somewhat more useful and is at least standardized. The loose version is that dynamic range is the range of signals over which dynamic effects (e.g., intermodulation, etc.) do not exceed the noise floor of the receiver. Dyer's recommendation for HF receivers is that the dynamic range is two-thirds of the difference between the noise floor and the third-order intercept point in a 3-kHz bandwidth. Dyer also states an alternative: dynamic range is the difference between the fundamental response input signal level and the third-order intercept point along the noise floor, measured with a 3-kHz bandwidth. For practical reasons, this measurement is sometimes made not at the actual noise floor (which is sometimes hard to ascertain) but rather at 3 dB above the noise floor.

Magne and Sherwood (1987) provide a measurement procedure that produces similar results (the same method is used for many amateur radio magazine product reviews). Two equal-strength signals are input to the receiver at the same time. The frequency difference has traditionally been 20 kHz for HF and 30 to 50 kHz for VHF receivers (modern band crowding might indicate a need for a specification at 5-kHz separation on HF). The amplitudes of these signals are raised until the third-order distortion products are raised to the noise floor level.

For 20-kHz spacing, using the two-signal approach, anything over 90 dB is an excellent receiver, and anything over 80 dB is at least decent.

The difference between the single-to-signal and two-signal (dynamic) performance is not merely an academic exercise. Besides the fact that the same receiver can show as much as a 40-dB difference between the two measures (favoring the single-to-signal measurement), the most severe effects of poor dynamic range show up most in the dynamic performance.

Blocking

The blocking specification refers to the ability of the receiver to withstand very strong off-tune signals that are at least 20 kHz (Dyer, 1993) away from the desired signal, although some use 100-kHz separation (Magne and Sherwood, 1987). When very strong signals appear at the input terminals of a receiver, they can desensitize the receiver (i.e., reduce the apparent strength of desired signals over what they would be if the interfering signal were not present).

Figure 9-20 shows the blocking behavior. When a strong signal is present, it takes up more of the receiver's resources than normal, so there is not enough of the output power budget to accommodate the weaker desired signals. But if the strong

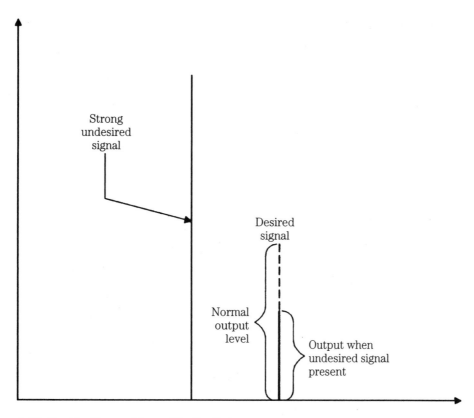

9-20 The blocking, or "desensitization," phenomenon.

undesired signal is turned off then the weaker signals receive a full measure of the unit's power budget.

The usual way to measure blocking behavior is to input two signals, a desired signal at 60 dBμV and another signal 20 (or 100) kHz away at a much stronger level. The strong signal is increased to the point where blocking desensitization causes a 3-dB drop in the output level of the desired signal. A good receiver will show ≥90 dBμV, with many being considerably better. An interesting note about modern receivers is that the blocking performance is so good that it's often necessary to specify the input level difference (dB) that causes a 1-dB drop, rather than a 3-dB drop, of the desired signal's amplitude (Dyer, 1993).

The phenomenon of blocking leads to an effect that is often seen as paradoxical on first blush. Many receivers are equipped with front-end attenuators that permit fixed attenuations of 1, 3, 6, 12, or 20 dB (or some subset of same) to be inserted into the signal path ahead of the active stages. When a strong signal that is capable of causing desensitization is present, adding attenuation often increases the level of the desired signals in the output—even though overall gain is reduced. This occurs because the overall signal that the receiver front end is asked to handle is below the threshold where desensitization occurs.

Cross-modulation

Cross-modulation is an effect in which amplitude modulation (AM) from a strong undesired signal is transferred to a weaker desired signal. Testing is usually done (in HF receivers) with a 20-kHz spacing between the desired and undesired signals, a 3-kHz IF bandwidth on the receiver, and the desired signal set to 1000 μV_{EMF} (-53 dBm). The undesired signal (20 kHz away) is amplitude-modulated to the 30% level. This undesired AM signal is increased in strength until an unwanted AM output 20 dB below the desired signal is produced.

A cross-modulation specification ≥ 100 dB would be considered decent performance. This figure is often not given for modern HF receivers, but if the receiver has a good third-order intercept point then it is likely to also have good cross-modulation performance.

Cross-modulation is also said to occur naturally—especially in transpolar and North Atlantic radio paths, where the effects of the aurora are strong. According to one (possibly apocryphal) legend there was something called the "Radio Luxembourg Effect" discovered in the 1930s. Modulation from very strong broadcasters appeared on the Radio Luxembourg signal received in North America. This effect was said to be an ionospheric cross-modulation phenomenon. If anyone has any direct experience with this effect, or a literature citation, then I would be interested in hearing from them.

Reciprocal mixing

Reciprocal mixing occurs when noise sidebands from the local oscillator (LO) signal in a superheterodyne receiver mix with a strong undesired signal that is close to the desired signal. Every oscillator signal produces noise, and that noise tends to amplitude modulate the oscillator's output signal. It will thus form sidebands on either side of the LO signal. The production of phase noise in all LOs is well-known, but in more-recent designs the digitally produced synthesized LOs are prone to additional noise elements as well. The noise is usually measured in $-$dBc (decibels below carrier, or, in this case, dB below the LO output level).

In a superheterodyne receiver, the LO beats with the desired signal to produce an intermediate frequency (IF) equal to either the sum (LO + RF) or difference (LO $-$ RF). If a strong unwanted signal is present then it might mix with the noise sidebands of the LO to reproduce the noise spectrum at the IF frequency (see Fig. 9-21). In the usual test scenario, the reciprocal mixing is defined as the level of the unwanted signal (dB) at 20 kHz required to produce noise sidebands 20 down from the desired IF signal in a specified bandwidth (usually 3 kHz on HF receivers). Figures of 90 dBc or better are considered good.

The importance of the reciprocal mixing specification is that it can seriously deteriorate the observed selectivity of the receiver, yet is not detected in the normal static measurements made of selectivity (it is a "dynamic selectivity" problem). When the LO noise sidebands appear in the IF, the distant frequency attenuation ($>$20 kHz off-center of a 3-kHz bandwidth filter) can deteriorate 20 to 40 dB.

The reciprocal mixing performance of receivers can be improved by eliminating the noise from the oscillator signal. Although this sounds simple, in practice it is

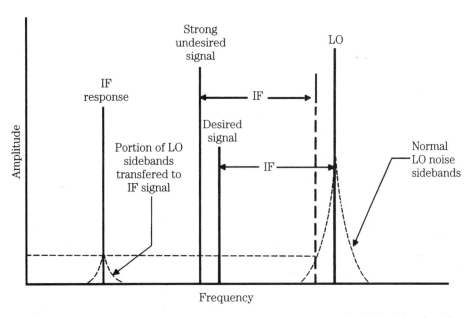

9-21 Reciprocal mixing phenomenon: (A) LO with noise sidebands, desired signal, and a strong undesired signal; (B) transfer of noise sidebands to IF signal.

often quite difficult. A tactic that will work well, at least for those designing their own receiver, is to add high-*Q* filtering between the LO output and the mixer input. The narrow bandwidth of the high-*Q* filter prevents excessive noise sidebands from getting to the mixer. Although this sounds like quite the easy solution, as they say "the devil's in the details."

Other important specifications

IF notch rejection

If two signals fall within the passband of a receiver they will both compete to be heard. They will also heterodyne together in the detector stage, producing an audio tone equal to their carrier frequency difference. For example, suppose we have an AM receiver with a 5-kHz bandwidth and a 455 kHz IF. If two signals appear on the band such that one appears at an IF of 456 kHz and the other is at 454 kHz, then both are within the receiver passband and both will be heard in the output. However, the 2-kHz difference in their carrier frequency will produce a 2-kHz heterodyne audio tone difference signal in the output of the AM detector.

In some receivers, a tunable, high-*Q* (narrow and deep) notch filter is in the IF amplifier circuit. This tunable filter can be turned on and then adjusted to attenuate the unwanted interfering signal, reducing the irritating heterodyne. Attenuation figures for good receivers vary from −35 to −65 dB or so (the more negative the better).

There are some tradeoffs in notch filter design. First, the notch filter *Q* is more easily achieved at low-IF frequencies (such as 50 to 500 kHz) than at high-IF frequencies (e.g., 9 MHz and up). Also, the higher the *Q* the better the attenuation of the undesired squeal but the touchier it is to tune. Some happy middle ground between the irritating squeal and the touchy tune is mandated here.

Some receivers use audio filters rather than IF filters to help reduce the heterodyne squeal. In the AM broadcast band, channel spacing is typically 10 kHz, and the transmitted audio bandwidth (hence the sidebands) are 5 kHz. Designers of AM BCB receivers usually insert an RC low-pass filter with a -3-dB point just above 4 or 5 kHz right after the detector in order to suppress the 10-kHz heterodyne. This RC filter is called a *tweet filter* in the slang of the electronic service/ repair trade.

Another audio approach is to sharply limit the bandpass of the audio amplifiers. For AM BCB reception, a 5-kHz bandpass is sufficient, so the frequencies higher can be rolled off at a fast rate in order to produce only a small response an octave higher (10 kHz). In shortwave receivers, this option is weaker because the station channels are typically 5 kHz, and many don't bother to honor the official channels anyway. And on the amateur radio bands frequency selection is a perpetually changing adhocracy, at best. Although the shortwave bands typically only need 3-kHz bandwidth for communications, and 5 kHz for broadcast, the tweet filter and audio roll-off might not be sufficient. In receivers that lack an effective IF notch filter, an audio notch filter can be provided. This accessory can even be added after the fact (as an outboard accessory) once you own the receiver.

Internal spurii

All receivers produce a number of internal spurious signals that sometimes interfere with the operation. Both old and modern receivers have spurious signals from assorted high-order mixer products, power-supply harmonics, parasitic oscillations, and a host of other sources. Newer receivers with either (or both) synthesized local oscillators and digital frequency readouts produce noise and spurious signals in abundance. (Note: low-power digital chips with slower rise times—CMOS, NMOS, etc.—are generally much cleaner than higher-power, fast-rise-time chips like TTL devices).

With appropriate filtering and shielding, it is possible to hold the "spurs" down to -100 dB relative to the main maximum signal output or within about 3 dB or the noise floor, whichever is lower.

A writer of shortwave books (Helms, 1994) has several high-quality receivers, including tube models and modern synthesized models. His comparisons of basic spur/noise level was something of a surprise. A high-quality tube receiver from the 1960s appeared to have a lower noise floor than the modern receivers. Helms attributed the difference to the internal spurs from the digital circuitry used in the modern receivers.

Receiver improvement strategies

There are several strategies that can improve the performance of a receiver. Some of them are available for after-market use with existing receivers and others are only practical when designing a new receiver.

One method for preventing dynamic effects is to reduce the signal level applied to the input of the receiver at all frequencies. A front-end attenuator will help here. Indeed, many modern receivers have a switchable attenuator built in to the front-end circuitry. This attenuator will reduce the level of all signals, backing the largest away from the compression point and, in the process, eliminating many of these dynamic problems.

Another method is to prevent the signal from ever reaching the receiver front end. This goal can be achieved by using any of several tunable filters, bandpass filters, high-pass filters, or low-pass filters (as appropriate) ahead of the receiver front end. These filters might not help if the intermodulation products fall close to the desired frequency but otherwise are quite useful.

When designing a receiver project there are some things that can be done. If an RF amplifier is used then use either field-effect transistors (MOSFETs are popular) or a relatively high-powered bipolar (NPN or PNP) transistor intended for cable-TV applications, such as the 2N5109 or MRF-586. Gain in the RF amplifier should be kept minimal (4 to 8 dB). It is also useful to design the RF amplifier with as high a dc power-supply potential as possible. That tactic allows a lot of "head room" as input signals become larger.

You might also wish to consider deleting the RF amplifier altogether. One design philosophy holds that one should use a high-dynamic range mixer with only input tuning or suboctave filters between the antenna connector and the mixer's RF input. It is possible to achieve noise figures of 10 dB or so using this approach, and that is sufficient for HF work (although it might be marginal for VHF and up). The tube-design 1960s vintage Squires-Sanders SS-1 receiver used this approach. The front-end mixer tube was the 7360 double-balanced mixer.

The mixer in a newly designed receiver project should be a high-dynamic range type regardless of whether an RF amplifier is used. Popular with some designers is the double-balanced switching-type mixer (Makhinson, 1993). Although examples can be fabricated from MOSFET transistors and MOS digital switches, there are also some IC versions on the market that are intended as MOSFET switching mixers. If you opt for diode mixers then consult a source, such as Mini-Circuits (1992) for DBMs, that operate up to RF levels of +20 dBm.

Anyone contemplating the design of a high dynamic-range receiver will want to consult Makhinson (1993) and DeMaw (1976) for design ideas before starting.

References

Carr, J. J. (1994). "Be a RadioScience Observer—Part 1." *Shortwave,* November, pp. 32–36.

Carr, J. J. (1994). "Be a RadioScience Observer—Part 2." *Shortwave,* December, pp. 36–44.

Carr, J. J. (1995). "Be a RadioScience Observer—Part 3." *Shortwave,* January, pp. 40–44.

DeMaw, D. (1976). "His Eminence—The Receiver Part 1." *QST,* June, American Radio Relay League (Newington, CT, USA), p. 27ff.

DeMaw, D. (1976). "His Eminence—The Receiver Part 2." *QST,* July, American Radio Relay League (Newington, CT, USA), p. 14ff.

Dyer, J. A. (1993). "Receiver Performance: Descriptions and Definitions of Each Performance Parameter." *Communications Quarterly,* Summer, pp. 73–88.

Gruber, M. (1992). "Synchronous Detection of AM Signals: What Is It and How Does It Work?" *QEX*, September, pp. 9–16.

Helms, H. (1994). Author of *All About Ham Radio* and *Shortwave Listeners Guide.* Solana Beach, CA: HighText Publications, Inc. Personal telephone conversation.

Kinley, R. H. (1985). *Standard Radio Communications Manual: With Instrumentation and Testing Techniques.* Englewood Cliffs, NJ, USA: Prentice-Hall.

Kleine, G. (1995). "Parameters of MMIC Wideband RF Amplifiers," *Elektor Electronics* (UK), February, pp. 58–60.

Magne, L. and Sherwood, J. R. (1987). *How To Interpret Receiver Specifications and Lab Tests, A Radio Database International White Paper,* Edition 2.0, 18 Feb, International Broadcasting Services, Ltd., Penn's Park, PA.

Makhinson, J. (1993). "A High Dynamic Range MF/HF Receiver Front-End." *QST,* February, pp. 23–28.

Mini-Circuits RF/IF Designers Handbook. Mini-Circuits (P.O. Box 350166, Brooklyn, NY 11235-0003, USA. In UK: Mini-Circuits/Dale, Ltd., Dale House, Wharf Road, Frimley Green, Camberley, Surry, GU16 6LF, England).

Poole, I. (1995). G3YWX, "Specification—The Mysteries Explained," *Practical Wireless,* April, p. 41.

Poole, I. (1995). G3YWX, "Specification—The Mysteries Explained," *Practical Wireless,* March, p. 57.

Rohde, U. L., and Bucher, T. T. N. (1998). *Communications Receivers: Principles & Design.* New York: McGraw-Hill.

Tsui, J. B. (1985). *Microwave Receivers and Related Components.* Los Altos, CA: Peninsula.

10
CHAPTER

Building signal-generator and oscillator circuits

Signal-generator circuits produce RF sine-wave (or other waveform) output for purposes of testing, troubleshooting, and alignment. Although the subject of designing top-notch signal generator circuits can be quite deep, it is possible to reduce the possibilities to a small number for people whose needs are less stringent. This chapter looks at RF signal generators that can be used in a wide variety of applications on the electronic and radio workbenches. But first, take a look at some generic types of oscillator circuits.

Types of oscillator circuits

The two major categories of oscillator circuits in electronics textbooks are relaxation oscillators and feedback oscillators. The relaxation oscillator uses some sort of voltage-breakdown device, such as a neon glow-lamp or unijunction transistor. The feedback oscillator (Fig. 10-1) uses an amplifier circuit and a feedback network to start and sustain oscillations on a particular frequency. Most oscillators that are useful for sine-wave signal generator circuits are of the feedback oscillator class.

The requirements for sustained oscillator, called *Nyquist's criteria*, are (1) the loop gain between feedback network losses and amplifier gain must be greater than or equal to one at the frequency of oscillation and (2) the feedback signal must be in-phase with the input signal at the frequency of oscillation. The second of these criteria means that the feedback signal must be phase-shifted 360°, of which 180° is usually obtained from the inversion of the amplifier and 180° from the frequency-selective feedback network.

Feedback oscillators can also be classified according to the nature of the feedback network. The three basic types of feedback oscillators (and many variations on the themes) are Armstrong, Colpitts or Clapp, and Hartley. Other forms of feedback network are used at audio frequencies, but these are the principal forms of

243

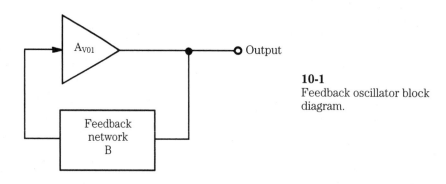

10-1
Feedback oscillator block diagram.

RF oscillators. You can tell which is which by looking at the feedback network (Fig. 10-2). The Armstrong oscillator (Fig. 10-2A) uses a separate tickler coil (L_2) to provide feedback to the main tuning coil (L_1). These coils are usually wound on the same coil form with each other. The Colpitts oscillator (Fig. 10-2B) uses a parallel resonant-tuned circuit and (this is the key) a tapped capacitor voltage divider (C_1 and C_2) to provide feedback. The voltage divider might be a direct part of the resonant circuit. The Clapp oscillator is a variant of the Colpitts circuit, in which a series-resonant tuning circuit is used. Finally, the Hartley oscillator uses a tapped inductor voltage divider (Fig. 10-2C) for the feedback network.

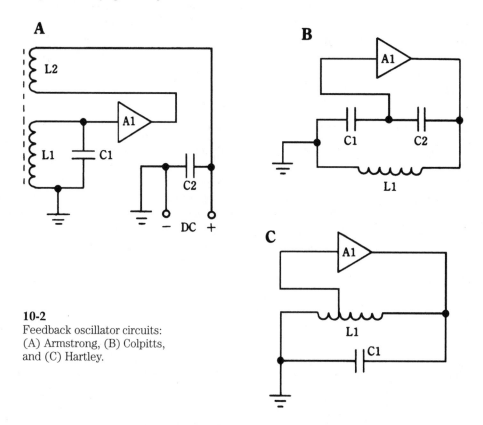

10-2
Feedback oscillator circuits:
(A) Armstrong, (B) Colpitts,
and (C) Hartley.

The resonator of the oscillator could be either an inductor-capacitor (LC) tuned circuit, as shown in the examples of Fig. 10-2, or a piezoelectric crystal resonator. The latter are single-channel resonators, but are more stable than LC-tuned circuits.

1- to 20-MHz crystal oscillator

Figure 10-3 shows a simple, nearly universal signal generator that can accommodate crystal frequencies between 1 and 20 MHz. This oscillator operates in the fundamental mode so the marked frequency of the crystal is the actual frequency that it operates on (as opposed to overtone crystals that operate on harmonics of the fundamental frequency). When specifying the crystal, if you are given a choice, ask for a crystal that operates into 32 pF of capacitance. Otherwise, the actual operating frequency might be a little different than the specified frequency.

The amplifier device is a simple bipolar NPN transistor. The transistor selected is not critical, but the 2N3904 and 2N2222 devices have been tried on numerous occasions in this circuit. Whatever transistor is selected, it must operate as an oscillator in the frequency range of interest. If you want to use a similarly rated PNP transistor (e.g., 2N3906), then simply reverse the polarity of the dc power supply. The transistor uses the simplest form of dc bias network: R_1 is directly connected between the transistor's collector and its base. The output signal is taken through a dc-blocking capacitor (C_5) across the emitter resistor (R_2). A bypass capacitor (0.01 to 0.1 μF) is connected between the collector and ground. This capacitor sets the collector at ground potential for ac potentials and keeps it at the rated dc power-supply potential (9 to 12 Vdc).

The feedback network consists of capacitors C_1 and C_2. The ratio of the values, C_1/C_2, is selected to achieve a reasonable tradeoff of output level and stability. These capacitors should be either silver mica or (preferably) NPO disk ceramic capacitors. Use of these types of capacitor will prevent frequency drift of the oscillator as a result of temperature-dependent changes in the capacitance of C_1 and C_2. For the same reason, capacitor C_3 should also be a silver mica or NPO disk ceramic.

The circuit as shown in Fig. 10-3 operates at a single frequency because it is crystal controlled. Although it's nice to think that the crystal always operates on the specified frequency, it's likely that differences in circuit capacitance will make it oscillate at a different frequency. If the exact frequency is important then replace C_3 with a variable capacitor. Alternatively, connect a 50-pF variable capacitor in parallel across the crystal.

The circuit of Fig. 10-3 is popular with people who have to troubleshoot FM radio receivers. Figure 10-4 shows a signal generator that can be used to align and test common FM radio receivers. The basic circuit is the same as Fig. 10-3, but the crystal is replaced with a SPDT switch and a pair of crystals. The 10.7-MHz crystal is used to test the FM IF amplifier stages, so it is on the standard FM IF frequency. The 9-MHz crystal is used to test the front-end and tuning dial accuracy of the receiver. A 9-MHz crystal will output sufficient harmonics at 90, 99, and 108 MHz to check the calibration of the low end, middle, and high end of the dial, respectively. Notice that this signal generator does not produce an audio tone in the output of the receiver but

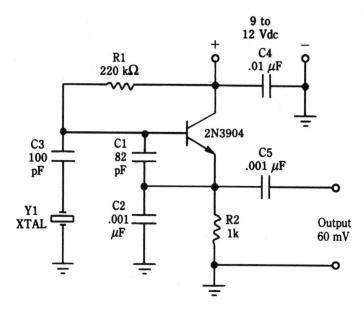

10-3 Colpitts crystal oscillator.

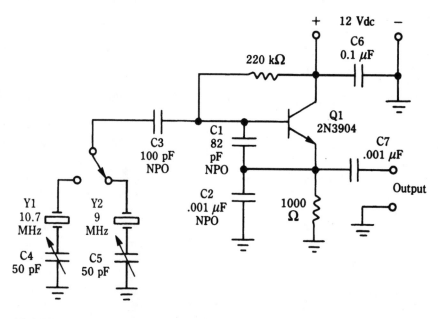

10-4 FM alignment oscillator. Use 10.7 MHz for IF alignment and 9-MHz to align local oscillator at 90, 99, or 108 MHz.

rather it is indicated by "quieting" of the receiver and an increase in the strength meter reading.

HF/VHF buffer amplifier

A buffer amplifier is used to isolate the output of the oscillator from the load, or circuits that follow. Variations in the load of an oscillator can "pull" the frequency incorrectly so the buffer is used to prevent that problem. Figure 10-5 shows the circuit for a buffer amplifier that can be used in the low-frequency (LF), high-frequency (HF), and the lower end of the VHF ranges.

The amplifier device in the buffer amplifier is a 40673 dual-gate MOSFET transistor (or NTE-222, which is a replacement). The MOSFET is connected in the standard "grounded source" (S) configuration. A small source bias is provided by the 100-Ω resistor (R_2) from source to ground. A 0.1-μF capacitor shunted across R_2 keeps the source resistor at ground potential for RF signals while also keeping it at the small positive dc potential caused by the voltage drop across the resistance.

The positive operating dc potential on the drain (D) is supplied through a 270-Ω collector load resistor (R_3) and a 1000-μH radio-frequency choke (RFC$_1$). The choke has an inductive reactance that increases linearly with frequency, so the combination of R_3 and RFC$_1$ provides a load impedance that increases with frequency, overcoming the tendency of the transistor to offer less gain at higher frequencies. The dc drain load network (R_3/RFC$_1$) is decoupled by a pair of 0.1-μF bypass capacitors. Use disk ceramic, polyester, or other forms of capacitor for this application, but be sure when ordering them that they will work as bypass capacitors at the low-VHF region (see the write-up in the catalog at the beginning of each section on capacitors).

The input is applied through a dc-blocking capacitor (C_1), across a gate resistor (R_1), to MOSFET gate G_1. The output signal is taken from the drain (D) through a 0.001-μF dc-blocking capacitor (C_3).

The second gate (G_2) of the 40673 MOSFET is biased to a potential of about 10 Vdc, which is set from the +12-Vdc power supply by voltage divider R_4/R_5. The G_2 terminal is set to ac ground by a bypass capacitor (C_4) selected to have a low capacitive reactance at the minimum operating frequency relative to R_5. In practical terms, this means that $X_{C_4} < R_5/10$.

A variation on the theme can be built by connecting the voltage divider network R_4/R_5 to a potentiometer or other variable voltage source instead of the V+ power supply (as shown). The variable voltage can then be used as an output level control. In some signal generators, the G_2 terminal of the MOSFET buffer amplifier is connected to an automatic gain control (AGC) to stabilize the output signal level.

Another use for the G_2 terminal is to amplitude modulate (AM) the output signal level. A potentiometer and capacitor (see inset to Fig. 10-5) is used to connect the G_2 circuit to an audio sine-wave source. Be sure that the sine-wave amplitude is high enough to make the amplifier nonlinear (otherwise modulation will not result) but not so much that the output waveform peaks and valleys do not "flat top" (as shown on an oscilloscope).

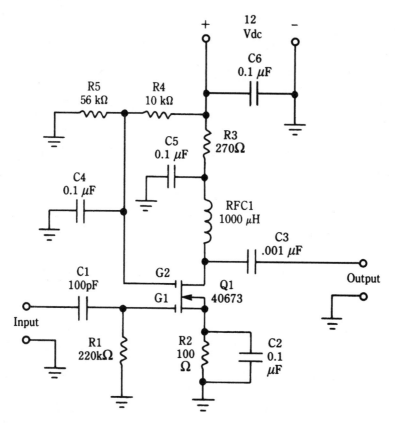

10-5 FR amplifier circuit.

455-kHz AM IF-amplifier test-and-alignment oscillator

The modern superheterodyne AM radio uses a 455-kHz IF amplifier (262.5 kHz is used on AM car radios). Figure 10-6 shows a simple signal-generator circuit that can be used to test, troubleshoot, or align the AM IF stage in common radios. The active element is a MPF-102 (or NTE-312) JFET.

As you can see by comparing Fig. 10-6 to Fig. 10-2, this circuit is of the basic Hartley oscillator class because it uses a tapped inductor (T_1) in the feedback network. The transformer used for this purpose is a standard 455-kHz IF transformer (Digi-Key Part No. TK-1301, or equivalent). This transformer has a tapped primary that is used to match the low collector impedance of bipolar (PNP or NPN) transistor amplifiers, but the tapped LC tank circuit can be pressed into service as a Hartley oscillator. The secondary of the transformer forms a handy output port to pass a 455-kHz signal to the following buffer stage (see Fig. 10-5 for a description of this circuit).

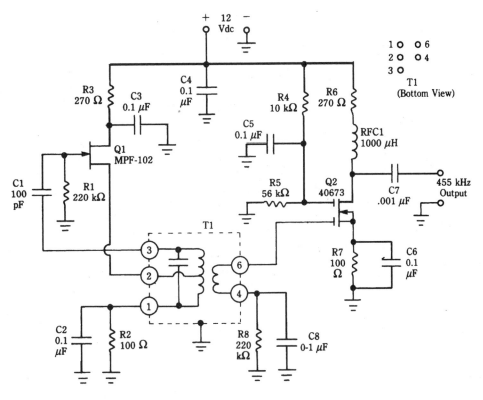

10-6 455-kHz AM IF alignment oscillator.

Signal generator for the AM and shortwave bands

The same general type of circuit can be used for the AM band (530 to 1610 kHz) or for the shortwave bands (1610 kHz to 30 MHz). The version shown in Fig. 10-7 uses a standard transformer that has a 217-µH inductance in the primary winding. This particular example uses a Toko coil from Digi-Key (Part No. TK-1903) in conjunction with a 365-pF broadcast variable capacitor to cover the AM broadcast band. For other frequencies:

$$F = \frac{1}{2\pi\sqrt{LC}} \qquad (10\text{-}1)$$

or if F and L are known but C is unknown:

$$C = \frac{1}{39.5\,F^2 L} \qquad (10\text{-}2)$$

or if F and C are known but L is unknown:

$$L = \frac{1}{39.5\,F^2 C}. \qquad (10\text{-}3)$$

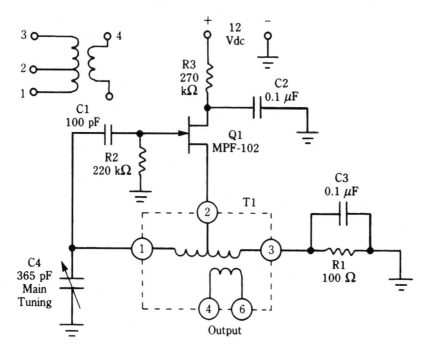

10-7 Variable alignment oscillator.

In all three cases, F is in hertz, C is in farads, and L is in henrys. For frequencies at which there is not an easily available coil, select a toroidal core and wind a coil yourself. In either case, allow about 20 pF for stray capacitances in the circuit (the actual amount might be more or less, depending on the layout).

A VHF voltage-tuned oscillator circuit is shown in Fig. 10-8A. I've used this circuit at frequencies from 20 to 150 MHz. The circuit uses a feedback capacitor (C_2) across the collector-emitter circuit of the transistor. This capacitor is critical, and the circuit probably won't oscillate without it. The tuning network consists of inductor, L_1; dc-blocking capacitor, C_1; and varactor, D_1. *Varactors* are voltage-variable capacitor diodes. The inductor is wound on either a VHF toroidal coil or an air form about $3/8''$ in diameter. If no. 20 or so solid wire is used for L_1 then the coil will be self-supporting after it is removed from the coil form.

The VHF oscillator of Fig. 10-8A is tuned with a dc voltage (V_t) applied through a 150-kΩ resistor (R_4). This voltage is positive in order to reverse bias D_1. A problem with this circuit is that the frequency is a nonlinear function of frequency because of the characteristics of D_1. Figure 10-8B shows a somewhat more linear version of this same circuit. In this case, the dc-blocking capacitor is replaced with a second varactor diode that is identical to the first. The total capacitance of the combination is half the capacitance of one diode, but the voltage-vs-frequency characteristic is more linear.

Both of these circuit variations can be fixed-tuned with a dc voltage, variable-tuned through a potentiometer, or swept across a band of frequencies with a sine-wave (FM) or sawtooth signal. In the case of the sawtooth, use a 10- to 60-Hz signal.

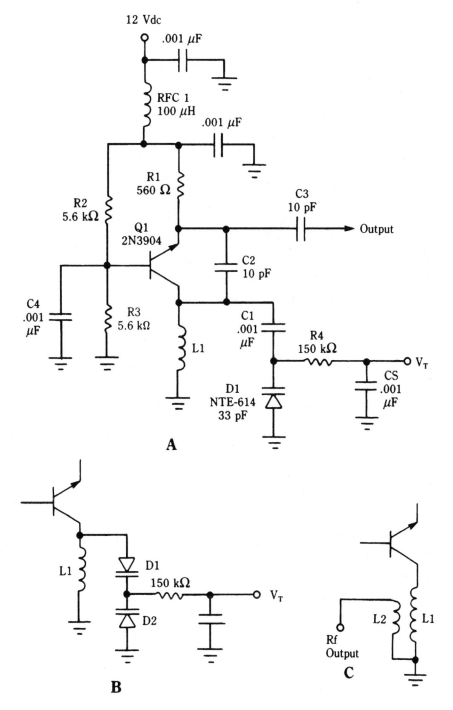

10-8 (A) Varactor tuned VHF variable frequency oscillator, (B) alternate control circuit, and (C) alternate output circuit.

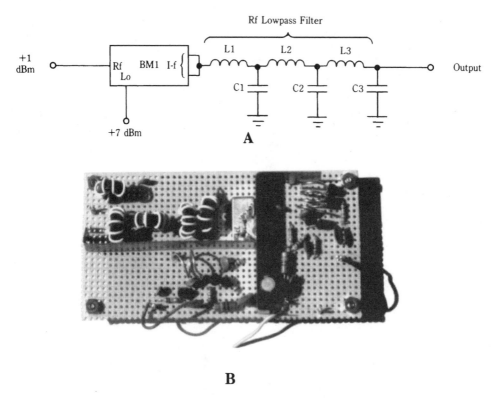

10-9 Balanced mixer with low-pass filter output for mixing two RF sources. (A) RF low-pass filter connected to balanced mixer; (B) physical implementation.

A variation on the theme is shown in Fig. 10-8C. Although the output in Fig. 10-8A is taken from the emitter of the transistor, the variation in Fig. 10-8C uses a secondary winding over the tuning inductor (L_1) to output the signal.

Another variation useful for a wide variety of circuits is shown in Fig. 10-9A, and an actual example prototyped by me is shown in Fig. 10-9B. This circuit uses a double-balanced mixer (DBM), such as the SRA-1 and related products by Mini-Circuits Laboratories, Inc. One signal is applied to the RF port of the DBM, and the other signal is applied to the LO port of the DBM (note the relative signal levels). The difference frequency is taken across the two shorted IF OUT ports. The low-pass filter is used to remove residual LO or RF signal. I used this scheme to produce a ham-band sweep generator. One oscillator was fixed tuned to 37 MHz then swept with a saw-tooth tuning voltage ±1 MHz. The other oscillator was tuned to frequencies that would heterodyne the 37-MHz signal to the middle of various HF ham bands. A bank of potentiometers selects the dc bias required to achieve the correct output frequency.

The RF signal generator circuit is generally well-behaved and thus is easy to build and align if proper technique is used. You can be quite successful making these circuits perform the tasks for which they were designed.

11
CHAPTER

RF directional couplers

Directional couplers are devices that will pass signal across one path while passing a much smaller signal along another path. One of the most common uses of the directional coupler is to sample a RF power signal either for controlling transmitter output power level or for measurement. An example of the latter use is to connect a digital frequency counter to the low-level port and the transmitter and antenna to the straight-through (high-power) ports.

The circuit symbol for a directional coupler is shown in Fig. 11-1. Note that there are three outputs and one input. The IN–OUT path is low-loss and is the principal path between the signal source and the load. The coupled output is a sample of the forward path while the isolated showed very low signal. If the IN and OUT are reversed then the roles of the coupled and isolated ports also reversed.

An implementation of this circuit using transmission line segments is shown in Fig. 11-2. Each transmission line segment (TL_1 and TL_2) has a characteristic impedance, Z_o, and is a quarter-wavelength long. The path from port 1 to port 2 is the low-loss signal direction. If power flows in this direction, then port 3 is the coupled port and port 4 is isolated. If the power flow direction reverses (port 2 to port 1) then the respective roles of port 3 and port 4 are reversed.

For a coupling ratio (port 3/port 4) ≤ -15 dB the value of coupling capacitance must be:

$$C_c < \frac{0.18}{\omega Z_o} \text{ (in farads).} \tag{11-1}$$

The coupling ratio is:

$$\text{C.R.} = 20 \log{(\omega C Z_o)} \text{ (in decibels)} \tag{11-2}$$

where

$$\omega = 2\pi f.$$

The bandwidth is about 12%.

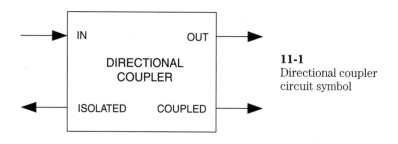

11-1
Directional coupler
circuit symbol

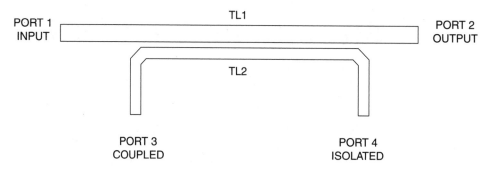

11-2 Transmission line directional coupler.

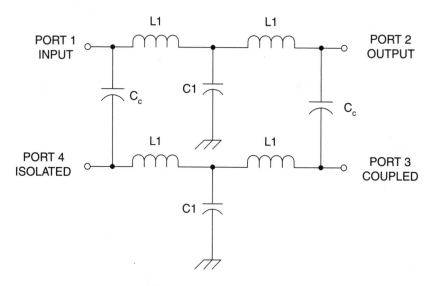

11-3 L-C version of directional coupler.

The circuit shown in Fig. 11-3 is an LC lumped constant version of the transmission lines. This network can be used to replace TL_1 and TL_2 in Fig. 11-2. The values of the components are:

$$L_1 = \frac{Z_0}{\omega_0} \qquad (11\text{-}3)$$

and

$$C_1 = \frac{1}{\omega_0 Z_0}. \qquad (11\text{-}4)$$

Figure 11-4 shows a directional coupler used in a lot of RF power meters and VSWR meters. The transmission lines are implemented as printed circuit board tracks. It consists of a main transmission line (TL_1) between Ports 1 and 2 (the low-loss path) and a coupled line (TL_2) to form the coupled and isolated ports. The coupling capacitance (in picofarads) is approximated by $9.399X$ (X is in meters) when implemented on G-10 Epoxy fiberglass printed circuit board.

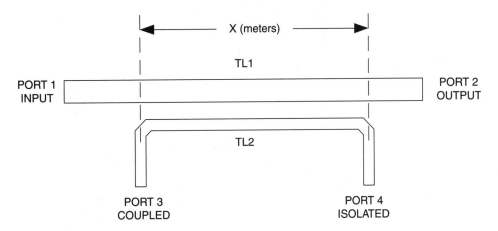

11-4 Transmission line directional coupler.

A reflectometer directional coupler is shown in Fig. 11-5A. This type of directional coupler is at the heart of many commercial VSWR meters and RF power meters used in the HF through low-VHF regions of the spectrum. This circuit is conceptually similar to the previous transmission line but is designed around a toroid transmission line transformer. It consists of a transformer in which the low-loss path is a single-turn primary winding and a secondary wound of enameled wire.

Details of the pick-up sensor are shown in Fig. 11-5B. The secondary is wound around the rim of the toroid in the normal manner, occupying not more than 330° of circumference. A rubber or plastic grommet is fitted into the center hole of the toroid core. The single-turn primary is formed by a single conductor passed once through the hole in the center of the grommet. It turns out the 3/16″ *o.d.* brass tubing (the kind sold in hobby shops that cater to model builders) will fit through several standard grommet sizes nicely and will slip-fit over the center conductor of SO-239 coaxial connectors.

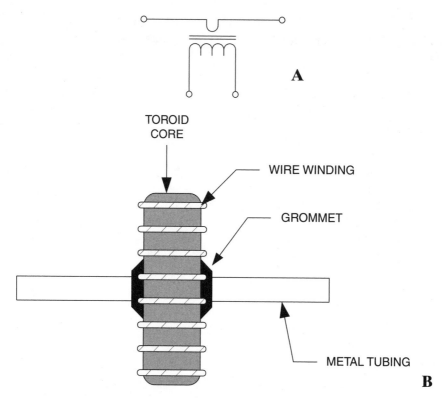

11-5 (A) Directional coupler suitable for use in an RF power meter; (B) Physical implementation.

Another transmission line directional coupler is shown in Fig. 11-6. Two short lengths of RG-58/U transmission line ($\approx$6″) are passed through a pair of toroid coils. Each coil is wound with 8 to 12 turns of wire. Note that the shields of the two transmission line segments are grounded only at one end.

Each combination of transmission line and toroid core form a transformer similar to the previous case. These two transformers are cross-coupled to form the network shown. The XMTR-ANTENNA path is the low-loss path, while (with the signal flow direction shown) the other two coupled ports are for forward and reflected power samples. These samples can be rectified and used to indicate the relative power levels flowing in the forward and reverse directions. Taken together these indications allow us to calculate VSWR.

Directional couplers are used for RF power sampling in measurement and transmitter control. They can also be used in receivers between the mixer or RF amplifier and the antenna input circuit. This arrangement can prevent the flow of LO signal and mixer products back toward the antenna where they could be radiated and cause electromagnetic interference (EMI) to other devices.

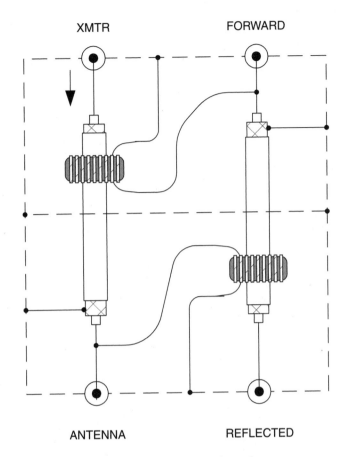

11-6 Dual transmission line directional coupler.

Conclusion

The passive diplexers and directional couplers discussed in this article are the mainstays of many electronic circuits used at higher frequencies.

References

ARRL Handbook for Radio Amateurs CD-ROM Version 1.0 (1996). Newington, CT: ARRL.

Carr, J. J. (1998). *Practical Antenna Handbook 3rd Edition.* New York: McGraw-Hill.

Carr, J. J. (1997). *Microwave and Wireless Communications Technology.* Boston: Newnes.

Carr, J. J. (1996). *Secrets of RF Circuit Design 2nd Edition.* New York: McGraw-Hill.

Hagen, J. B. (1996). *Radio-Frequency Electronics: Circuits and Applications.* Cambridge (UK): Cambridge Univ. Press.

Hardy, J. (1979). *High Frequency Circuit Design.* Reston, VA: Reston (Division of Prentice-Hall).

Kinley, R. H. (1985). *Standard Radio Communications Manual: With Instrumentation and Testing Techniques.* Englewood Cliffs, NJ: Prentice-Hall.

Laverghetta, T. S. (1984). *Practical Microwaves.* Indianapolis, IN: Howard W. Sams.

Liao, S. Y. (1990). *Microwave Devices & Circuits.* Englewood Cliffs, NJ: Prentice-Hall.

Sabin, W. E., and Schoenike, E. O. (Eds.) (1998). *HF Radio Systems & Circuits 2nd Edition.* Atlanta: Noble.

Shrader, R. L. (1975). *Electronic Communication 3rd Edition.* New York: McGraw-Hill.

Vizmuller, P. (1995). *RF Design Guide.* Boston/London: Artech House.

12
CHAPTER

The RF hybrid coupler

The hybrid coupler (Fig. 12-1) is an AF or RF device that will either (a) split a signal source into two directions or (b) combine two signal sources into a common path. The circuit symbol shown in Fig. 12-1 is essentially a signal path schematic. Consider the situation where an RF signal is applied to port 1. This signal is divided equally, flowing to both ports 2 and 3.

Because the power is divided equally the hybrid is called a 3-dB divider, i.e., the power level at each adjacent port is one-half (-3 dB) of the power applied to the input port.

If the ports are properly terminated in the system impedance then all power is absorbed in the loads connected to the ports adjacent to the injection port. None travels to the opposite port. The termination of the opposite port is required, but it does not dissipate power because the power level is zero.

The one general rule to remember about hybrids is that *opposite ports cancel.* That is, power applied to one port in a properly terminated hybrid will not appear at the opposite port. In the case cited above, the power was applied to port 1 so no power appeared at port 4.

One of the incredibly useful features of the hybrid is that it accomplishes this task while allowing all devices connected to it to see the system impedance, R_0. For example, if the output impedance of the signal source connected to port 1 is 50 Ω, the loads of ports and port 3 are 50 Ω, and the dummy load attached to port 4 is 50 Ω then all devices are either looking into, or driven by, the 50-Ω system impedance.

One source of reasonably priced hybrid devices is Mini-Circuits Laboratories (13 Neptune Avenue, Brooklyn, NY, 11235, USA. Web site: http://www.minicircuits.com). They have a large selection of 0°, 90°, and 180° hybrid combiners and splitters.

Applications of hybrids

The hybrid can be used for a variety of applications where either combining or splitting signals is required.

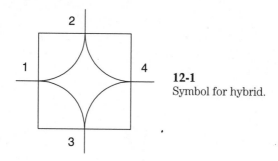

12-1
Symbol for hybrid.

Combining signal sources

In Fig. 12-2 there are two signal generators connected to opposite ports of a hybrid (ports 2 and 3). Power at port 2 from signal generator 1 is therefore canceled at port 3, and power from signal generator 2 (port 3) is canceled at port 2. Therefore, the signals from the two signal generators will not interfere with each other.

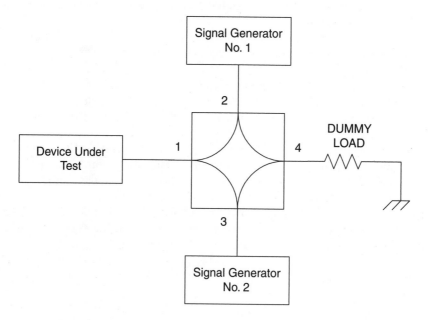

12-2 Combining two signal sources.

In both cases, the power splits two ways. For example, the power from signal generator 1 flows into port 2 and splits two ways. Half of it (3 dB) flows from port 2 to port 1 while the other half flows from port 2 to port 4. Similarly with the power from signal generator 2 applied to port 3. It splits into two equal portions, with one flowing to port 1 and the device under test and half flowing to the dummy load.

Bidirectional amplifiers

A number of different applications exists for bidirectional amplifiers, i.e., amplifiers that can handle signals from two opposing directions on a single line. The telecommunications industry, for example, uses such systems to send full duplex signals over the same lines.

Similarly, cable TV systems that use two-way (e.g., cable MODEM) require two-way amplifiers. Figure 12-3 shows how the hybrid coupler can be used to make such an amplifier. In some telecommunications textbooks the two directions are called "east" and "west", so this amplifier is occasionally called an east-west (E-W) amplifier. At other times this circuit is called a *repeater.*

In the bidirectional E-W amplifier of Fig. 12-3, amplifier A_1 amplifies the signals traveling west to east while A_2 amplifies signals traveling east to west. In each case, the amplifiers are connected to hybrids HB_1 and HB_2 via opposite ports so will not interfere with each other. Otherwise, connecting two amplifiers input-to-output-to-input-to-output is a recipe for disaster—if only a large amount of destructive feedback.

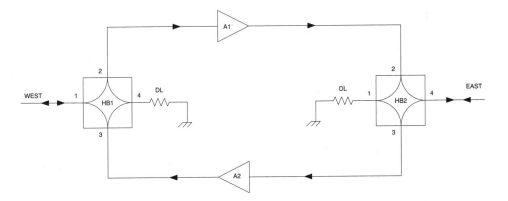

12-3 Bidirectional "repeater" amplifier.

Transmitter/receiver isolation

One of the problems that exists when using a transmitter and receiver together on the same antenna is isolating the receiver input from the transmitter input. Even a weak transmitter will burn out the receiver input if its power were allowed to reach the receiver input circuits. One solution is to use one form of transmit/receive (T/R) relay. But that solution relies on an electromechanical device, which adds problems of its own (not the least of which is reliability).

A solution to the T/R problem using a hybrid is shown in Fig. 12-4. In this circuit the transmitter output and receiver input are connected to opposite ports of a hybrid device. Thus, the transmitter power does not reach the receiver input.

The antenna is connected to the adjacent port between the transmitter port and the receiver port. Signal from the antenna will flow over the path from port 1 to port 2 to reach the receiver input. Transmitter power, on the other hand, enters at port 3 and is split into two equal portions. Half the power flows to the antenna from port 3 to port 1 while the other half flows to a dummy load from port 3 to port 4.

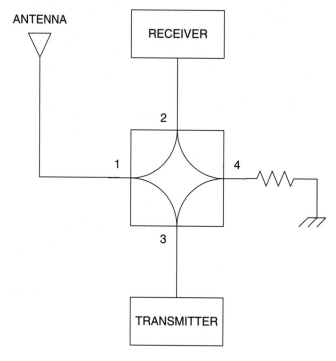

12-4 Use of hybrid as a T/R switch.

There is a problem with this configuration. Because half the power is routed to a dummy load, there is a 3-dB reduction in the power available to the antenna. A solution is shown in Fig. 12-5. In this configuration a second antenna is connected in place of the dummy load. Depending on the spacing (S), and the phasing, various directivity patterns can be created using two identical antennas.

If the hybrid produces no phase shift of its own then the relative phase shift of the signals exciting the antennas is determined by the length of the transmission line between the hybrid and that antenna. A 0° phase shift is created when both transmission lines are the same length. Making one transmission line a half-wavelength longer than the other results in a 180° phase shift. These two relative phase relationships are the basis for two popular configurations of phased array antenna. Consult an antenna book for other options.

Phase-shifted hybrids

The hybrids discussed thus far split the power half to each adjacent port but the signals at those ports are in-phase with each other. That is, there is a 0° phase shift over the paths from the input to the two output ports. There are, however, two forms of phase-shifted hybrids. The form shown in Fig. 12-6A is a 0°–180° hybrid. The sig-

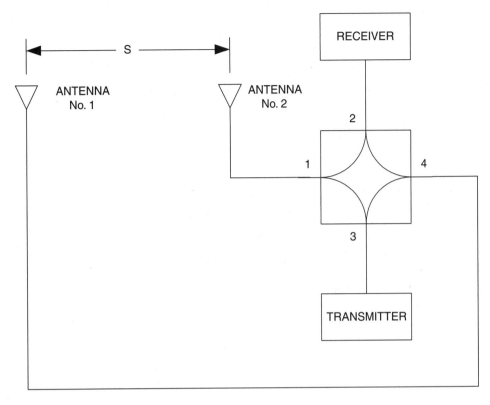

12-5 Combining two antennas in T/R switch.

nal over the path from port 1 to port 2 is not phase-shifted (0°) while that between port 1 and port 3 is phase-shifted 180°. Most transformer based hybrids are inherently 0°–180° hybrids.

A 0°–90° hybrid is shown in Fig. 12-6B. This hybrid shows a 90° phase shift over the port 1/port 2 path and a 0° phase shift over the port 1/port 3 path. This type of hybrid is also called a *quadrature hybrid.*

One application for the quadrature hybrid is the balanced amplifier shown in Fig. 12-7. Two amplifiers, A_1 and A_2, are used to process the same input signal arriving via hybrid HB_1. The signal splits in HB_1 and so becomes the inputs to both A_1 and A_2. If the input impedances the amplifiers are not matched to the system impedance then signal will be reflected from the inputs back toward HB_1. The reflected signal from A_2 arrives back at the input in-phase (0°) but that reflected from A_1 has to pass through the 90° phase shift arm twice and so has a total phase shift of 180°. Thus, the reflections caused by mismatching the amplifier inputs are cancelled out.

The output signals of A_1 and A_2 are combined in hybrid HB_2. The phase balance is restored by the fact that the output of A_1 passes through the 0° leg of HB_2, while the output of A_2 passes through the 90° leg. Thus, both signals have undergone a 90° phase shift, so are now restored to the in-phase condition.

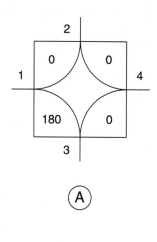

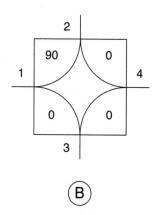

12-6
(A) 180 degree hybrid;
(B) 90 degree hybrid.

Use with receive antennas

Examples given above combine a receiver and transmitter on a single antenna or antenna system. It is also possible to use the hybrid for antenna arrays intended for receivers. Antenna spaced some distance X apart will have different patterns and gains depending on the value of X and the relative phase of the currents in the two antennas. One can, therefore, connect the antennas to ports 2 and 3 and the receiver antenna input to port 1. A terminating resistor would be used at port 4. You can use 0°, 90°, or 180° hybrids depending on the particular antenna system.

Conclusion

The hybrid coupler–combiner splitter is a remarkably useful passive RF component that will solve a lot of practical problems.

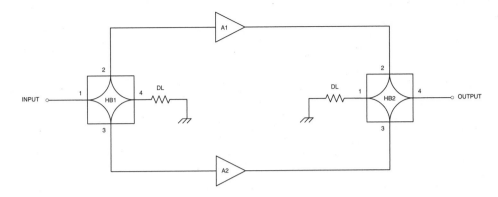

12-7 Balanced amplifier.

References

Carr, J. J. (1998). *Practical Antenna Handbook 3rd Edition*. New York: McGraw-Hill.

Carr, J. J. (1997). *Microwave and Wireless Communications Technology*. Boston: Newnes.

Carr, J. J. (1996). *Secrets of RF Circuit Design 2nd Edition*. New York: McGraw-Hill.

Hagen, J. B. (1996). *Radio-Frequency Electronics: Circuits and Applications*. Cambridge (UK): Cambridge Univ. Press.

Hardy, J. (1979). *High Frequency Circuit Design*. Reston, VA: Reston (Division of Prentice-Hall).

Kinley, R. H. (1985). *Standard Radio Communications Manual: With Instrumentation and Testing Techniques*. Englewood Cliffs, NJ: Prentice-Hall.

Laverghetta, T. S. (1984). *Practical Microwaves*. Indianapolis, IN: Howard W. Sams.

Liao, S. Y. (1990). *Microwave Devices & Circuits*. Englewood Cliffs, NJ: Prentice-Hall.

Sabin, W. E., and Schoenike, E. O. (Eds.) (1998). *HF Radio Systems & Circuits 2nd Edition*. Atlanta: Noble.

Shrader, R. L. (1975). *Electronic Communication 3rd Edition*. New York: McGraw-Hill.

Vizmuller, P. (1995). *RF Design Guide*. Boston/London: Artech House.

13
CHAPTER

Building simple VLF radio receivers

The very low frequencies (VLF) are located between a few kilohertz and around 300 kHz, depending on whose definition is used. For purposes of this chapter VLF represents the 5- to 100-kHz region. The reason for this seemingly arbitrary designation is that many ham-band and SWL communications receivers operate down to 100 kHz and only a few operate below that limit.

A lot of radio stations are active in the region below 100 kHz. Perhaps the best-known station is WWVB on 60 kHz. This station is operated from Colorado by the National Institutes for Standards and Technology (NIST). WWVB is a very accurate time and frequency station and for many purposes is preferred over the high-frequency WWV and WWVH transmissions. The U.S. Navy operates submarine communications stations in the VLF region, NSS on 21.4 kHz from Annapolis, Maryland (400 kW) and NAA on 24 kHz from Cutler, Maine (1000 kW) being two most commonly heard. Other stations, in both the United States and abroad, are found throughout the VLF region.

But DX-ing in the VLF band is not all that easy. Besides the fact that propagation doesn't support "skip" the way the 20-m ham band does, huge noise signals are in the VLF region. Two sources seem to predominate. First, the 60-Hz power lines are terrible offenders. Although it might seem counterintuitive that a 60-Hz signal could be of much concern at, for example, 30 kHz, it nonetheless is. The high harmonics are caused by harmonic-laden alternating current from the power lines. In addition, the large amounts of power carried by normal residential power lines makes even the higher harmonics at least strong enough to interfere with sensitive receivers.

The second form of interference is the neighborhood television sets. The horizontal oscillator in TV receivers operates at 15.734 kHz, and it is a pulse. As a result, harmonics from television sets can be found up and down the VLF spectrum. And furthermore, the TV horizontal pulse produces its own "sidebands," so each harmonic actually wipes out a lot of spectrum space on either side of the integer multiple of 15.734 kHz. Listening to VLF allows you to identify the evenings when a popular TV presentation is on the air. As a result of TV interference, it is common to find VLFers listening during daylight hours and during the period between midnight and daybreak.

Amateur scientists use VLF receivers in two different types of activity. Some are active in monitoring solar activity that affects radio propagation. Sudden ionospheric disturbances (SIDs) can be noted by sudden increases in VLF signal levels (Taylor and Stokes, 1991; Taylor, 1993). The SID monitoring activity occurs in the 20- to 30-kHz region, although some articles cite as high as 60 kHz.

The other VLF amateur science activity involves looking for naturally created radio signals called *whistlers*. These signals are believed to be created by lightning storms and are propagated at long distances. They occur in the 1- to 10-kHz region (Mideke, 1992 and Eggleston, 1993). At least one project under the sponsorship of NASA engaged amateurs to look at whistlers (Pine, 1991).

Receiver types

Virtually all common designs are used for VLF receivers, except possibly the crystal set. This chapter covers the superheterodyne, the direct conversion, the tuned radio frequency, and the tuned-input gain block methods. It also includes the use of a converter to translate the VLF bands to the HF bands so that an ordinary ham band or SWL receiver can be used.

Superheterodynes

The "superhet" (Fig. 13-1) is the basic receiver design used in communications and broadcast receivers. It dates from the 1920s and is the most successful receiver design. In the superhet receiver, the incoming RF signal (at frequency F_1) is filtered by a tuned RF-resonant circuit or a bandpass filter then applied to a mixer circuit. In most cases, the RF signal is amplified in an RF amplifier (Fig. 13-1), but that is not a necessary requirement. The mixer nonlinearly combines F_1 with the signal from a local oscillator (at frequency F_2) to produce an output spectrum of $F_3 = mF_1 \pm nF_2$. In our simplified case, $m = n = 1$, so the output will consist of the two original signals (F_1 and F_2), the sum signal ($F_1 + F_2$), and the difference signal ($F_1 - F_2$). A filter at the output of the mixer selects either sum or difference signal as F_3 and this is called the *intermediate frequency* (IF). Most of the receiver's gain and selectivity are provided in the IF amplifier. The output of the IF amplifier is fed to a detector that will demodulate the type of signal being received. For an AM signal, a simple envelope detector is used; for CW and SSB, a product detector is used.

In older radios, only the difference signal was used for the IF frequency, but in modern receivers, either the sum or the difference is used. In a VLF receiver, it is possible to use a 10- to 100-kHz RF range, a local oscillator range of 465 to 555 kHz, to produce an IF of 455 kHz (one of the common "traditional" frequencies).

Converters

A subclass of superheterodyne receivers is the converter technique (Fig. 13-2), in which the VLF band is frequency translated to the high-frequency (HF) bands. The typical converter circuit is rather simple. An input filter (either bandpass or

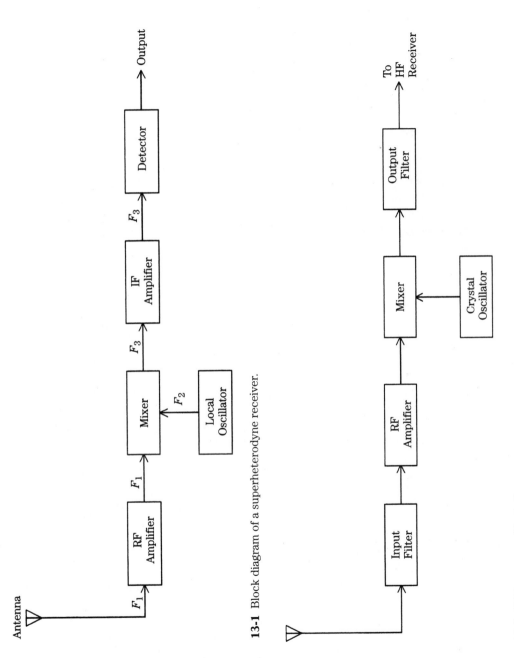

13-1 Block diagram of a superheterodyne receiver.

13-2 VLF converter block diagram.

tuned to a specific frequency) feeds an optional RF amplifier and then a mixer. A fixed-frequency crystal oscillator mixes with the RF signal to produce an output on an IF that is in the ham band or some other shortwave band. For example, to receive 10 to 100 kHz, the input filter (Fig. 13-2) could be a bandpass filter with −3-dB points at 10 and 100 kHz. The local oscillator could be a 3600-kHz crystal oscillator. The output filter is a 3610- to 3710-kHz bandpass filter, the output of which is fed to an 80-m ham-band receiver. However, the feedthrough of the 3600-kHz crystal oscillator signal must be reduced to the receiver. A combination of using a double-balanced mixer and proper termination of the mixer will generally cure this problem.

Tuned radio frequency (TRF) receivers

The TRF receiver (Fig. 13-3) uses a cascade chain of tuned RF amplifiers (A_1, A_2, and A_3) to amplify the radio signal. The TRF was the first really sensitive design in the early 1920s and was eclipsed by the superhet in popular commercial receivers. But in the VLF range, the TRF is still popular—especially among homebrewers. A problem with the TRF receiver is the possibility of unwanted oscillations, which are common in tuned triode devices, such as NPN bipolar transistors.

Peter Taylor's column in the Spring 1993 *Communications Quarterly* gave details of a 20- to 30-kHz TRF receiver for SID monitoring that was designed by Art Stokes (Taylor, 1993).

Tuned gain block (TGB) receivers

The TGB receiver (Fig. 13-4A) is basically a variant of the TRF receiver, except that the tuning circuits are all up front, ahead of the gain. Two benefits are realized. First, the oscillation problem of conventional TRF receivers is avoided because the gain block is untuned. Second, the tuning up front eradicates much of the unwanted noise prior to it being amplified. The actual circuit is shown in Fig. 13-4B. The three amplifier stages consist of high-gain NPN transistors, such as the 2N4416. Tuning is accomplished by connecting a fixed or variable capacitor across input coil L_1. The output circuit consists of a voltage doubler detector (D_1/D_2) and an integrator capacitor (C_{10}). This combination converts the output of the receiver to a dc voltage used to detect the signal. This dc level is recorded on a 1-mA recorder or is input to a computer via an A/D converter. If the latter approach is taken, connect a 10-kΩ resistor across capacitor C_{10} to convert the output current to an output voltage.

Tuning circuit problems

The principal problem to solve in designing and building VLF receivers is the tuning circuits. The capacitance and inductance values tend to be rather large. If you want to use a standard "broadcast variable" capacitor, which is typically 10 to 365 pF in capacitance value, then a 20- to 30-kHz receiver needs an inductor in the 88-mH range. The calculated value of capacitance needed to resonate 88 mH at 20 to 30 kHz is 720 pF. The reason why a smaller variable capacitance is used is that the distributed capacitance of large coils, such as 88 mH, is typically quite large.

The Stokes TGB receiver used a tuned circuit (Fig. 13-4B). Originally, a J. W. Miller 6319 inductor was used, but these apparently are no longer available. The 1993

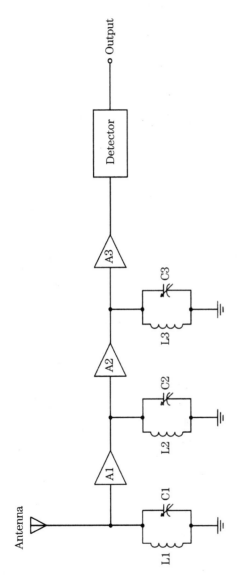

13-3 Tuned radio frequency (TRF) receiver.

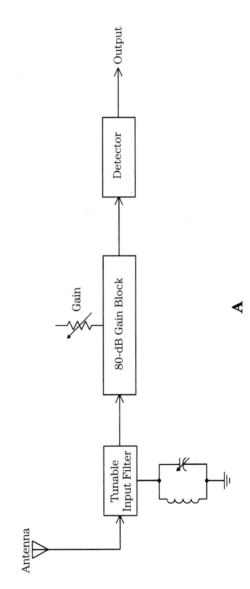

13-4 (A) Block diagram of the VLF SID hunter's receiver and (B) schematic diagram of circuit.

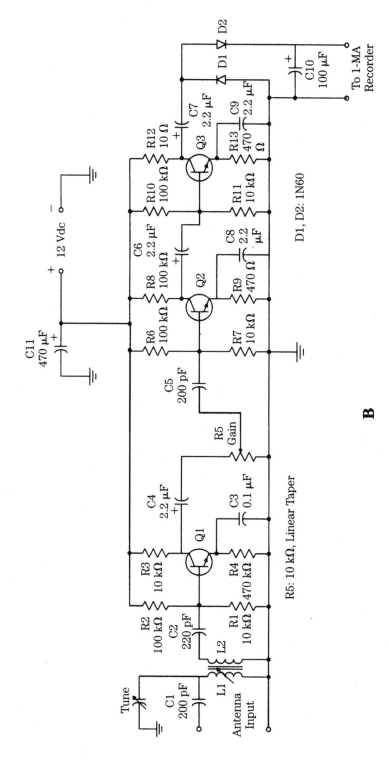

B

13-4 *Continued.*

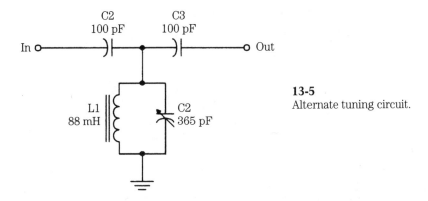

13-5
Alternate tuning circuit.

TRF design used 88-mH "telephone" toroid inductors. The TRF circuit (Fig. 13-5) used the 88-mH inductor paralleled by a 365-pF variable capacitor and isolated it for dc by a pair of 100-pF capacitors (C_2 and C_3 in Fig. 13-5).

A problem with the simple circuit of Fig. 13-5 is that it is not easy to adjust the tuning range (i.e., the receiver cannot be aligned). A variant on the theme, shown in Fig. 13-6, adds a small variable "trimmer" capacitor (C_4) shunted across the main tuning capacitor (C_1), and a trimmer inductor (L_{1B}) is in series with the main fixed inductor (L_{1A}). It would be nice to obtain a single inductor that has the tuning range, but they are hard to obtain these days. As shown, the circuit can be adjusted over about 10% of the inductance range. In one version, a 100-μH fixed inductor was connected in series with a 56-μH variable inductor.

When using the TGB design, all of the tuning is done ahead of the gain block. This poses certain problems for the tuning circuit—especially if more than one tuned LC circuit is desired for selectivity purposes. Figures 13-7 and 13-8 show several methods for combining LC-tuned circuits; both of them fall into the "mutual reactance" method but with different approaches.

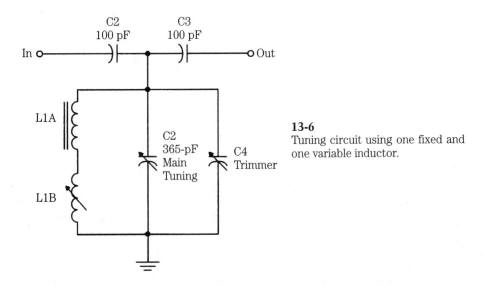

13-6
Tuning circuit using one fixed and one variable inductor.

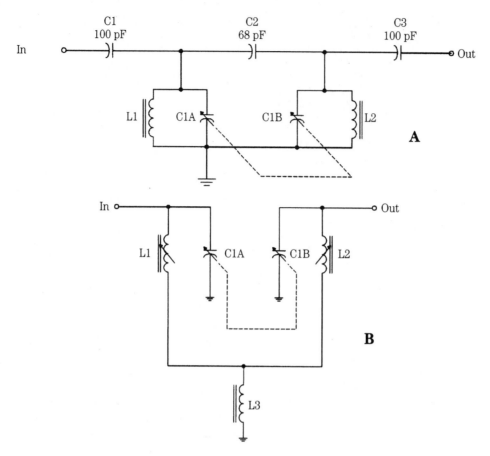

13-7 Mutual reactance coupling: (A) capacitance and (B) inductance.

In Fig. 13-7A, a small capacitance (33 to 120 pF) is used to couple the two LC resonant circuits (L_1/C_{1A} and L_2/C_{1B}). In Fig. 13-7B, a common inductor (L_3) is used for the same purpose. Experiments show that a value of 150 to 700 μH is needed for L_3. *The ARRL Handbook for Radio Amateurs* (all recent editions) provides details on selecting values for the components in these circuits.

A different approach to the design of receiver front ends is shown in Fig. 13-8. One of the problems in VLF receiver design is providing low-impedance link coupling into and out of the LC tank circuit. Large inductance coils are not very often available with low impedance transformer windings. For example, in the Digi-Key catalog, the largest coil with such a winding is a 220-μH unit, which is intended for the AM broadcast band. The solution was the series inductor method (Fig. 13-6). In this case, however, a 56-mH variable inductor was connected in series with a xenon flash-tube trigger transformer. The version selected was intended to fire a 6000-V fast-rise-time pulse into a xenon flash tube, and it has a nominal inductance of 6.96 mH $\pm$20%. The measured inductance was 7.1 μH, and the self-resonance frequency was 140 kHz. Without the trimmer capacitors, the tuning range was 30 to 60 kHz when a 2 $\times$ 380-pF variable capacitor was used. Adding a 600-pF trimmer across each sec-

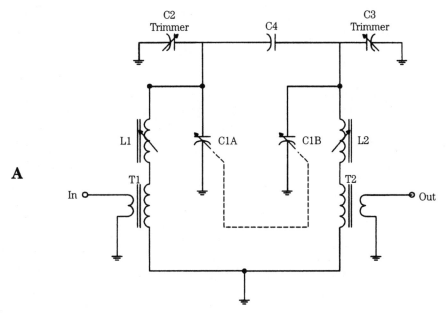

A

T1, T2:: 7-μH Trigger Transformer

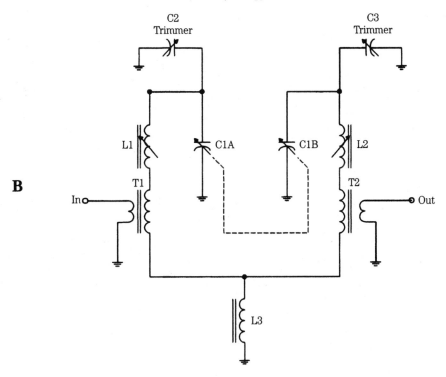

B

13-8 (A) Using a flash-tube trigger transformer as a circuit element and (B) using custom wound coils.

tion of the main tuning capacitor reduced the tuning range to 20 to 30 kHz. The approach in Fig. 13-8A uses a capacitor (C_4) for the mutual reactance, and in Fig. 13-8B, an inductor is used (L_3).

The trigger transformers are widely available from mail-order sources. However, I ordered several from an English source, Maplin Electronics (P.O. Box 3, Rayleigh, Essex SS6 8LR, England). They have variable capacitors, coil forms, and a number of other things of interest to amateur radio constructors. The U.S. credit cards accepted include VISA and American Express and the currency exchange is automatic.

Another alternative, although I've not tested it, is to use pulse transformers. Unfortunately, these transformers typically have limited turns ratios (2:1:1).

A VLF receiver project

The circuit for a modified VLF receiver is shown in Fig. 13-9. It uses the same basic circuit as the Stokes design but with these modified tuning circuits. The 82-mH fixed inductors (L_{1A}, L_{2A}, and L_{3A}) are the Toko 181LY-823J, which are available from Digi-Key (P.O. Box 677, Thief River Falls, MN 56701-0677; 1-800-344-4539) under catalog number TK-4424. The 56-mH variable inductors are Toko CLNS-T1039Z,

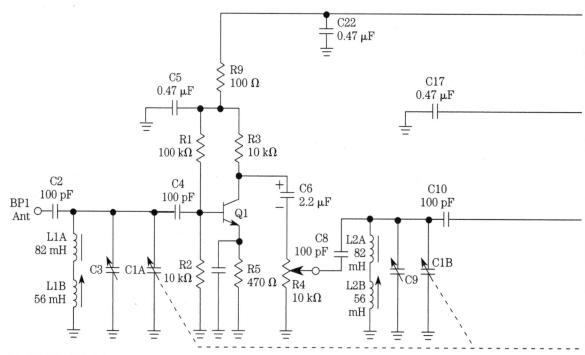

Q1, Q2, Q3: 2N4401
C1: 3 × 365 pF (see text)
C3, C9, C14: 8–80 pF trimmer capacitor (Sprague-Goodman GZC80000) Digi-key SG3010

13-9 Schematic of the VLF receiver project.

available under Digi-Key TK-1724. An additional degree of adjustment is provided by using an 8- to 80-pF trimmer capacitor across each section of the 3 × 365-pF variable main tuning capacitor. These capacitors are Sprague–Goodman GZC8000 units (Digi-Key SG3010).

The output circuit for this receiver reflects the fact that it is a SID monitor receiver. The detector is a voltage doubler (D_1/D_2) made from germanium diodes. The original diodes specified were 1N34 but 1N60s work well also. If you cannot find these diodes (RadioShack and Jim-Pak sells them) then try using replacements from the "universal" service shop replacement lines, such as SK, NTE, and ECG. The NTE-109s and ECG-109s will work well. The output of the detector is heavily integrated by a 470-μF electrolytic capacitor. The output as shown is designed to feed a current-input recorder or a microammeter. If a voltage output is desired then connect a resistor (3.3 to 10 kΩ) across capacitor C_{19}.

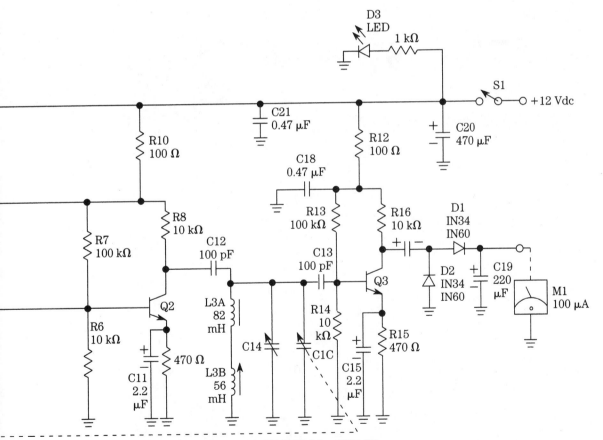

L1A, L2A, L3A: 82 mH (Toko 181LY-823J) Digi-key TK4424
L1B, L2B, L3B: 56 mH (Toko CLNS-T1039Z) Digi-key TK1724

Figure 13-10A shows a printed circuit board for use with this circuit, and Fig. 13-10B is the components' placement "roadmap." The board is designed for these specific components (Fig. 13-9). Variations on the theme can be accommodated by using different value inductors from the same Toko series (see Digi-Key catalog; the L_1-series coils are Toko size 10RB and the L_2-series coils are size 10PA). Also, if you don't want

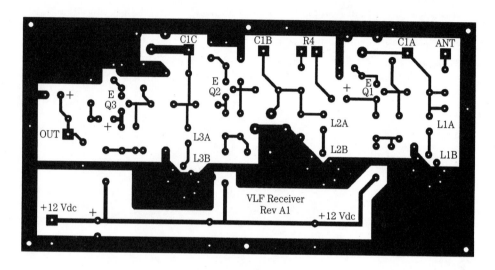

A

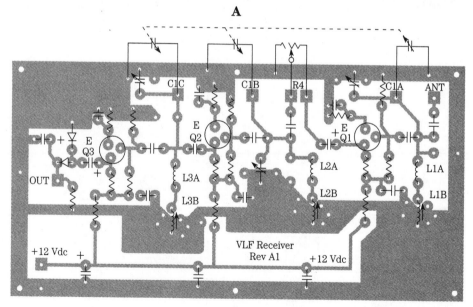

B

13-10 PC board for the VLF receiver. (A) PCB pattern for VLF receiver; (B) parts layout.

to use two coils in each tuning circuit, then short out the holes for L_1 positions and use a coil in the L_2 positions with the required inductance.

The final receiver is shown in Fig. 13-11. A front-panel view is in Fig. 13-11A, and an internal view is shown in Fig. 13-11B. The tuning capacitor was a three-

A

B

13-11 Completed receiver.

section model purchased from Antique Electronic Supply, Tempe, AZ. Although I first thought it was a 3×365-pF unit, it measured at 550 pF (which is better for VLF anyway). A $2''$ vernier dial (Ocean State Electronics, P.O. Box 1458, Westerly, RI 02891; 1-800-866-6626) was used to drive the variable capacitor (notice: Ocean State also sells multisection variable capacitors). The final receiver tuned from 16.5 to 31 kHz.

A later VLF receiver design by Stokes used the gyrator concept to simulate the inductance. Recall that the inductance is the hardest problem in the design of VLF receivers. A gyrator is an operational amplifier circuit (A_1 and A_2 in Fig. 13-11) that acts as if it were an inductor. The value of the inductance is set by the values of the resistors and of capacitor C_3. Op amp A_3 is a noninverting follower with a gain of 100 and provides half of the RF gain and most of the dc gain for the circuit.

After being amplified in A_3, the signal is split into two paths. The original Stokes design went directly to a precision rectifier/integrator, but I needed an RF output as well. When I built the original design, I found that the RF output waveform was distorted by the action of the precision rectifier. In order to overcome that problem, I changed the design to that of Fig. 13-11. Amplifier A_5 buffers the input of the precision rectifier (A_6) to isolate it from the RF output circuit. The RF output could be taken directly from the output of A_3, but I opted instead to use a gain-of-2 noninverting follower (A_4) to buffer the RF output and provide some additional gain.

This receiver is a single-channel model and will tune the 17- to 30-kHz range preferred by SID hunters. Because it is single-channel, I used 15-turn PC-mounted trimmer potentiometers for R_3 and R_7. The receiver was built into a SESCOM SB-7 RF box.

The operational amplifiers used in this circuit must have a gain-bandwidth product high enough to support amplification at the frequencies involved (741-family op amps won't cut it). The original Stokes design used four op amps, so he used the 4136 device. The 4136 is a quad op amp. Other alternatives include the 5534, CA-3140, or CA-3160 (all single op amps) or the CA-3240 (dual op amp).

The dc power supply is extremely important in this circuit. The circuit will operate from any potential over ± 9 to ± 12 Vdc. It is critical that the dc power supply lines be well-bypassed. In Fig. 13-12, only a single op amp is shown in the power connections inset but it serves as a reference for all stages. Each package (whether quad, dual, or single op amp) will have $V-$ and $V+$ terminals. Each of these terminals must be bypassed to ground through 0.1-μF capacitors. These capacitors should be placed as close to the body of the op amp as possible (otherwise oscillation might occur). The $V-$ and $V+$ lines should be bypassed with electrolytic capacitors (be mindful of polarity!) in the 470- to 2000-μF range.

A multiband variant is shown in Fig. 13-13. This receiver is identical to the version of Fig. 13-12, except that the band-set capacitor (C_3) is replaced by a bank of four switched capacitors.

The inductance simulated by the gyrator is calculated from:

$$L_{\mu H} = C_{3\mu F} \times R_5 \times (R_2 + R_3). \tag{13-1}$$

The inductance required for any given frequency (in hertz) is calculated from:

$$L_{\mu H} = \frac{1}{39.5 \, F_{HZ}^2 \, C_{\mu F}}. \tag{13-2}$$

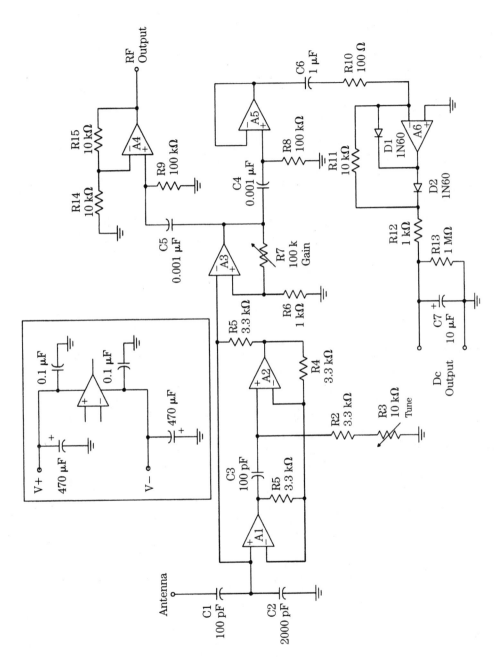

13-12 VLF receiver using "virtual inductor" gyrator circuit (after a design by Art Stokes).

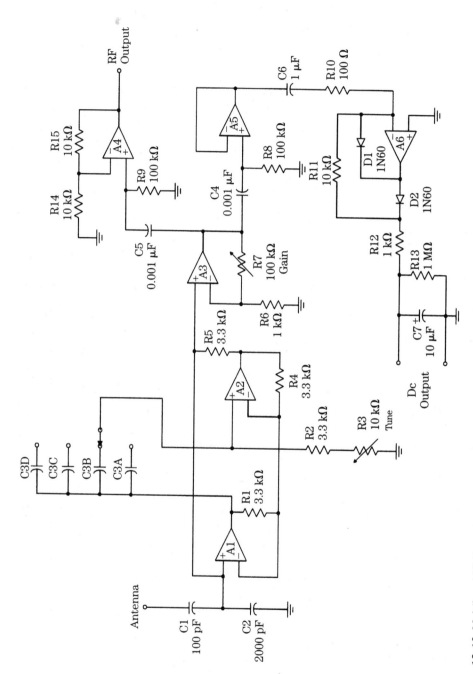

13-13 Multiband VLF receiver.

The use of a VLF converter is shown in Fig. 13-14A, and an actual converter circuit is shown in Fig. 13-14B. The converter is placed in the signal path between the antenna and the antenna input of the receiver. The converter circuit in Fig. 13-14B is based on the NE-602 chip. The oscillator section operates at 4000 kHz (i.e., 4 MHz), and the input is tunable over whatever range needed in the range 10 to 500 kHz. Tuning is accomplished by resonating the secondary winding of T_1 with capacitor C_3. The difference frequency will be either 3500 to 4000 kHz (difference IF), or 4000 to 4500 (sum IF), so the VLF signals will appear in either range (the sense of the tuning will be low to high if the sum IF is selected).

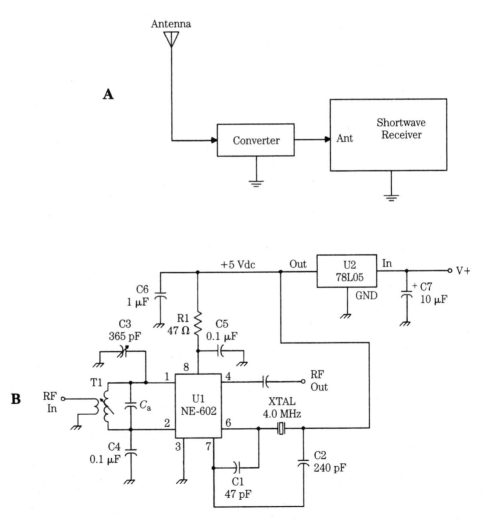

13-14 (A) Converter connection to SW receiver and (B) typical VLF converter circuit.

References and notes

Eggleston, G. (1993). "The Sky Chorus," *Popular Electronics*, July, p. 46.

Mideke, M. (1992). "Listening to Nature's Radio," *Science Probe!*, July, p. 87.

Pine, B. "High School Support for Space Physics Research." Unpublished paper sent to participants in Project INSPIRE.

Taylor, P., and Stokes, A. (1991). "Recording Solar Flares Indirectly." *Communications Quarterly*, Summer, p. 29.

Taylor, P. (1993). "The Solar Spectrum: Update on the VLF Receiver." *Communications Quarterly*, Spring, p. 51.

14
CHAPTER

What's that mess coming from my receiver?

AM-band, shortwave, and VHF monitor receivers often produce output signals that are a terrible mess of weird squawks, squeals, and other assorted objectionable noises. Shortwave receivers in particular show a lot of weird signals, but the growing number of local electronic products are making the cacophony louder and messier than ever before. Fortunately, many of the interference problems that you'll see on your receiver can be fixed or at least reduced significantly in severity. Interference problems fit into two major categories: unwanted local radio station signals and unwanted radiation from other electronic devices.

Radio station interference

Signals from AM, FM, and TV stations are quite powerful, so they can produce a number of problems for receiver owners in their immediate vicinity. Unfortunately, in many areas of the country, residences are located quite close to these types of stations (in a few areas, even shortwave stations are located near homes).

Typical problems from local transmitters include blanketing, desensitization, harmonic generation, and intermodulation. Let's look at each of these in turn.

Blanketing

The blanketing problem occurs when a very strong local signal completely washes out radio reception all across the band. It generally affects the band that the signal is found in, but not other bands. The offending signal can be heard all across the band or at numerous discrete points.

Desensitization

The desensitization problem is a severely reduced receiver sensitivity caused by the presence of a strong local signal. Desensitization can occur across a wide frequency spectrum, not only in the band where the offending signal is located. The strong offending signal might not be heard, except when the receiver is tuned to or near its frequency.

Harmonic generation

If the signal is strong enough to drive the RF amplifier of the receiver into non-linearity then it might generate harmonics (integer multiples) of the strong signal. For example, if you live close to a 780-kHz AM broadcast band signal and the receiver overloads from that signal then you might be able to pick up the signal at $2f$ (1560 kHz), $3f$ (2340 kHz), $4f$ (3120 kHz), and so on up throughout the shortwave bands. Notice that this problem is not the fault of the radio station (which must be essentially clean of harmonics because of FCC regulations) but rather is an inappropriate response on the part of your receiver.

Intermodulation

This problem occurs when two signals of different frequencies (for example, F_1 and F_2) mix together in a nonlinear element to produce a third frequency (F_3). The third signal is sometimes called a *phantom signal*. The frequencies involved might be (and probably are) the assigned fundamental frequency of a legitimate station or their harmonics. The nonlinearity can be caused by receiver overload, improper receiver design, or some other source (legend has it that rusted downspouts and corroded antenna connections can serve as a nonlinear PN junction). As they say in the new science of Chaos Theory, ". . . nonlinearity can arise throughout nature in subtle ways."

The possible new frequency (F_3) will be found from $F_3 = mF_1 \pm nF_2$, where m and n are integers, although not all possibilities are likely to occur in any given situation. Suppose one local ham is operating on 10,120 kHz (in the new 30-m band) and another is operating on 21,390 kHz (in the 15-m band). Both stations are operating normally, but at least one is close enough to overload your receiver and produce nonlinearity in the RF amplifier. When the second harmonic of 10,120 kHz (i.e., 20,240 kHz) combines with 21,390 kHz, the result is a third frequency at 21,390 kHz − 20,240 kHz = 1,150 kHz—right in the middle of the AM broadcast band.

With all of the signals that might exist in your locality, an extremely large number of possible "intermod" combinations can arise. A hill in my hometown is right in the middle of a densely populated residential neighborhood. On that hill are two 50-kW FM broadcast stations; a 5-kW AM broadcast station; scores of VHF and UHF landmobile radio base stations or repeater transmitters; and an assortment of paging systems, ham operators, medical telemetry systems, and the microwave towers of an AT&T long-lines relay station—all within a city block or two. Only a few radio receivers work well unassisted in that neighborhood!

There are two approaches to overcoming these problems. First, either reject or somehow selectively attenuate the offending signal (or one of the signals, in the case of an intermod problem). This is done using a wavetrap. Second, add a passive preselector to the front end of the radio receiver between the antenna line and the antenna input of the receiver.

Figure 14-1 shows three wavetrap circuits that can be used from the AM band

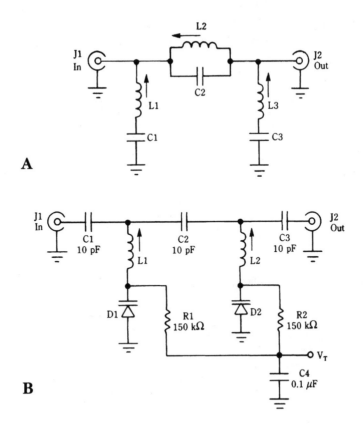

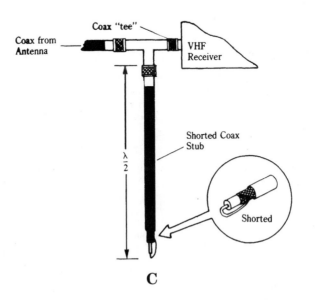

14-1 (A) Series-parallel wave trap, (B) voltage tunable filter/trap, and (C) VHF/UHF half-wave shorted stub trap.

up through VHF/UHF frequencies. The circuit in Fig. 14-1A is an LC wavetrap based on inductors and capacitors. Two LC resonant circuits are shown in Fig. 14-1A. In series with the signal path is a parallel resonant circuit (L_2/C_2). These circuits have a very high impedance at the resonant frequency so they will attenuate signals of that frequency trying to pass through the line between J_1 and J_2. At the same time, the impedance at all other frequencies is high, so those frequencies will pass easily from J_1 to J_2. Shunted across the line on both ends of the signal path are a pair of series resonant circuits. These LC circuits have a low impedance at the resonant frequency and a high impedance at all other frequencies so they will shunt only the offending signals to ground.

The wavetrap of Fig. 14-1A can be built using either fixed inductors and variable capacitors or vice versa. It can be built either for one fixed frequency or as a variable wavetrap that can be tuned from a front-panel knob to attenuate the offending signal.

Figure 14-1B shows the circuit of a similar type of wavetrap, but it is built with voltage-variable capacitance diodes, also called *varactors*. These diodes produce a capacitance across their terminals that is proportional to the applied reverse bias tuning voltage (V_T). In the circuit of Fig. 14-1B, only series resonant circuits are used.

Wavetraps built from LC circuit elements are useful well into the VHF region. In fact, many video and other electronics stores sell wavetraps. (Fig. 14-1A) built especially for the FM broadcast band (88 to 108 MHz) because nearby FM broadcasting stations are frequent sources of interference to VHF television receivers.

Another VHF/UHF wavetrap is shown in Fig. 14-1C. This type of wavetrap is called a *half-wavelength shorted stub*. One of the properties of a transmission line is that it will reflect the load impedance every half-wavelength along the line. When the end of the stub is shorted, therefore, a "virtual" short-circuit will also appear every half-wavelength along the line. This length is frequency-sensitive so the physical length of the line is a tuning factor for this wavetrap.

The length of the line is found from $L_{ft} = 492\ V/F_{MHz}$, where L is in feet and F is in megahertz. The V term is the velocity factor of the transmission line. For common coaxial lines, the value of V ranges from 0.66 (polyethylene) to 0.82 (polyfoam). For example, assume that you need to eliminate the signal from a local FM broadcaster on 88.5 MHz. The shorted stub is made from ordinary coax ($V = 0.66$) and must have a length of $[(492)(0.66)/88.5] = 3.669$ ft (or 44").

The half-wavelength shorted stub is connected in parallel with the antenna input on the receiver. In the case shown in Fig. 14-1C, the stub is connected to the receiver and antenna transmission line through a coaxial "tee" connector. Although UHF connectors are shown, the actual connector that you would select must match your receiver and antenna. Also, it is possible to use 300-Ω twinlead transmission line, rather than coax, as long as that type of line is used on the receiver and antenna.

The other approach to solving the problem is to use a passive preselector ahead of the receiver. Again, the preselector is inserted directly in the transmission line between the receiver antenna input connector and the antenna transmission line. Figure 14-2 shows a typical circuit for this type of preselector. The tuning is controlled by C_{1A}/L_2 and C_{1B}/L_3 and are trimmed by C_2 and C_3 to permit tracking of the two LC circuits. The circuit should be built in a closed, shielded metal box.

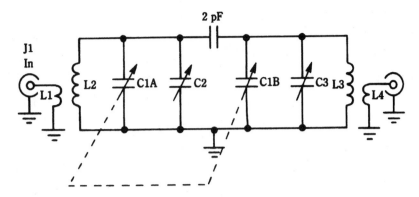

14-2 Passive tunable preselector circuit.

Other interference

The world is a terrible place for sensitive radio receivers: a lot of pure crud (other than the programming material!) comes over the airwaves from a large variety of electrical and electronic devices. There are also some renegade transmitters out there. For example, in the HF shortwave spectrum, you will occasionally hear a "beka-beka-beka" pulsed signal that seems to hop around quite a bit. It will set down on your favorite listening frequency and then go to another. Unfortunately, the nature of a pulsed signal is to spread out over several megahertz—wiping out large segments of the spectrum. That signal is an over-the-horizon backscatter (OTHB) radar in the USSR called by North American SWLs and hams the "Russian Woodpecker." That's one that we can do little about, unfortunately, so let's look at some interfering signals that we can affect.

Light dimmers

A lot of homes are equipped with dimmers instead of switches to control the lighting. These devices are based on silicon-controlled rectifiers (SCRs) that cut the ac sine wave off over part of its cycle, producing a harmonic-rich sharp waveform. These devices can produce a sound described as "frying eggs" well into the shortwave spectrum. Diagnosis is simple: turn off the light. If the noise stops when the light is turned off, then the dimmer is at fault. Although it is possible to install LC line noise filters at the dimmer, that approach is not usually feasible. It would be better to (a) remove the dimmer and replace it with an on/off switch, (b) replace the dimmer with a special model that is designed to suppress radio noise, or (c) keep the light turned off when using the receiver.

Videocassette recorders (VCR)

The VCR is a magnificent entertainment product, and I own one. But I also own SWL and ham radio receivers, and I can always tell when a popular movie is on TV. Because lots of people videotape the movie (legalities notwithstanding), their VCRs are in operation for a couple of hours at a time. The VCR contains a number of radiation-producing circuits, including a 3.58-MHz color subcarrier oscillator. On popular TV nights, I can hear a load of trash around that frequency, which is right in the middle of the 80-m amateur radio band.

TV receivers

The TV set in your home contains at least two major interference producers: the 60-Hz vertical deflection system and the 15,734-kHz horizontal deflection system. The horizontal system includes a high-powered amplifier driving a high-voltage "flyback" transformer and a deflection yoke. As you tune up and down the shortwave band, you will hear "birdies" of the horizontal deflection signal every 15,734 kHz; these are the harmonic signals of the TV horizontal signal.

One quick solution to many interference problems from VCRs and TVs is the line-noise filters shown in Fig. 14-3. These "EMI filters" are placed in the power line

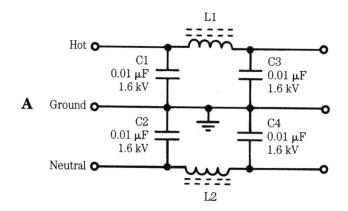

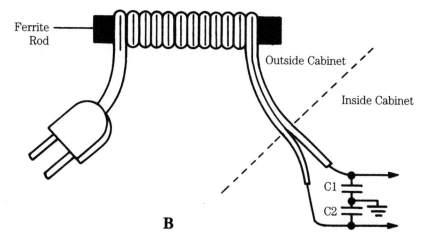

14-3 (A) Ac line EMI filter and (B) power cord on ferrite rod also reduces EMI by acting as a RF choke.

coming from the offending device. It works much of the time because the principal source of the noise signal is radiation from the power line of the VCR or TV. The EMI filter should be installed as close as possible to the body of the offending device.

The filter in Fig. 14-3A uses LC elements to form a low-pass filter network that is placed in series with the ac power line. Homebrew filters should be built inside of a quality, heavy-duty, shielded metal box. If you make your own filter, be sure to use capacitors and inductors that are rated for continuous use across ac power lines. Most readers should consider buying a ready-made LC line filter from a distributor (even RadioShack offers models suitable for most applications). Homebrew EMI filters have a potential for danger if built incorrectly—they can electrocute you or cause a fire in your house!

The filter shown in Fig. 14-3B can be made for any appliance that uses ordinary zipcord for the power line. The cord is wrapped around a $3/8''$ to $1/2''$ ferrite rod (Amidon Associates), which is then taped to keep it in place. This filter should be mounted as close as possible to the TV or VCR chassis. The optional bypass capacitors (C_1 and C_2) inside the equipment cabinet should be 0.01-μF 1600-WVdc disk ceramics that are rated for continuous ac service.

Microwave ovens

Most American homes are equipped with microwave ovens. These devices use a magnetron tube to produce several hundred watts of microwave power on a frequency of approximately 2450 MHz. The high voltage applied to the "maggie" is typically pulsating dc [i.e., it is rectified ac, but not ripple-filtered (as are true dc power supplies)]. This pulsating dc causes "hash" in radio receivers. Although better-quality microwave ovens are equipped with EMI filters, many are not. However, most manufacturers or servicers of microwave ovens can install EMI filters inside the oven. Alternatively, one of the EMI filters shown in Fig. 14-3 can be used.

Personal computer noise

The proliferation of personal computers has greatly increased the amount of noise in the radio spectrum. The noise is caused by the internal circuits that operate using digital clock pulses. Older machines, which use internal clock frequencies of 1 to 4.77 MHz (original IBM PC), wipe out large portions of the AM and shortwave bands. If you doubt this, try using an AM radio near the computer! Later computers (XT-turbo and AT-class machines) used higher clock frequencies (e.g., 8, 10, 12, 16, 25, or 33 MHz) and these can wipe out the VHF bands as well—including the FM broadcast band. I've noticed that the XT-turbo machine that I use for wordprocessing my technical articles and books wipes out the AM band in the regular mode and the FM band in the turbo mode—but the turbo mode produces little interference on AM (a result of the higher clock frequency, I suspect).

Most of the noise produced by the computer is radiated from the power line or from the keyboard cable. In the latter case, be sure that a shielded keyboard cable is used. The power line noise is often susceptible to the same kinds of EMI filters.

Another source of noise is the printer (or more commonly, the printer cable). If the cable between the computer and the printer is unshielded then replace it with a shielded type. Otherwise, you might want to consider a ferrite clamp-on filter bar, such as the Amidon 2X-43 or equivalent.

Cable TV noise

Many communities today are wired for cable TV. These systems transmit a large number of TV and FM broadcast and special-service signals along a coaxial transmission line. They operate on frequencies of 54 to 300 MHz in 36-channel systems and 54 to 440 MHz in the 55-channel systems. Whenever a large number of signals get together in one system, intermodulation is a possibility—and that means signals outside of the official spectrum.

The problem is that signals leak out of the cable TV system and interfere with your receiver. The only thing that you can do legally is to complain to the cable operator and insist he or she eliminate the interference. Fortunately, the FCC is your ally: by law, the operator must keep the signals home!

EMI interference to shortwave and monitor receivers is terribly annoying and sometimes gets so bad that it wipes out reception altogether. Some of these "fixes" might well eliminate the problem and make your radio hobby a lot more fun.

15
CHAPTER

Filtering circuits against EMI

Electronic pollution is all around us. Radio and noise waves impinge us all of the time. Never has electromagnetic interference (EMI) been so great as it is today. One source defines EMI thusly: "... electromagnetic interference is a degradation in performance of an electronic system caused by an electromagnetic disturbance." At worst, EMI can cause a loss of human life, as when it interferes with an aircraft or automobile electronic system. At best, it will pass unnoticed or will interfere with the electronic system on a subaudible basis.

The European Community has issued regulations pertaining to EMI in all manner of electrical and electronic equipment. Electrical and electronic products sold in Europe must exhibit that it neither emits nor is affected by radiation and conduction of EMI. In other words, it must be electromagnetically compatible (EMC).

Means of EMI transmission

EMI is transmitted from the source to the victim system in two basic ways: *conduction* and *radiation*. The difference is that the EMI travels along a wire in conduction and travels by air in radiation. In general (but not always), radiation (Fig. 15-1A) occurs at high frequencies (>30 MHz) and conduction (Fig. 15-1B) occurs at low frequencies (<30 MHz). In some cases, both radiation and conduction can occur. In those cases, either radiation occurs first and then the wave is conducted into the equipment on a line (Fig. 15-1C) or the radiation occurs after conduction (Fig. 15-1D).

In general, the existence of EMI can occur only if (1) there is a source of energy, (2) there is a receiver of that energy, and (3) there is a transmission path between the two.

If any of the three do not exist, EMI cannot occur. What we do about EMI depends on the situation. For example, in the case of some noise sources we can turn

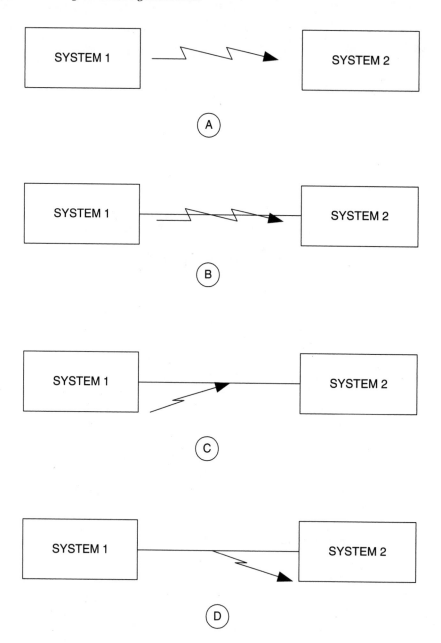

15-1 (A) Radiation, (B) conduction, (C) conduction from radiation, and (D) radiation from conduction.

it off or otherwise suppress it. Otherwise, we might have to live with the effects of EMI as best we can.

Electronic noise

Noise comes in two different types: *continuous* and *transient*. By definition, noise has been standardized as "transient" if it lasts less than 1/60th of a second (16.67 ms) and continuous if it is of longer duration.

Continuous noise

The low-frequency noise sources include fluorescent lights, electric motors, and switching-mode dc power supplies. High-frequency noise is mostly radio frequency interference (RFI), and it can originate in radio transmitters, computer clocks, or other sources.

In the typical RFI environment, signals levels can be between a few microvolts/meter (μV/m) and 300 V/m. While the latter field strengths are only found close to transmitting antennas for high-power radio and radar stations, anything in excess of 1 V/m can cause damage to unprotected circuits. The test specifications for commercial systems may call for protection to 10 V/m, while automotive and medical environments can call for up to 200 V/m and military environments up to 300–400 V/m. Analog circuitry tends to be more influenced by RFI than digital circuits because they can be interfered with by lower-voltage fields.

Transient noise

A "transient" is any temporary (<16.6 ms) overvoltage or overpower condition. Transients are either *repeatable* or *random* in nature. An example of the repeatable type of transient is the discharge of an inductor or capacitor. Examples of random-type transients include electrostatic discharge (ESD), lightning, and the nuclear electromagnetic pulse. In the case of lightning, Fig. 15-2 shows the exposure of the U.S. electrical power system to lightning strikes. Clearly, high-voltage lines are struck more frequently than low-voltage lines. Unfortunately, that exposure transmits along the lines to your home or business to disrupt electronic circuits. In fact, the lightning doesn't have to actually strike the line but by induction can cause disruptive currents to flow in the power system by striking something close!

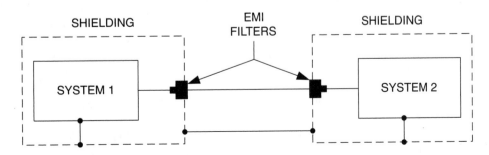

15-2 Average lightning strikes against U.S. power systems.

Counters to EMI

There are two effective ways to counter EMI: *shielding* and *filtering* (Fig. 15-3). The shielding is used to guard against radiation interference, while the filter is used to guard against conduction interference. The filters have the advantage of being bidirectional, so they also prevent interference from flowing out of the system as well as prevent it from flowing in.

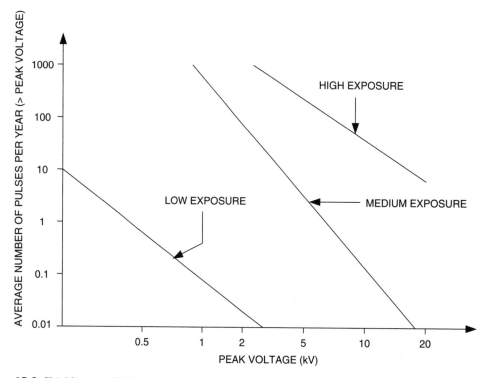

15-3 Shielding and EMI filtering are suitable defenses.

Shielding

Shielding is used to attenuate the interfering RF signal before it reaches the protected circuitry. Very frequently, the hidden difference between a higher-priced appliance and a lower-priced offering is in the internal shielding that one gets. For example, consider computer monitors. The principal difference between a high-priced model and a cheap model is in the shielding that is provided.

Unfortunately, all shielded enclosures for electronic projects are not created equal. Some enclosures are butt-fitted and have dimples or notches to hold the half-shells together. The minimum requirement, as far as I am concerned, for low-frequency radiation noise is shown in Fig. 15-4A. Note that the flange of the lower half-shell overlaps the upper half-shell by 4 to 6 mm (at least). There are four screws (two shown) to hold the assembly together. For high frequencies even this box is insufficient but can be made sufficient by the addition of more screws (Fig. 15-4B).

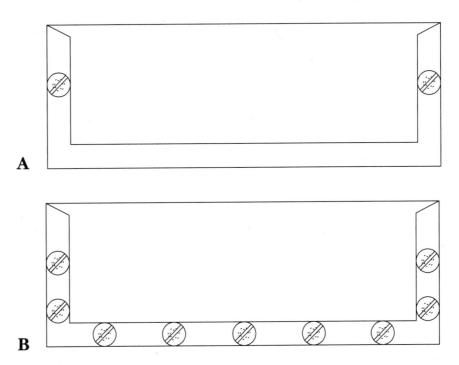

15-4 (A) Simple but effective low-frequency shielding and (B) high-frequency shielding.

The rules are that the screws should be not more than a $\frac{1}{2}$-wavelength apart and $\frac{1}{8}$th-wavelength is better.

Filtering

Filtering can take on different meanings for different situations. In general, most EMI filters are *low-pass* filters, although *high-pass* and *bandpass filters* exist. In some cases of a particular frequency being the cause of interference, a *notch filter* may be used. In general, a perfect, ideal single component filter (either a capacitor or inductor) has a theoretical roll-off or gain of -20 dB/decade, with a practical maximum of something between -60 and -120 dB. In fact, real components do not achieve that theoretical goal. Capacitors are more useful in high-impedance circuits, whereas inductors are more useful in lower-impedance circuits.

Perhaps the simplest form of single component filter is the *feedthrough capacitor* (a.k.a. "EMI filter"). When combined with good shielding, the use of such a capacitor can be quite sufficient. Figure 15-5 shows two methods of passing a feedthrough capacitor through a shielded panel. In Fig. 15-5A we see the screw in variety. The threaded nut is cinched tight against the chassis or panel. In Fig. 15-5B we see the installation of a solder-in type of feedthrough capacitor. A small fillet of solder is used to hold the capacitor against the chassis or panel. This type of capacitor assumes a solderable chassis or panel, thus it eliminates the use of aluminum.

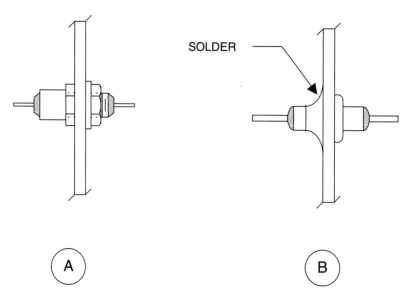

15-5 Two types of feedthrough capacitor.

Where greater suppression is needed, a combination of L and C elements is needed. A two-component L-section filter is shown in Fig. 15-6A. This filter produces a theoretical gain of −40 dB/decade, which means 100:1 per decade between input and output signals. This filter can be used at any frequency, although the values will tend to differ between, say, LF and VHF. The ideal is to keep the lead lengths as short as possible to prevent radiation of the signal. As an alternative, a higher-order filter can be realized by replacing the grommet in Fig. 15-6A with a second feedthrough capacitor. In that case, a theoretical gain of −60 dB/decade is realized.

In Fig. 15-6B we see a case where the opposite situation occurs, i.e., the inductor input L-section filter. In this case the inductor or RF choke is mounted external to the chassis and directly drives the feedthrough capacitor.

Figure 15-7 shows a graphic of which filters to use when the source impedances are known. For example, in the case where the input impedance and the output impedance are both low then use either a single inductor or RF choke or a T-filter consisting of two inductors and a capacitor. When the source impedance is high and the output impedance is low then use a single capacitor input L-section filter. Similarly, when the input impedance is low and the output impedance is high use an inductor input L-section filter. Finally, when the input and output impedances are both high use either a single capacitor or a pi-section filter as shown.

Common mode and differential currents

Noise currents can flow in two modes: *differential* and *common mode*. These are defined as follows.

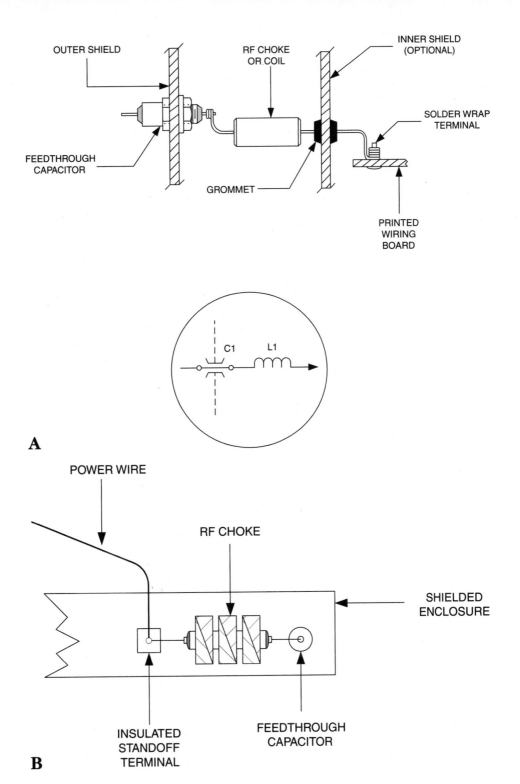

15-6 Simple L-section filters: (A) internal capacitor input type and (B) external inductor input type.

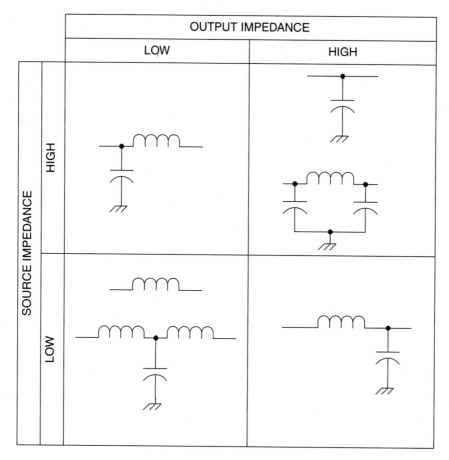

15-7 Several situations and the filters to implement them.

Differential mode

As shown in Fig. 15-8A there is a conducted signal on a signal line that is returned along the return line or via a grounded connection. The noise current is characterized by the arrows flowing in different directions.

Common mode

In common mode conduction noise appears in multiple conductors flowing in the same direction.

The filtering necessary for differential and common mode filtering is different. In differential mode filtering it is necessary to place a filter in series with the "hot" lead, with suitable bypassing to ground. In the common mode case the same filtering must be applied to all affected leads. Figure 15-9 shows a filter that is suitable for both differential and common mode forms of EMI.

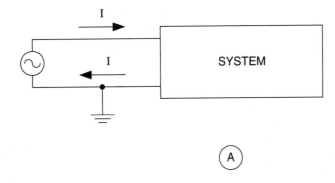

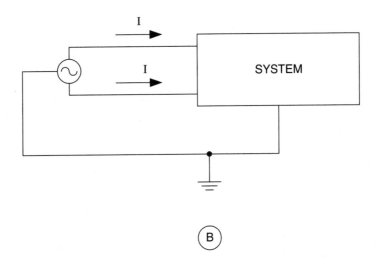

15-8 Common mode and differential signals.

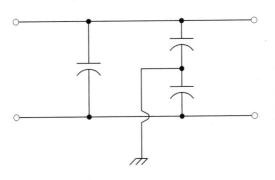

15-9
Filter for both differential and common mode signals.

AC power line filtering

The ac power lines are a source of conducted EMI and must be filtered extensively to make them really clean. Not only should the ac lines be filtered against RFI but also against lightning strikes at a distance. The lightning and other high-voltage transients can be handled per Fig. 15-10 with *metal oxide varistors* (MOV). These devices are made by various suppliers and can act like a pair of zener diodes back-to-back. In other words, they snip the high-voltage transient to a lower level, regardless of polarity. Basically, the way they work is that they act like an insulator at lower voltages but when a certain critical threshold voltage is exceeded the varistors develop a low resistance, shunting the offending voltage transient to a low value. Varistors are used primarily for transient voltage spikes on the ac power line.

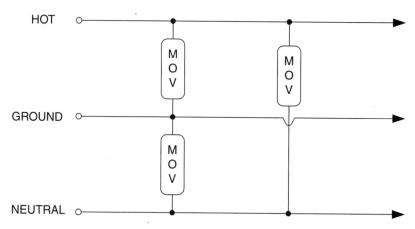

15-10 Metal oxide varistors (MOV) use on AC power mains.

For RFI a filter must be provided to the ac power line. Figure 15-11 shows a suitable filter that affects both common mode and differential RFI on the line. When the values are high enough, it will also protect somewhat against lightning and other transients because those transients have a high-frequency component as well as the fundamental frequency. Suitable filters have been molded into ac power line sockets.

Special medical EMI problems

EMI problems exist in medical electronic devices such as electrocardiograph (ECG) or electroencephalograph (EEG) machines. These machines have to process signals on the order of a few microvolts (μV) to about 1 millivolt (mV) in the presence of strong interfering signals. In addition to regular RFI problems, there are two additional problems. First, there is the problem of the defibrillator. This is a high-voltage (several kilovolts!) capacitive discharge device used to "jump start" the heart of a patient undergoing resuscitation. The second problem is in the application of 500- to 2500-kHz high-powered RF electrosurgery units just a few centimeters from

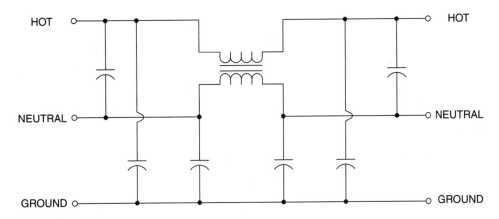

15-11 AC power mains filter.

the site of the EEG or ECG electrodes. As much as 400 or 500 W of RF could be applied to the "circuit" only a few centimeters from the pick-up electrodes!

Figure 15-12 shows the input to a bioelectric amplifier suitable for EEG or ECG use in the presence of high electrical or electromagnetic fields. Both defibrillator and electrosurgery units can be accommodated by the filtering shown. Both ECG and EEG signals consists of signals below 100 Hz, so an RC filter will suffice in this case. The RC filter consists of resistors R_1 through R_6 and capacitors C_1 through C_4. It is a low-pass filter.

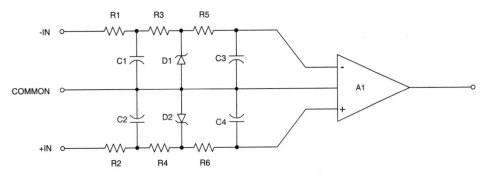

15-12 ECG/EEG preamplifier filtering and defibrillator protection.

The defibrillator protection is the zener diodes and the series resistors (R_1–R_3 and R_2–R_4) between the source and the amplifier (A_1). Sometimes, in older machines neon glow lamps (e.g., NE-2) are used instead of the zener diodes. The disadvantage of the neon lamps is their relatively high protection voltage.

Computer EMI

The case of computer EMI is very serious. Just place an AM radio anywhere close to a modern computer and you will hear lots of hash. In fact, with computer clock speeds reaching several hundred megahertz, the interference to FM radios can be tremendous. Figure 15-13 shows a method for connecting a digital connector pin that can carry EMI to a printed wiring board. The ferrite beads act like little RF chokes so will eliminate RFI in the VHF/UHF region. Because the filtering is bidirectional, it will attenuate the noise going out as well as that coming into the computer.

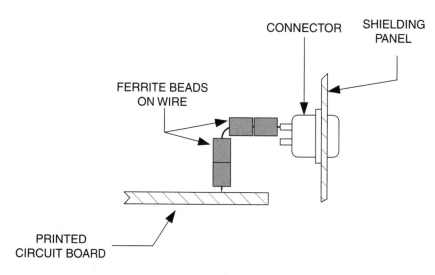

15-13 Computer connector fitted with ferrite beads.

The principal offender with respect to noise from computer systems is the monitor. This is because of two factors. First, the deflection circuits tend to operate in frequency ranges (under 40 kHz) that are below many other systems and they have lots of harmonics. Second, those deflection circuits tend to be high-power. The answer to the problem is to place shielding around the circuits and a common mode choke in the signal line.

Conclusion

EMI protection is often an afterthought in the design of electronic equipment. It should be a first requirement but, unfortunately, this isn't often the case. The methods shown herein will go a long way toward suppressing the RFI or transient conditions on the signal or power lines.

16
CHAPTER

Measuring inductors and capacitors at RF frequencies

The measurement of the values of inductors (L) and capacitors (C) at radio frequencies differs somewhat from the same measurements at low frequencies. Although similarities exist, the RF measurement is a bit more complicated. One of the reasons for this situation is that stray or "distributed" inductance and capacitance values of the test set-up will affect the results. Another reason is that capacitors and inductors are not ideal components but rather all capacitors have some inductance and all inductors have capacitance. In this chapter we will take a look at several methods for making such measurements.

VSWR method

When a load impedance ($R + jX$) is connected across an RF source the maximum power transfer occurs when the load impedance (Z_L) and source (Z_S) impedances are equal ($Z_L = Z_S$). If these impedances are not equal, then the *voltage standing wave ratio* (VSWR) will indicate the degree of mismatch. We can use this phenomenon to measure values of inductance and capacitance using the scheme shown in Fig. 16-1A. The instrumentation required includes a signal generator or other signal source and a VSWR meter or VSWR analyzer.

Some VSWR instruments require a transmitter for excitation, but others will accept the lower signal levels that can be produced by a signal generator. An alternative device is the SWR analyzer type of instrument. It contains the signal generator and VSWR meter along with a frequency counter to be sure of the actual test frequency. Whatever signal source is used, however, it must have a variable output frequency. Further, the frequency read-out must be accurate (the accuracy of the method depends on knowing the actual frequency).

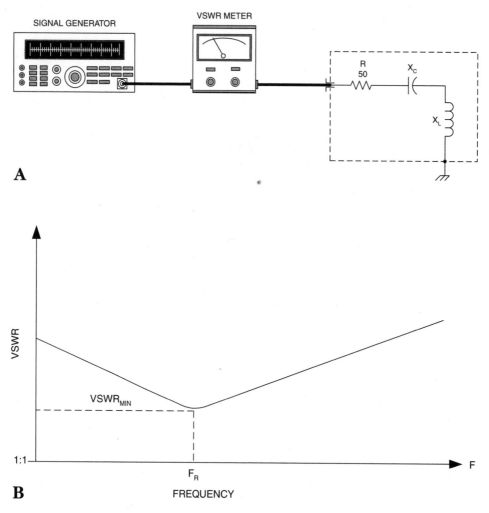

16-1 (A) VSWR method for measuring L and C; (B) VSWR-vs-Frequency curve.

The load impedance inside the shielded enclosure consists of a noninductive re-sistor (R_1) that has a resistance equal to the desired system impedance resistive component (50 Ω in most RF applications and 75 Ω in television and video). An in-ductive reactance (X_L) and a capacitive reactance (X_C) are connected in series with the load. The circuit containing a resistor, capacitor, and inductor simulates an an-tenna feedpoint impedance. The overall impedance is given by:

$$Z_L = \sqrt{R^2 + (X_L - X_C)^2}. \qquad (16\text{-}1)$$

Note the reactive portion of Eq. (16-1). When the condition $|X_L| = |X_C|$ exists, the series network is at resonance and VSWR is minimum (Fig. 16-1B). This gives us a means for measuring the values of the capacitor or inductor, provided that the other is known. That is, if you want to measure a capacitance then use an inductor of known value. Alternatively, if you want to know the value of an unknown inductor use a capacitor of known value.

Using the test set-up in Fig. 16-1A, adjust the frequency of the signal source to produce minimum VSWR.

For finding an inductance from a known capacitance:

$$L_{\mu H} = \frac{10^{12}}{4\pi^2 f^2 C_{pF}},$$ (16-2)

where

$L_{\mu H}$ = inductance in microhenrys (μH)
C_{pF} is the capacitance in picofarads (pF)
f is the frequency in hertz (Hz)
For finding a capacitance from a known inductance:

$$C_{pF} = \frac{10^{12}}{4\pi^2 f^2 L_{\mu H}}.$$ (16-3)

The accuracy of this approach depends on how accurately the frequency and the known reactance are known and how accurately the minimum VSWR frequency can be found.

Voltage divider method

A resistive voltage divider is shown in Fig. 16-2A. This circuit consists of two resistors (R_1 and R_2) in series across a voltage source, V. The voltage drops across R_1 and R_2 are V_1 and V_2, respectively. We know that either voltage drop is found from:

$$V_X = \frac{V R_X}{R_1 + R_2},$$ (16-4)

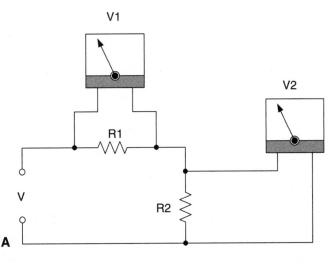

16-2 (A) Simple voltage divider; (B) reactance voltage divider.

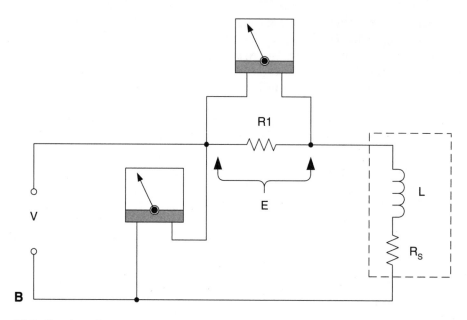

B

16-2 *Continued.*

where V_X is V_1 and R_X is R_1 or V_X is V_2 and R_X is R_2, depending on which voltage drop is being measured.

We can use the voltage divider concept to find either inductance or capacitance by replacing R_2 with the unknown reactance. Consider first the inductive case. In Fig. 16-2B resistor R_2 has been replaced by an inductor (L). The resistor R_1 is the inductor series resistance. If we measure the voltage drop across R_1 (i.e., "E" in Fig. 16-2B) then we can calculate the inductance from:

$$L = \frac{R}{2\pi f} \times \sqrt{\left(\frac{V}{E}\right)^2 - \left(1 + \frac{R_S}{R_1}\right)^2}. \tag{16-5}$$

As can be noted in Eq. (16-5) if $R_1 \gg R_S$, then the quotient R_S/R_1 becomes negligible.

In capacitors the series resistance is typically too small to be of consequence. We can replace L in the model of Fig. 16-2B with a capacitor and again measure voltage E. The value of the capacitor will be:

$$C = \frac{2\pi f \times 10^6}{R \times \sqrt{\left(\frac{V}{E}\right)^2 - 1}}. \tag{16-6}$$

The value of resistance selected for R_1 should be approximately the same order of magnitude as the expected reactance of the capacitor or inductor being measured. For example, if you expect the reactance to be, say, between 1K and 10K at some frequency then select a resistance for R_1 in this same range. This will keep the voltage values manageable.

Signal generator method

If the frequency of a signal generator is accurately known then we can use a known inductance to find an unknown capacitance or use a known capacitance to find an unknown inductance. Figure 16-3 shows the test set-up for this option. The known and unknown components (L and C) are connected together inside a shielded enclosure. The parallel tuned circuit is lightly coupled to the signal source and the display through very low value capacitors (C_1 and C_2). The rule is that the reactance of C_1 and C_2 should be very high compared with the reactances of L and C.

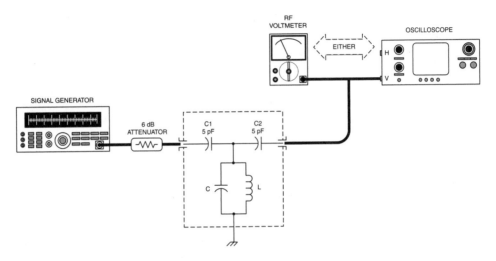

16-3 Signal generator method.

The signal generator is equipped with a 6-dB resistive attenuator in order to keep its output impedance stable. The output indicators should be any instrument that will read the RF voltage at the frequency of resonance. For example, you could use either an RF voltmeter or an oscilloscope.

The procedure requires tuning the frequency of the signal source to provide a peak output voltage reading on the voltmeter or 'scope. If the value of one of the components (L or C) is known, then the value of the other can be calculated using Eq. (16-2) or Eq. (16-3), as appropriate.

Alternate forms of coupling are shown in Fig. 16-4. In either case, the idea is to isolate the instruments from the L and C elements. In Fig. 16-4A the isolation is provided by a pair of high-value (10K to 1 Meg) resistors, R_1 and R_2. In Fig. 16-4B the coupling and isolation is provided by a one- or two-turn link winding over the inductor. The links and the main inductor are lightly coupled to each other.

Frequency-shifted oscillator method

The frequency of a variable-frequency oscillator (VFO) is set by the combined action of an inductor and a capacitor. We know that a change in either capacitance

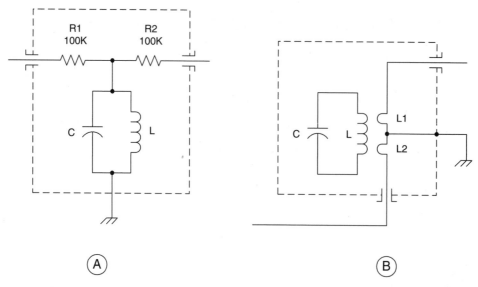

16-4 Alternative coupling methods.

or inductance produces a frequency change equal to the square of the component ratio. For example, for an inductance change:

$$L_2 = L_1 \times \left[\left(\frac{F_1}{F_2} \right)^2 - 1 \right],$$

(16-7)

where

    L_1 is the original inductance
    L_2 is the new inductance
    F_1 is the original frequency
    F_2 is the new frequency.

From this equation we can construct an inductance meter such as Fig. 16-5. This circuit is a Clapp oscillator designed to oscillate in the high-frequency (HF) range up to about 12 MHz. The components L_1, C_2, and C_3 are selected to resonate at some frequency. Inductor L_1 should be of the same order of magnitude as L_1. The idea is to connect the unknown inductor across the test fixture terminals. Switch S_1 is set to position "b" and the frequency (F_1) is measured on a digital frequency counter. The switch is then set to position "a" in order to put the unknown inductance (L_2) in series with the known inductance (L_1). The oscillator output frequency will shift to F_2. When we know L_1, F_1, and F_2 we can apply Eq. (16-7) to calculate L_2.

If we need to find a capacitance, then modify the circuit to permit a capacitance to be switched into the circuit across C_1 instead of an inductance as shown in Fig. 16-5. Replace the "L" terms in Eq. (16-7) with the corresponding "C" terms.

Using RF bridges

Most RF bridges are based on the dc *Wheatstone bridge* circuit (Fig. 16-6). In use since 1843, the Wheatstone bridge has formed the basis for many different mea-

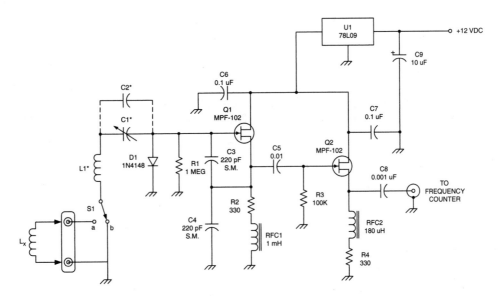

16-5 Frequency shift method.

surement instruments. The *null condition* of the Wheatstone bridge exists when the voltage drop of R_1/R_2 is equal to the voltage drop of R_3/R_4. When the condition $R_1/R_2 = R_3/R_4$ is true, then the voltmeter (M_1) will read zero. The basic measurement scheme is to know the values of three of the resistors and use them to measure the

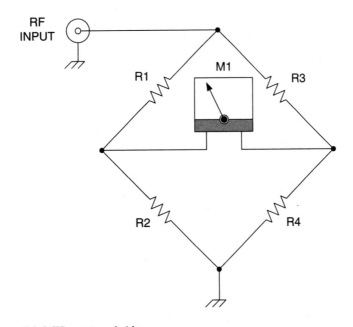

16-6 Wheatstone bridge

value of the fourth. For example, one common scheme is to connect the unknown resistor in place of R_4 and make R_1 and R_3 fixed resistors of known value and R_2 a calibrated potentiometer marked in ohms. By adjusting R_2 for the null condition and then reading its value, we can use the ratio $(R_2 \times R_3)/R_1 = R_4$.

The Wheatstone bridge works well for finding unknown resistances from dc to some relatively low RF frequencies, but to measure L and C values at higher frequencies we need to modify the bridges. Three basic versions are used: Maxwell's bridge (Fig. 16-7), Hay's bridge (Fig. 16-8), and Schering's bridge (Fig. 16-9).

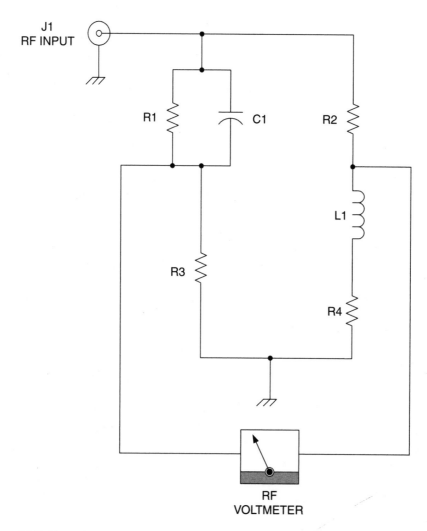

16-7 Maxwell bridge.

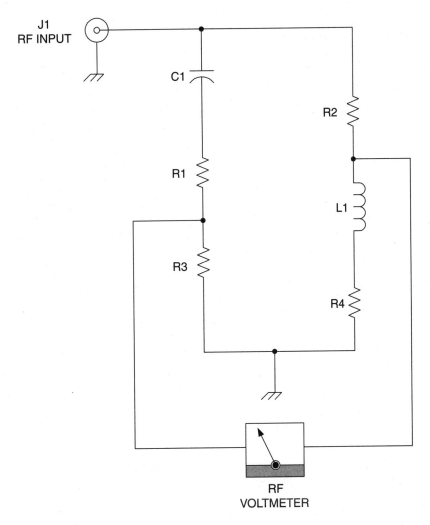

16-8 Hay bridge.

Maxwell bridge

The Maxwell bridge is shown in Fig. 16-7. The null condition for this bridge occurs when:

$$L_1 = R_2 \times R_3 \times C_1 \tag{16-8}$$

and

$$R_4 = \frac{R_2 \times R_3}{R_1}. \tag{16-9}$$

The Maxwell bridge is often used to measure unknown values of inductance (e.g., L_1) because the balance equations are totally independent of frequency. The bridge is also not too sensitive to resistive losses in the inductor (a failing of some other methods). Additionally, it is much easier to obtain calibrated standard capaci-

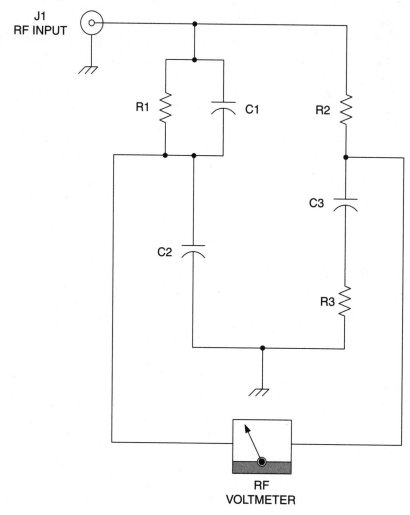

16-9 Schering bridge.

tors for C_1 than it is to obtain standard inductors for L_1. As a result, one of the principal uses of this bridge is inductance measurements.

Maxwell bridge circuits are often used in measurement instruments called *Q-meters,* which measure the quality factor (Q) of inductors. The equation for Q is, however, frequency sensitive:

$$Q = 2 \times \pi \times F \times R_1 \times C_1, \qquad (16\text{-}10)$$

where
   F is in hertz
   R_1 is in ohms
   C_1 is in farads.

Hay bridge

The Hay bridge (Fig. 16-8) is physically similar to the Maxwell bridge except that the R_1/C_1 combination is connected in series rather than in parallel. The Hay bridge is, unlike the Maxwell bridge, frequency-sensitive. The balance equations for the null condition are also a little more complex:

$$L_1 = \frac{R_2 \times R_3 \times C_1}{1 + \left[\dfrac{1}{Q}\right]^2} \tag{16-11}$$

$$R_4 = \left[\frac{R_2 \times R_3}{R_1}\right] \times \left[\frac{1}{Q^2 + 1}\right] \tag{16-12}$$

where

$$Q = \frac{1}{\omega \times R_1 \times C_1}. \tag{16-13}$$

The Hay bridge is used for measuring inductances with high Q figures while the Maxwell bridge is best with inductors that have a low Q value.

Note: A frequency-independent version of Eq. (16-11) is possible when Q is very large (i.e., >100):

$$L_1 = R_2 \times R_3 \times C_1. \tag{16-14}$$

Schering bridge

The Schering bridge circuit is shown in Fig. 16-9. The balance equation for the null condition is:

$$C_3 = \frac{C_2 \times R_1}{R_2} \tag{16-15}$$

$$R_3 = \frac{C_2 \times R_1}{R_2}. \tag{16-16}$$

The Schering bridge is used primarily for finding the capacitance and the power factor of capacitors. In the latter applications no actual R_3 is connected into the circuit, making the series resistance of the capacitor being tested (e.g., C_3) the only resistance in that arm of the bridge. The capacitor's Q factor is found from:

$$Q_{C_3} = \frac{1}{\omega \times R_1 \times C_1}. \tag{16-17}$$

Finding parasitic capacitances and inductances

Capacitors and inductors are not ideal components. A capacitor will have a certain amount of series inductance (called "parasitic inductance"). This inductance is created by the conductors in the capacitor, especially the leads. In older forms of ca-

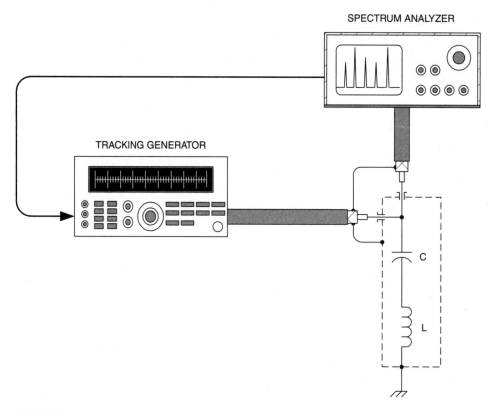

16-10 Measuring series inductance in a capacitor.

pacitor, such as the wax-paper dielectric devices used prior to about 1960, the series inductance was very large. Because the inductance is in series with the capacitance of the capacitor, it forms a series resonant circuit.

Figure 16-10 shows a test set-up for finding the series resonant frequency. A *tracking generator* is a special form of sweep generator that is synchronized to the frequency sweep of a spectrum analyzer. They are used with spectrum analyzers in order to perform stimulus response measurements such as Fig. 16-10.

The nature of a series resonant circuit is to present a low impedance at the resonant frequency and a high impedance at all frequencies removed from resonance. In this case (Fig. 16-10), that impedance is across the signal line. The display on the spectrum analyzer will show a pronounced, sharp dip at the frequency where the capacitance and the parasitic inductance are resonant.

The value of the parasitic series inductance is:

$$L = \frac{1}{2^2 \pi^2 f^2 C}. \qquad (16\text{-}18)$$

Inductors are also less than ideal. The adjacent turns of wire form small capacitors which when summed up can result in a relatively large capacitance value. Figure 16-11 shows a method for measuring the parallel capacitance of an inductor.

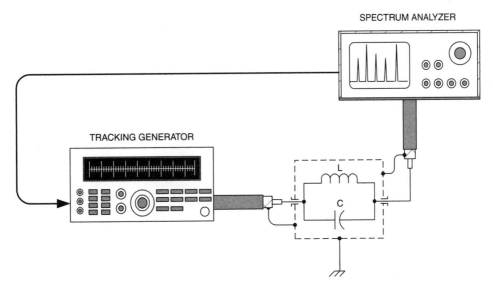

SPECTRUM ANALYZER

TRACKING GENERATOR

16-11 Measuring parallel capacitance in an inductor.

Because the capacitance is in parallel with the inductance, it forms a parallel resonant circuit. These circuits will produce an impedance that is very high at the resonant frequency and very low at frequencies removed from resonance. In Fig. 16-11 the inductor and its parasitic parallel capacitance are in series with the signal line, so will (like the other circuit) produce a pronounced dip in the signal at the resonant frequency. The value of the parasitic inductance is:

$$C = \frac{1}{2^2 \pi^2 f^2 L}. \tag{16-19}$$

Conclusion

There are other forms of bridge, and other methods, for measuring L and C elements in RF circuits, but those discussed above are very practical, especially in the absence of specialist instrumentation.

17
CHAPTER

Building and using the RF noise bridge

This chapter explores a device that has applications in general RF electronics as well as in antenna work: the RF noise bridge. It is one of the most useful, low-cost, and over-looked test instruments in the servicer's armamentarium.

Several companies have produced low-cost noise bridges: Omega-T, Palomar Engineers, and the now out-of-the-kit-business Heath Company. The Omega-T device (Fig. 17-1A) is a small cube with minimal dials and a pair of BNC coax connectors (marked *antenna* and *receiver*). The dial is calibrated in ohms and measures only the resistive component of impedance. The Palomar Engineers device (Fig. 17-1B) is a little less eye-appealing but does everything the Omega-T does, plus it allows you to make a rough measurement of the reactive component of impedance.

The Heath Company had their Model HD-1422 in the line-up. Over the years, I have found the noise bridge terribly useful for a variety of test and measurement applications—especially in the HF and low-VHF regions, and those applications are not limited to the testing of antennas (which is the main job of the noise bridge). In fact, the two-way technician (including CB) will measure antennas with the device, but consumer technicians will find other applications.

Figure 17-2 shows the block diagram of this instrument. The bridge consists of four arms. The inductive arms (L_{1b} and L_{1c}) form a trifilar-wound transformer over a ferrite core with L_{1a} so signals applied to L_{1a} are injected into the bridge circuit. The measurement consists of a series circuit with a 200-Ω potentiometer and a 120-pF variable capacitor. The potentiometer sets the range (0 to 200 Ω) of the resistive component of measured impedance and the capacitor sets the reactance component. Capacitor C_2 in the UNKNOWN arm of the bridge is used to balance the measurement capacitor. With C_2 in the circuit, the bridge is balanced when C is approximately in the center of its range. This arrangement accommodates both inductive and capacitive reactances, which appear on either side of the "zero" point (i.e., the mid-range capacitance of C). When the bridge is in balance, the settings of R and C reveal the impedance across the UNKNOWN terminal.

A

B

17-1 (A) Omega-T noise bridge and (B) Palomar noise bridge.

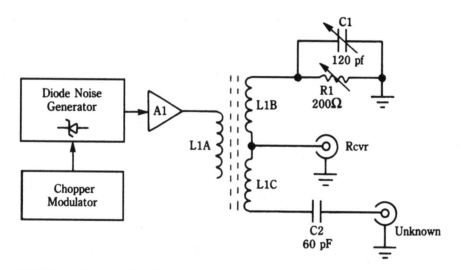

17-2 Block diagram of an RF noise bridge.

A reverse-biased zener diode (zeners normally operate in the reverse bias mode) produces a large amount of noise because of the avalanche process inherent in zener operation. Although this noise is a problem in many applications, in a noise bridge it is highly desirable: the richer the noise spectrum the better. The spectrum is enhanced somewhat in some models because of a 1-kHz square-wave modulator that chops the noise signal. An amplifier boosts the noise signal to the level needed in the bridge circuit.

The detector used in the noise bridge is a HF receiver. The preferable receiver uses an AM demodulator, but both CW (morse code) and SSB receivers will do in a pinch. The quality of the receiver depends entirely on the precision with which you need to know the operating frequency of the device under test.

Adjusting antennas

Perhaps the most common use for the antenna noise bridge is finding the impedance and resonant points of a HF antenna. Connect the RECEIVER terminal of the bridge to the ANTENNA input of the HF receiver through a short length of coaxial cable as shown in Fig. 17-3. The length should be as short as possible, and the

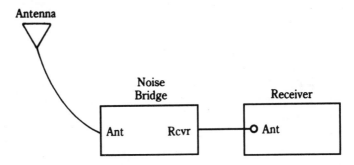

17-3 Connection of the noise bridge to the receiver and antenna

characteristic impedance should match that of the antenna feedline. Next, connect the coaxial feedline from the antenna to the ANTENNA terminals on the bridge. You are now ready to test the antenna.

Finding impedance

Set the noise bridge resistance control to the antenna feedline impedance (usually 50 or 75 Ω for most amateur antennas). Set the reactance control to mid-range (zero). Next, tune the receiver to the expected resonant frequency (F_{exp}) of the antenna. Turn the noise bridge on and look for a noise signal of about S9 (will vary on different receivers and if in the unlikely event that the antenna is resonant on the expected frequency).

Adjust the resistance control (R) on the bridge for a null (i.e., minimum noise, as indicated by the S meter). Next, adjust the reactance control (C) for a null. Repeat the adjustments of the R and C controls for the deepest possible null, as indicated by the lowest noise output on the S meter (there is some interaction between the two controls).

A perfectly resonant antenna will have a reactance reading of zero Ω and a resistance of 50 to 75 Ω. Real antennas might have some reactance (the less the better) and a resistance that is different from 50 or 75 Ω. Impedance-matching methods can be used to transform the actual resistive component to the 50- or 75-Ω characteristic impedance of the transmission line.

If the resistance is close to zero, suspect that there is a short circuit on the transmission line and an open circuit if the resistance is close to 200 Ω.

A reactance reading on the X_L side of zero indicates that the antenna is too long and a reading on the X_c side of zero indicates an antenna that is too short.

An antenna that is too long or too short should be adjusted to the correct length. To determine the correct length, we must find the actual resonant frequency, F_r. To do this, reset the reactance control to zero and then slowly tune the receiver in the proper direction—downband for too long and upband for too short—until the null is found. On a high-Q antenna, the null is easy to miss if you tune too fast. Don't be surprised if that null is out of band by quite a bit. The percentage of change is given by dividing the expected resonant frequency (F_{exp}) by the actual resonant frequency (F_r) and multiplying by 100:

$$\text{Change} = \frac{(F_{exp})\,(100\%)}{F_r}. \tag{17-1}$$

Resonant frequency

Connect the antenna, noise bridge, and the receiver in the same manner as above. Set the receiver to the expected resonant frequency (i.e., 468/F for half-wavelength types and 234/F for quarter-wavelength types). Set the resistance control to 50 or 75 Ω, as appropriate for the normal antenna impedance and the transmission-line impedance. Set the reactance control to zero. Turn the bridge on and listen for the noise signal.

Slowly rock the reactance control back and forth to find on which side of zero the null appears. Once the direction of the null is determined, set the reactance control to zero then tune the receiver toward the null direction (downband if null is on the X_L side and upband if on the X_c side of zero).

A less-than-ideal antenna will not have exactly 50 or 75 Ω of impedance so you must adjust R and C to find the deepest null. You will be surprised how far off some dipoles and other forms of antennas can be if they are not in "free space" (i.e., if they are close to the Earth's surface).

Nonresonant antenna adjustment

We can operate antennas on frequencies other than their resonant frequency if we know the impedance. For the antenna to radiate properly, however, it is necessary to match the impedance of the antenna feedpoint to the source (e.g., a transmission line from a transmitter). We can find the feedpoint resistance from setting the potentiometer in the noise bridge. The reactances can be calculated from the reactance measurement on the bridge by looking at the capacitor setting—and using a little arithmetic:

$$X_c = X = \left(\frac{55}{68 - C}\right) - 2340 \tag{17-2}$$

or

$$X_L = X = 2340 - \left(\frac{159{,}155}{68 + C}\right). \tag{17-3}$$

Now, plug "X" calculated from one of the previous equations into $X_f = X/F$, where F is the desired frequency in megahertz.

Other RF jobs for the noise bridge

The noise bridge can be used for a variety of jobs. We can find the values of capacitors and inductors, the characteristics of series and parallel-tuned resonant circuits, and the adjustment transmission lines.

Transmission line length

Some antennas and (nonnoise) measurements require antenna feedlines that are either a quarter-wavelength or a half-wavelength at some specific frequency. In other cases, a piece of coaxial cable of specified length is required for other purposes: for instance, the dummy load used to service-depth sounders is nothing, but a long piece of shorted coax that returns the echo at a time interval that corresponds to a specific depth. We can use the bridge to find these lengths as follows:

1. Connect a short-circuit across the UNKNOWN and adjust R and X for the best null at the frequency of interest (note: both will be near zero).
2. Remove the short-circuit.
3. Connect the length of transmission line to the unknown terminal—it should be longer than the expected length.
4. For quarter-wavelength lines, shorten the line until the null is very close to the desired frequency. For half-wavelength lines, do the same thing, except the line must be shorted at the far end for each trial length.

Transmission-line velocity factor

The velocity factor of a transmission line (usually designated by V in equations) is a decimal fraction that tells us how fast the radiowave propagates along the line relative to the speed of light in free space. For example, foam dielectric coaxial cable is said to have a velocity factor of $V = 0.80$. This number means that the signals in the line travel at a speed 0.80 (or 80%) of the speed of light.

Because all radio-wavelength formulas are based on the velocity of light, you need the V value to calculate the physical length needed to equal any given electrical length. For example, a half wavelength piece of coax has a physical length of $[(492)(V)/F_{MHz}]$ ft. Unfortunately, the real value of V is often a bit different from the published value. You can use the noise bridge to find the actual value of V for any sample of coaxial cable:

1. Select a convenient length of the coax (more than 12 ft in length) and install a PL-259 RF connector (or other connector compatible with your instrument) on one end and short-circuit the other end.
2. Accurately measure the physical length of the coax in feet; convert the "remainder" inches to a decimal fraction of 1 ft by dividing by 12 (e.g., 32' 8" = 32.67' because 8"/12" = 0.67). Alternatively, cut off the cable to the nearest foot and reconnect the short circuit.
3. Set the resistance and reactance controls to zero.

4. Adjust the monitor receiver for deepest null. Use the null frequency to find the velocity factor, $V = FL/492$, where V is the velocity factor (a decimal fraction), F is the frequency in magahertz, and L is the cable length in feet.

Tuned circuit measurements

An inductor/capacitor (LC) tuned "tank" circuit is the circuit equivalent of a resonant antenna so there is some similarity between the two measurements. You can measure resonant frequency with the noise bridge to within ±20% or better if care is taken. This accuracy might seem poor, but it is better than one can usually get with low-cost signal generators, dip meters, absorption wavemeters, and so on.

Series-tuned circuits

A series-tuned circuit exhibits a low impedance at the resonant frequency and a high impedance at all other frequencies. Start the measurement by connecting the series-tuned circuit under test across the unknown terminals of the bridge. Set the resistance control to a low resistance value, close to 0 Ω. Set the reactance control at mid-scale (zero mark). Next, tune the receiver to the expected null frequency then tune for the null. Be sure that the null is at its deepest point by rocking the R and X controls for best null. At this point, the receiver frequency is the resonant frequency of the tank circuit.

Parallel resonant-tuned circuits

A parallel resonant circuit exhibits a high impedance at resonance and a low impedance at all other frequencies. The measurement is made in exactly the same manner, as for the series resonant circuits, except that the connection is different. Figure 17-4 two-turn link coupling is needed to inject the noise signal into the parallel resonant tank circuit. If the inductor is the toroidal type then the link must go through the hole in the doughnut-shaped core and then connect to the UNKNOWN terminals on the bridge. After this, do exactly as you would for the series-tuned tank measurement.

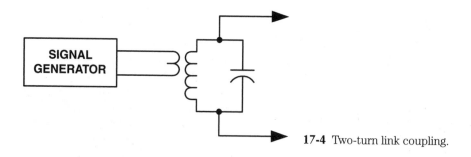

17-4 Two-turn link coupling.

Capacitance and inductance measurements

The Heathkit Model HD-1422 (a noise bridge similar to those mentioned in this chapter) comes with a calibrated 100-pF silver mica test capacitor (called C_{TEST} in the Heath literature) and a calibrated 4.7-μH test inductor (called L_{TEST}), which are used to measure inductance and capacitance, respectively. The idea is to use the test components to form a series-tuned resonant circuit with an unknown component. If you find the resonant frequency then you can calculate the unknown value. In both cases, the series-tuned circuit is connected across the UNKNOWN terminals of the HD-1422 and the series-tuned procedure is followed.

Inductance

To measure inductance, connect the 100-pF C_{TEST} capacitor in series with the unknown coil across the UNKNOWN terminals of the HD-1422. When the null frequency is found, find the inductance from $L = 253/F_2$; L is the inductance in microhenrys (μH) and F is the frequency in megahertz.

Capacitance

Connect L_{TEST} across the UNKNOWN terminals in series with the unknown capacitance. Set the RESISTANCE control to zero, tune the receiver to 2 MHz, and readjust the REACTANCE control for null. Without readjusting the noise bridge control connect L_{TEST} in series with the unknown capacitance and retune the receiver for a null. Capacitance can now be calculated from $C = 5389/F_2$; C is in picofarads and F is in megahertz.

18
CHAPTER

Vectors for RF circuits

A *vector* (Fig. 18-1A) is a graphical device that is used to define the *magnitude* and *direction* (both are needed) of a quantity or physical phenomena. The *length* of the arrow defines the magnitude of the quantity, while the direction in which it is pointing defines the direction of action of the quantity being represented.

Vectors can be used in combination with each other. For example, in Fig. 18-1B we see a pair of displacement vectors that define a starting position (P_1) and a final position (P_2) for a person who traveled from point P_1 12 miles north and then 8 miles east to arrive at point P_2. The *displacement* in this system is the hypotenuse of the right triangle formed by the "north" vector and the "east" vector. This concept was once illustrated vividly by a university bumper sticker's directions to get to a rival school: "North 'til you smell it, east 'til you step in it."

Figure 18-1C shows a calculations trick with vectors that is used a lot in engineering, science, and especially electronics. We can *translate* a vector parallel to its original direction and still treat it as valid. The "east" vector (E) has been translated parallel to its original position so that its tail is at the same point as the tail of the "north" vector (N). This allows us to use the *Pythagorean theorem* to define the vector. The magnitude of the displacement vector to P_2 is given by:

$$P_2 = \sqrt{N^2 + E^2}. \tag{18-1}$$

But recall that the magnitude only describes part of the vector's attributes. The other part is the *direction* of the vector. In the case of Fig. 18-1C the direction can be defined as the angle between the "east" vector and the displacement vector. This angle (θ) is given by:

$$\theta = \arccos\left(\frac{E_1}{P}\right). \tag{18-2}$$

In generic vector notation there is no "natural" or "standard" frame of reference so the vector can be drawn in any direction so long as the user understands what it means. In the system above, we have adopted—by convention—a method that is basically the same as the old-fashioned *Cartesian coordinate system* X–Y graph. In

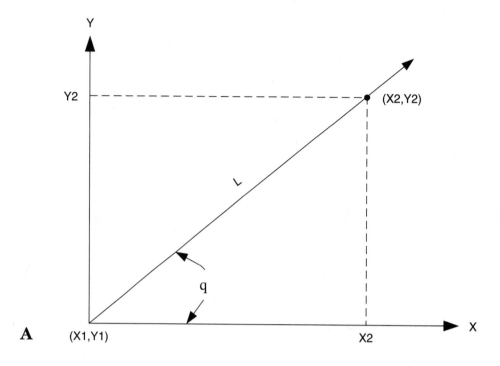

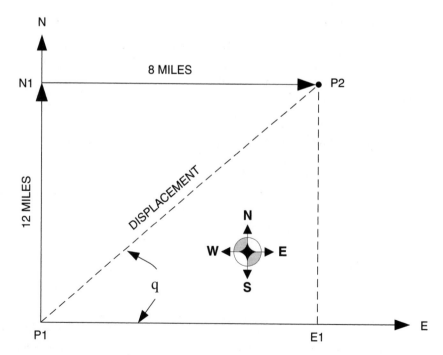

18-1 (A) Simple vector system; (B) N-E vector system; (C) Displacement vector.

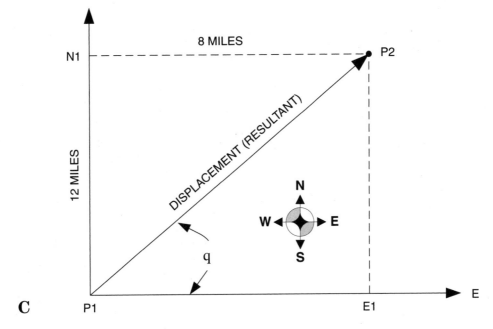

C

18-1 *Continued.*

the example of Fig. 18-1B the X axis is the "east" vector while the Y axis is the "north" vector.

In electronics, the vectors are used to describe voltages and currents in ac circuits are standardized (Fig. 18-2) on this same kind of Cartesian system in which the inductive reactance (X_L), i.e., the opposition to ac exhibited by inductors, is graphed in the "north" direction, the capacitive reactance (X_C) is graphed in the "south" direction, and the resistance (R) is graphed in the "east" direction. Negative resistance ("west" direction) is sometimes seen in electronics. It is a phenomenon in which the current *decreases* when the voltage increases. RF examples of negative resistance include tunnel diodes and Gunn diodes.

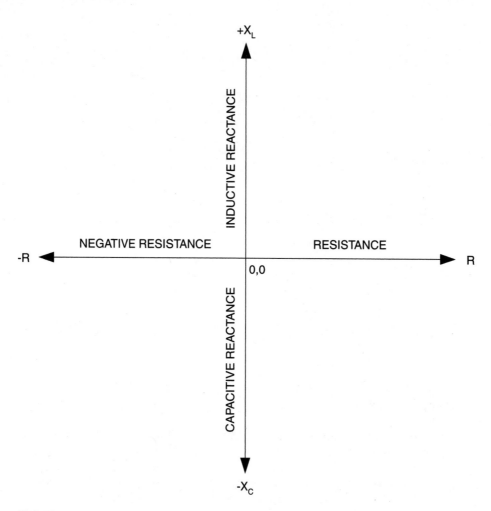

18-2 Vectors used in L-C circuits.

19
CHAPTER

Impedance matching: methods and circuits

One of the first things that you learn in radio communications and broadcasting is that antenna impedance must be matched to the transmission line impedance and that the transmission line impedance must be matched to the output impedance of the transmitter. The reason for this requirement is that maximum power transfer between a source and a load always occurs when the system impedances are matched. In other words, more power is transmitted from the system when the load impedance (the antenna), the transmission line impedance, and the transmitter output impedance are all matched to each other.

Of course, the trivial case is where all three sections of the system have the same impedance. For example, an antenna could have a simple 75-Ω resistive feedpoint impedance (typical of a half-wave dipole in free space) and a transmitter with an output impedance that will match 75 Ω. In that case, you only need to connect a 75-Ω standard-impedance length of coaxial cable between the transmitter and the antenna. Job done! Or so it seems. . . .

But, in other cases, the job is not so simple. In the case of the "standard" antenna, for example, the feedpoint impedance is rarely what the books say it should be. That ubiquitous dipole, for example, is nominally rated at 75 Ω but even the simplest antenna books show that value is an approximation of the theoretical free-space impedance. At locations closer to the earth's surface, the impedance could vary over the approximate range of 30 to 130 Ω and it might have a substantial reactive component.

But there is a way out of this situation. You can construct an impedance-matching system that will marry the source impedance to the load impedance. This chapter examines several matching systems that are useful in a number of antenna situations.

Impedance matching approaches

Antenna impedance might contain both reactive and resistive components. In most practical applications, we are searching for a purely resistive impedance ($Z = R$), but that ideal is rarely achieved. A dipole antenna, for example, has a theoretical free-space impedance of 73 Ω at resonance. But as the frequency applied to the dipole is varied away from resonance, however, a reactive component ($\pm jX$) appears. When the frequency is greater than resonance, then the antenna looks like an inductive reactance so the impedance is $Z = R + jX$. Similarly, when the frequency is less than the resonance frequency, the antenna looks like a capacitive reactance so the impedance is $Z = R - jX$. Also, at distances closer to the earth's surface the resistive component might not be exactly 73 Ω, but might vary from about 30 to 130 Ω. Clearly, whatever impedance coaxial cable is selected to feed the dipole, it stands a good chance of being wrong.

The method used for matching a complex load impedance (such as an antenna) to a resistive source, the most frequently encountered situation in practical radio work, is to interpose a matching network between the load and the source (Fig. 19-1). The matching network must have an impedance that is the complex conjugate of the complex load impedance. For example, if the load impedance is $R + jX$, then the matching network must have an impedance of $R - jX$; similarly, if the load is $R - jX$, then the matching network must be $R + jX$. The sections that follow show some of the more popular networks that accomplish this job.

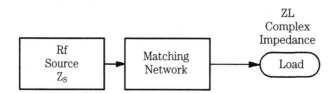

19-1 Block diagram of an antenna impedancing-matching scheme.

L-section network

The L-section network is one of the most used, or at least most published, antenna-matching networks in existence: it rivals even the pi-network. A circuit for the L-section network is shown in Fig. 19-2A. The two resistors represent the source (R_1) and load (R_2) impedances. The elementary assumption of this network is that $R_1 < R_2$. The design equations are:

$$(R_1 < R_2, \text{ and } 1 < Q < 5)$$

$$X_L = 2\pi FL = QR_1 \tag{19-1}$$

$$X_C = \frac{1}{2\pi FC} \tag{19-2}$$

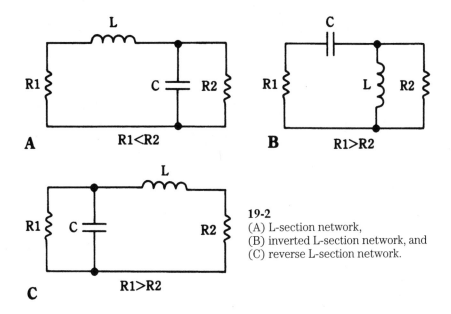

19-2
(A) L-section network,
(B) inverted L-section network, and
(C) reverse L-section network.

$$Q > \sqrt{\frac{R_2}{R_1} - 1} \qquad (19\text{-}3)$$

and

$$Q = \frac{X_L}{R_1} = \frac{R_L}{X_C}. \qquad (19\text{-}4)$$

You will probably most often see this network published in conjunction with less-than-quarter-wavelength longwire antennas. One common fault of those books and articles is that they typically call for a "good ground" in order to make the antenna work properly. But they don't tell you what a "good ground" is and how you can obtain it. Unfortunately, at most locations a good ground means burying a lot of copper conductor—something that most of us cannot afford. In addition, often the person who is forced to use a longwire instead of a better antenna cannot construct a "good ground" under any circumstances because of landlords and/or logistics problems. The very factors that prompt the use of a longwire antenna in the first place also prohibit any form of practically obtainable "good ground." But there is a way out: radials. A good ground can be simulated with a counterpoise ground constructed of quarter-wavelength radials. These radials have a length in feet equal to $246/F_{\text{MHz}}$, and as few as two of them will work wonders.

Another form of L-section network is shown in Figure 19-2B. This circuit differs from the previous circuit in that the roles of the L and C components are reversed. As you might suspect, this role reversal brings about a reversal of the impedance relationships: in this circuit, the assumption is that driving source impedance R_1 is larger than load impedance R_2 (i.e., $R_1 > R_2$). The equations are:

$$(R_2 > R_1)$$

$$X_L = R_2 \sqrt{\frac{R_1}{R_2 - R_1}} \qquad (19\text{-}5)$$

$$X_C = \frac{R_1 R_2}{X_L}. \qquad (19\text{-}6)$$

Still another form of L-section network is shown in Figure 19-2C. Again, we are assuming that the driving source impedance, R_1, is larger than load impedance, R_2 (i.e., $R_1 > R_2$). In this circuit, the elements are arranged similarly to those in Figure 19-2A, with the exception that the capacitor is at the input, rather than the output, of the network. The equations governing this network are:

$$(R_1 > R_2 \text{ and } 1 < Q < 5)$$

$$X_L = 2\pi\, FL = \sqrt{(R_1 R_2) - (R_3)^2} \qquad (19\text{-}7)$$

$$X_C = \frac{1}{2\,\pi\,FC} = \frac{R_1 R_2}{X_L} \qquad (19\text{-}8)$$

$$C = \frac{1}{2\,\pi\,FX_C} \qquad (19\text{-}9)$$

$$L = \frac{X_L}{2\,\pi\,F}. \qquad (19\text{-}10)$$

Thus far, only matching networks that are based on inductor and capacitor circuits have been considered. Also, transmission-line segments can be used as impedance-matching devices. Two basic forms are available: quarter-wave sections and series-matching section.

Pi- (π) networks

The pi-network shown in Fig. 19-3 is used to match a high source impedance to a low load impedance. These circuits are typically used in vacuum-tube RF power amplifiers that need to match low antenna impedances. The name of the circuit comes from its resemblance to the greek letter "pi" (π). The equations for the pi-network are:

$$(R_1 > R_2 \text{ and } 5 < Q < 15)$$

$$Q > \sqrt{\frac{R_2}{R_1} - 1} \qquad (19\text{-}11)$$

$$X_{C2} = \frac{R_2}{\sqrt{\dfrac{R_2}{1 + Q^2} - 1}} \qquad (19\text{-}12)$$

$$X_{C1} = \frac{R_1}{Q} \qquad (19\text{-}13)$$

and

$$X_L = \frac{R_1 \left(Q + \dfrac{R_2}{X_{C2}} \right)}{Q^2 + 1}. \qquad (19\text{-}14)$$

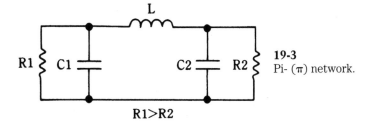

19-3
Pi- (π) network.

Split-capacitor network

The split capacitor network shown in Fig. 19-4 is used to transform a source impedance that is less than the load impedance. In addition to matching antennas, this circuit is also used for interstage impedance matching inside communications equipment. The equations for design are:

$$(R_1 < R_2)$$

$$Q > \sqrt{\frac{R_2}{R_1} - 1} \qquad (19\text{-}15)$$

and

$$X_L = \frac{R_2}{Q}. \qquad (19\text{-}16)$$

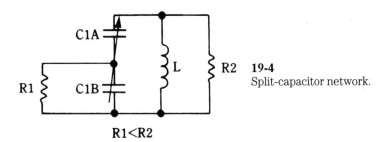

19-4
Split-capacitor network.

Transmatch circuit

One version of the transmatch is shown in Fig. 19-5. This circuit is basically a combination of the split-capacitor network and an output tuning capacitor (C_2). For the HF bands, the capacitors are on the order of 150 pF per section for C_1 and 250 pF for C_2. The roller inductor should be 28 μH. The transmatch is essentially a coax-to-coax impedance matcher and is used to trim the mismatch from a line before it affects the transmitter.

Perhaps the most common form of transmatch circuit is the tee-network shown in Fig. 19-6. This network is lower in cost than some of the others, but suffers a problem. Although it does, in fact, match impedance (and thereby, in a naive sense, "tunes out" VSWR on coaxial lines), it also suffers a high-pass characteristic. This network, therefore, does not reduce the harmonic output of the transmitter. The simple tee-network does not serve one of the main purposes of the antenna tuner:

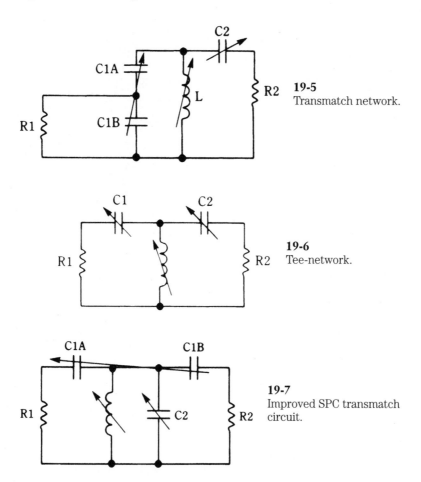

19-5
Transmatch network.

19-6 Tee-network.

19-7
Improved SPC transmatch
circuit.

harmonic reduction. An alternative network, called the *SPC transmatch,* is shown in Fig. 19-7. This version of the circuit offers harmonic attenuation as well as matching impedance.

Figure 19-8 shows commercially available antenna tuners based on the transmatch design. The unit shown in Fig. 19-8A is manufactured by MFJ. It contains the usual three tuning controls, here labeled Transmitter, Antenna, and Inductor. Included in this instrument is an antenna selector switch that allows the operator to select a coax cable antenna through the tuner to connect input to output (coax) without regard to the tuner and select a balanced antenna or an internal dummy load. The instrument also contains a multifunction meter that can measure 200 or 2000 W (full-scale) in either forward or reverse directions. In addition, the meter operates as a VSWR meter.

Figures 19-8 shows a tuner from the United Kingdom. This instrument, the Nevada model, is low-cost but contains the three basic controls. For proper operation, an external RF power meter or VSWR meter is required. This tuner and a Heathkit transmatch antenna tuner are shown in Fig. 19-8B. There are SO-239 coax-

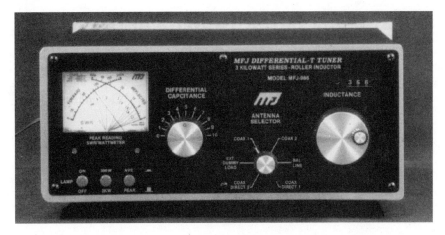

A

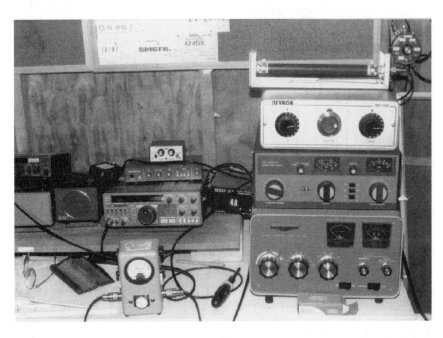

B

19-8 Commercial HF antenna-matching networks.

ial connectors for input and unbalanced output along with a pair of posts for the parallel line output. A three-post panel is used to select which antenna the RF goes to: unbalanced (coax) or parallel. The roller inductor is in the center and it allows the user to set the tuner to a wide range of impedances over the entire 3- to 30-MHz HF band.

Coaxial cable balun transformers

A balun is a transformer that matches an *un*balanced resistive source impedance (such as a coaxial cable) and a *bal*anced load (such as a dipole antenna). With
the circuit of Fig. 19-9, you can make a balun that will transform impedance at a 4:1
ratio, with $R_2 = 4 \times R_1$. The length of the balun section coaxial cable is:

$$L_{\text{ft}} = \frac{492\,V}{F_{\text{MHz}}},$$ (19-17)

where

    L_{ft} = the length in feet
    V = the velocity factor of the coaxial cable (a decimal fraction)
    F_{MHz} = the operating frequency in megahertz.

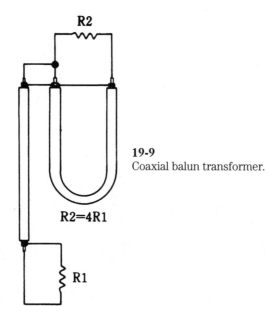

19-9
Coaxial balun transformer.

Matching stubs

A shorted stub can be built to produce almost any value of reactance. This can
be used to make an impedance-matching device that cancels the reactive portion of
a complex impedance. If you have an impedance of, for example, $Z = R + j30\ \Omega$, you
need to make a stub with a reactance of $-j30\ \Omega$ to match it. Two forms of matching
stubs are shown in Figs. 19-10A and 19-10B. These stubs are connected exactly at
the feedpoint of the complex load impedance, although they are sometimes placed
further back on the line at a (perhaps) more convenient point. In that case, however,
the reactance required will be transformed by the transmission line between the load
and the stub.

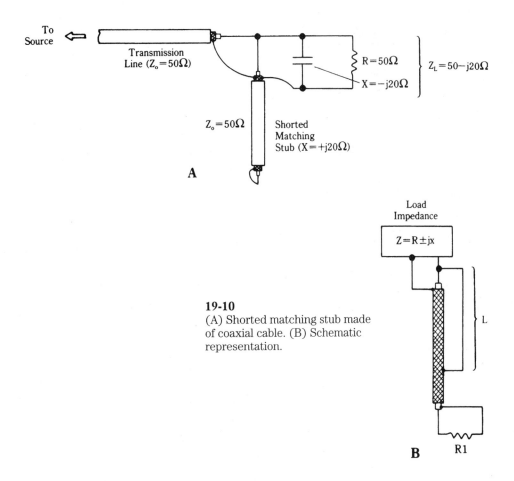

19-10
(A) Shorted matching stub made of coaxial cable. (B) Schematic representation.

Quarter-wavelength matching sections

Figure 19-11 shows the elementary quarter-wavelength transformer section connected between the transmission line and the antenna load. This transformer is also sometimes called a *Q-section*. When designed correctly, this transmission-line transformer is capable of matching the normal feedline impedance (Z_S) to the antenna feedpoint impedance (Z_R). The key factor is to have available a piece of transmission line that has an impedance Z_o of:

$$Z_o = \sqrt{Z_S Z_R}. \tag{19-18}$$

Most texts show this circuit for use with coaxial cable. Although it is certainly possible and even practical in some cases, for the most part, there is a serious flaw in using coax for this project. It seems that the normal range of antenna feedpoint impedances, coupled with the rigidly fixed values of coaxial-cable surge impedance available on the market, combines to yield unavailable values of Z_o. Although there are certainly situations that yield to this requirement, many times the quarter-wave section is not usable on coaxial-cable antenna systems using standard impedance values.

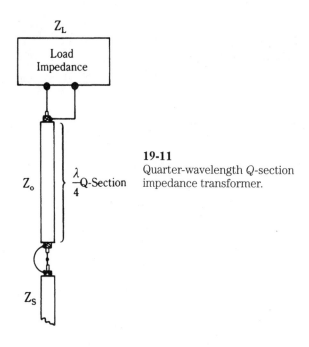

19-11
Quarter-wavelength Q-section
impedance transformer.

On parallel transmission-line systems, on the other hand, it is quite easy to achieve the correct impedance for the matching section. Use Eq.(19-18) to find a value for Z_o then calculate the dimensions of the parallel feeders. Because you know the impedance, and can more often than not select the conductor diameter from available wire supplies, you can use the following equation to calculate conductor spacing:

$$S = 10\left(\frac{Z}{276}\right),$$ (19-19)

where
   S = the spacing
   D = the conductor diameter (D and S in the same units)
   Z = the desired surge impedance.

From there we can calculate the length of the quarter-wave section from the familiar $246/F_{MHz}$.

Series-matching section

The quarter-wavelength section discussed above suffers from several drawbacks: it must be located at the antenna feedpoint, it must be a quarter-wavelength, and it must use a specified (often nonstandard) value of impedance. The series-matching section is a generalized case of the same idea and permits us to build an impedance transformer that overcomes most of these faults. According to *The ARRL Antenna Book,* this form of transformer is capable of matching any load resistance between about 5 and 1200 Ω. In addition, the transformer section is not located at the antenna feedpoint.

Figure 19-12 shows the basic form of the series-matching section. There are three lengths of coaxial cable: L_1, L_2, and the line to the transmitter. Length L_1 and the line to the transmitter (which is any convenient length) have the same characteristic impedance, usually 75 Ω. Section L_2 has a different impedance from L_1 and the line to the transmitter, usually 75 Ω. Notice that only standard, easily obtainable values of impedance are used here.

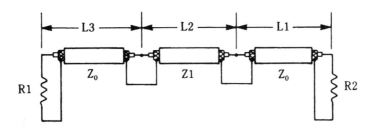

19-12 Matching-section impedance transformer.

The design of this transformer consists of finding the correct lengths for L_1 and L_2. You must know the characteristic impedance of the two lines (50 and 75 Ω are given as examples) and the complex antenna impedance. If the antenna is non-resonant, this impedance is of the form: $Z = R \pm jX$, where R is the resistive portion, X is the reactive portion (inductive or capacitive) and j is the so-called "imaginary" operator (i.e., $\sqrt{-1}$). If the antenna is resonant, then $X = 0$ and the impedance is simply R.

The first chore in designing the transformer is to normalize the impedances:

$$N = \frac{Z_{L_1}}{Z_0} \tag{19-20}$$

$$R = \frac{R_L}{Z_0} \tag{19-21}$$

$$X = \frac{X_L}{Z_0}. \tag{19-22}$$

The lengths are determined in electrical degrees, and from that determination you can find length in feet or meters. If you adopt ARRL notation and define A = tan (L_1), and B = tan (L_2) then the following equations can be written:

If

$$Z_L = R_L \pm jX_L \tag{19-23}$$

$$\tan L_2 = B = \frac{(r-1)^2 + X^2}{r\left(N - \left(\frac{1}{N}\right)^2 - (r-1)^2 - X^2\right)} \tag{19-24}$$

$$\tan L_1 = A = \frac{\left(N - \left(\frac{r}{R}\right)B\right) + X}{r + (XNB) - 1}, \tag{19-25}$$

where

$N = Z_1/Z_0$

$r = R_L/Z_0$

$Z = X_L/Z_0$.

Constraints:

$$Z_1 = Z_0 \sqrt{VSWR} \tag{19-26}$$

or

$$Z_1 < \frac{Z_0}{\sqrt{VSWR}}. \tag{19-27}$$

If $L_1 < 0$, then add 180°

If $B < 0$, then Z_1 is too close to Z_0

Z_1 not equal to Z_0.

$$Z_0 \sqrt{VSWR} < Z_1 < \frac{Z_0}{\sqrt{VSWR}}. \tag{19-28}$$

Physical length in feet:

$L_1' = (L_1 \lambda)/360$

$L_2' = (L_2 \lambda)/360$

where

$$\lambda = \frac{984 \times \text{velocity factor}}{\text{frequency in MHz}}. \tag{19-29}$$

The physical length is determined from arctan (A) and arctan (B) divided by 360 and multiplied by the wavelength along the line and the velocity factor. Although the sign of B might be selected as either − or +, the use of + is preferred because a shorter section is obtained. In the event that the sign of A turns out negative, add 180 degrees to the result.

There are constraints on the design of this transformer. The impedances of the two sections (L_1 and L_2) cannot be too close together. In general, the following relationships must be observed:

Either

$$Z_{L_1} = > Z_0 VSWR \tag{19-30}$$

or

$$Z_{L_1} < \frac{Z_0}{VSWR}. \tag{19-31}$$

20
CHAPTER

Using the double-balanced mixer (DBM)

In radio-frequency electronics, a mixer is a nonlinear circuit or device that permits frequency conversion by the process of heterodyning. Mixers are used in the "front end" of the most common form of radio, the superheterodyne (regardless of wave band), in certain electronic instruments and in certain measurement schemes (receiver dynamic range, oscillator phase noise, etc.).

The block diagram for a basic mixer system is shown in Fig. 20-1; this diagram is generic in form, but it also represents the front end of superheterodyne radio receivers. The mixer has three ports: F_1 receives a low-level signal and would correspond to the RF input from the aerial in radio receivers, F_2 is a high-level signal and corresponds to the local oscillator (LO) in superhet radios; and F_3 is the resultant mixer product (corresponding to the intermediate frequency or "IF" in superhet radios). These frequencies are related by:

$$F_3 = mF_1 \pm nF_2, \qquad (20\text{-}1)$$

where

F_1, F_2, and F_3 are as described

m and n are counting numbers (zero plus integers 0, 1, 2, 3, . . .).

In any given system, m and n can be zero, or any integer, but in practical circuits, it is common to consider only the first-, second-, and third-order products. For sake of simplicity, let's consider a first-order circuit ($m = n = 1$). Such a mixer would output four frequencies: F_1, F_2, $F_{3a} = F_1 + F_2$, and $F_{3b} = F_1 - F_2$. In terms of a radio receiver, these frequencies represent the RF input signal, the local oscillator signal, the sum IF, and the difference IF. In radios, it is common practice to select either sum or difference IF by filtering and rejecting all others.

There are a number of different types of mixer circuits, but only a few different classes: single-ended, singly balanced (or simply "balanced"), and doubly balanced. Most low-cost superheterodyne radio receivers use single-ended mixers, although a few of the more costly "communications receiver" models use singly or doubly

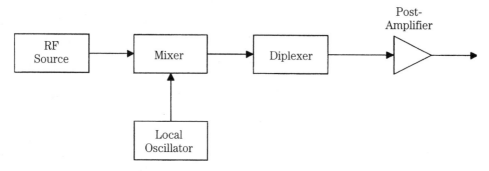

20-1 Block diagram of a mixer circuit.

balanced mixers for improved performance. This chapter focuses on the doubly balanced mixer (DBM) because it offers superior performance over the other forms but is not as well known in electronic hobby circles.

One of the advantages of the DBM over the other forms of mixer is that it suppresses F_1 and F_2 components of the output signal, passing only the sum and difference signals. In a radio receiver using a DBM, the IF filtering and amplifier would only have to contend with sum and difference IF frequencies and not bother with the LO and RF signals. This effect is seen in DBM specifications as the port-to-port isolation figure, which can reach 30 to 60 dB, depending on the DBM model.

A diplexer stage is shown in Fig. 20-1 and is used to absorb unwanted mixer products and pass desired frequencies. A postamplifier stage is typically included because the insertion loss of most passive DBMs is considerable (5 to 12 dB). The purpose of the amplifier is to make up for the loss of signal level in the mixing process. Some active DBMs, incidentally, have a conversion gain figure, not a loss. For example, the popular Signetics NE-602 device offers 20 dB of conversion gain (Carr, 1992).

Diplexer circuits

The RF mixer is like most RF circuits in that it wants to be terminated in its characteristic impedance. Otherwise, a standing-wave ratio (SWR) problem will result, causing signal loss and other problems. In addition, certain passive DBMs will not work well if improperly terminated. A number of different diplexer circuits are known, but two of the most popular are shown in Figs. 20-2 and 20-3.

A diplexer has two jobs: (1) it absorbs undesired mixer output signals so that they are not reflected back into the mixer and (2) it transmits desired signals to the output. In Fig. 20-2, these goals are met with two different LC networks; a high-pass filter and a low-pass filter. The assumption in this circuit is that the difference IF is desired so a high-pass filter with a cutoff above the difference IF is used to shunt the sum IF (plus and LO and RF signals that survived the DBM process) to a dummy load (R_1). The dummy load shown in Fig. 20-2 is set to 50 Ω because that is the most common system impedance for RF circuits (in practice, a 51-Ω resistor might be used).

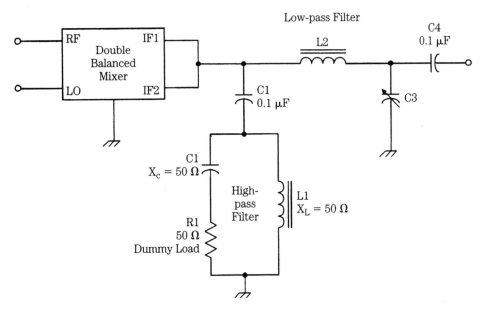

20-2 Diplexer circuit.

The dummy-load resistor can be a $^{1}/_{4}$-W unit in most low-level cases, but regardless of power level, it must be a noninductive type (e.g., carbon composition or metal film).

The inductance (L_1) and capacitance (C_1) values in the high-pass filter are designed to have a 50-Ω reactance at the IF frequency. These values can be calculated from:

$$L_{\text{henrys}} = \frac{50\,\Omega}{2\pi F3_{\text{Hz}}} \tag{20-2}$$

$$C_{\text{farads}} = \frac{1}{2\pi F3_{\text{Hz}}(50\,\Omega)}. \tag{20-3}$$

The low-pass filter transmits the desired difference IF frequency to the output, rejecting everything else. Like the high-pass filter, the L and C elements of this filter are designed to have reactances of 50 Ω at the difference IF frequency.

I built a sweep generator to facilitate a high-performance AM band (540 to 1700 kHz) receiver that I designed. The sweeper circuit is a varactor-tuned VCO that is driven with a 45-Hz sawtooth wave. The swept oscillator is heterodyned against a 14-MHz crystal oscillator, producing difference frequencies that can be tuned to either 455 kHz (my selected IF) or any frequency in the AM band. My diplexer unit consists of two 3000-kHz Butterworth filters, one high-pass and the other low-pass; each filter consists of five reactance elements. The high-pass filter was terminated in a 50-Ω dummy load made by paralleling two 100-Ω metal film resistors (that were hand-selected for the correct value). The low-pass filter was terminated in a postamplifier with a 50-Ω input impedance.

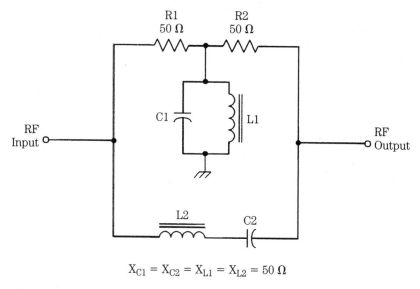

$$X_{C1} = X_{C2} = X_{L1} = X_{L2} = 50 \ \Omega$$

20-3 Alternate diplexer.

Another popular diplexer circuit is shown in Fig. 20-3. This circuit consists of a parallel-resonant 50-Ω tank circuit (C_1/L_1) and a series-resonant 50-Ω tank circuit (C_2/L_2). The series-resonant circuit passes its resonant frequency while rejecting all others because its impedance is low at resonance and high at other frequencies. Alternatively, the parallel-resonant tank circuit offers a high impedance to its resonant frequency and a low impedance to all other frequencies. Because C_1/L_1 are shunted across the signal line, it will short out all but the resonant frequency.

In direct-conversion receivers, a mixer is used to down convert CW and SSB signals directly to audio by setting F_2 close to F_1, without using an IF amplifier chain. In those receivers, it is common practice to build a diplexer similar to Fig. 20-4. Capacitor C_1 passes high frequencies (which are RF), and they are absorbed in a 50-Ω

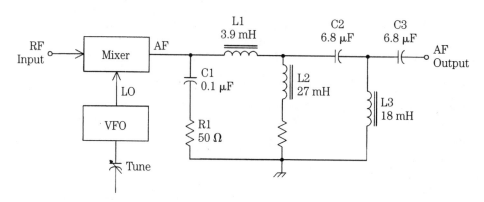

20-4 Diplexer for direct-conversion receiver.

dummy load. The low-frequency audio signals (300 to 3000 Hz) are passed by a 50-Ω low-pass filter consisting of L_1, L_2, L_3, R_2, and C_2.

JFET and MOSFET doubly balanced mixer circuits

Junction field-effect transistors (JFETs) and metal-oxide semiconductor transistors (MOSFETs) can be arranged in a ring circuit that provides good doubly balanced mixer operation. Figure 20-5 shows a circuit that is based on JFET devices. Although discrete JFETs such as a MPF-102 or its equivalent can be used in this circuit with success, performance is generally better if the devices are matched or are part of a single IC device (e.g., the U350 IC). The circuit is capable of better than 30-dB port-to-port isolation over an octave (2:1) frequency change.

The inputs and output of this circuit are based on broadband RF transformers. These transformers are bifilar and trifilar wound on toroidal cores. The input circuit consists of two bifilar-wound impedance transformers (T_1 and T_2). The LO and output circuits are trifilar-wound RF transformers. Part of the output circuit includes a

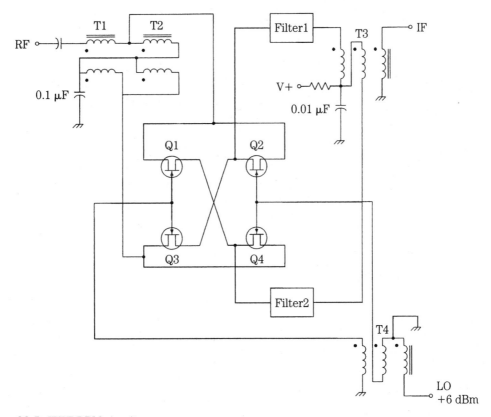

20-5 JFET DBM circuit.

pair of low-pass filters that also transform the 1.5- to 2-kΩ impedance of the JFET devices to 50 Ω. As a result of the needed impedance transformation, the filters must be designed with different R_{in} and R_{out} characteristics, and that complicates the use of loop-up tables (which would be permitted if the input and output resistances were equal).

Figure 20-6 shows an IC DBM that is based on MOSFET transistors. This device was first introduced as the Siliconix Si8901, but they no longer make it. Today, the same device is made by Calogic Corporation [237 Whitney Place, Fremont, CA 94539, USA; phones: 510-656-2900 (voice) and 510-651-3026 (fax)] under the part number SD8901. The SD8901 comes in a seven-pin metal can package. The specifications data sheet for the SD8901 claims that it provides as much as a 10-dBm improvement in third-order intercept point over the U350 JFET design or the diode-ring DBM.

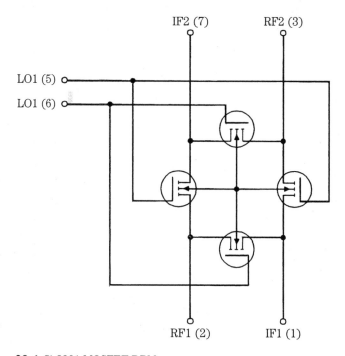

20-6 Si-8901 MOSFET DBM.

A basic circuit for the SD8901 is shown in Fig. 20-7. The input and output terminals are connected to center-tapped, 4:1 impedance ratio RF transformers. Although these transformers can be homemade (using toroidal cores), the Mini-Circuits type T4-1 transformers were used successfully in an amateur radio construction project (Makhinson, 1993). As was true in other DBM circuits, a diplexer is used at the output of the SD8901 circuit.

The local oscillator inputs are driven in push–pull by fast-rise-time square waves. This requirement can be met by generating a pair of complementary square

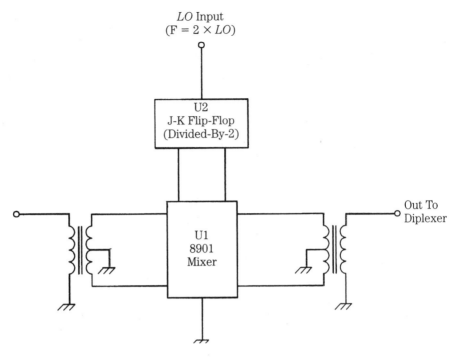

LO Input
(F = 2 × LO)

U2
J-K Flip-Flop
(Divided-By-2)

U1
8901
Mixer

Out To
Diplexer

20-7 Connection of the 8901 mixer.

waves from the same source. One circuit uses a pair of high-speed TTL J-K flip-flops connected with their clock inputs in parallel, driven from a variable-frequency oscillator that operated at twice the required LO frequency. The complementary requirement was met by using the Q-output of one J-K flip-flop and the NOT-Q output of its parallel twin.

The SD8901 device is capable of very good performance—especially in the dynamic range that is achievable. In the design by Makhinson (cited previously), a +35-dBm third-order intercept point was achieved, along with a +16-dBm 1-dB output compression point and a 1-dB output blocking desensitization of +15 dBm. Insertion loss was measured at 7 dB.

One problem with the SD8901 device is its general unavailability to amateur and hobbyist builders. Although low in cost, Calogic has a minimum order quantity of 100 and that makes it a little too rich for most hobbyists to consider. A compromise that also works well is the MOS electronic switch IC devices on the market from several companies, including Calogic. Figure 20-8 shows a typical MOS switch, and it is easy to see how it can be wired into a circuit, such as Fig. 20-7. At least one top-of-the-line amateur radio transceiver uses a quad MOS switch for the DBM in the receiver. It would be interesting to see how well low-cost MOS switches (such as the CMOS 4066) would work. I've seen that chip work well as a doubly balanced phase-sensitive detector in medical blood-pressure amplifiers, and those circuits are closely related to the DBM.

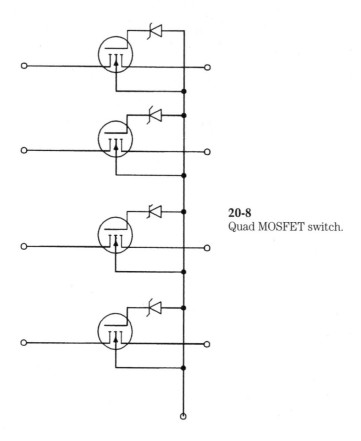

20-8
Quad MOSFET switch.

Doubly balanced diode mixer circuits

One of the easiest doubly balanced circuits, whether homebrew or commercial, is the circuit of Fig. 20-9. This circuit uses a diode ring mixer and balanced input, output, and LO ports. It is capable of 30 to 60 dB of port-to-port isolation yet is reasonably well-behaved in practical circuits. DBMs such as Fig. 20-9 have been used by electronic hobbyists and radio amateurs in a wide variety of projects from direct-conversion receivers to single-sideband transmitters to high-performance shortwave receivers. With proper design, a single DBM can be made to operate over an extremely wide frequency range; several models claim operation from 1 to 500 MHz, with IF outputs from dc to 500 MHz.

The diodes (D_1 through D_4) can be ordinary silicon VHF/UHF diodes, such as 1N914 or 1N4148. However, superior performance is expected when Schottky hot carrier diodes, such as 1N5820 through 1N5822, are used instead. Whatever diode is selected, all four devices should be matched. The best matching of silicon diodes is achieved by comparison on a curve tracer, but failing that, there should at least be a matched forward/reverse resistance reading. Schottky hot-carrier diodes can be matched by ensuring that the selected diodes have the same forward voltage drop when biased to a forward current of 5 to 10 mA.

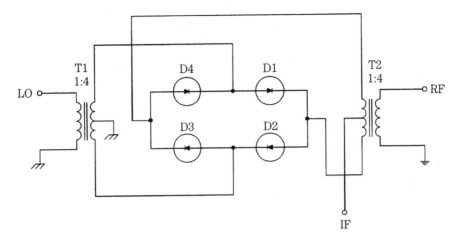

20-9 Diode DBM.

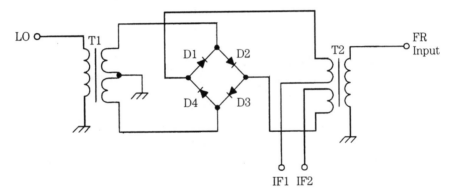

20-10 SBL/SBA mixer circuit.

Figure 20-10 shows the internal circuitry for a very popular commercial diode DBM device, the Mini-Circuits SRA-1 and SBL-1 series. A typical SRA/SBL package is shown in Fig. 20-11. These devices offer good performance and are widely available to hobbyist and radio amateur builders. Some parts houses sell them at retail, as does Mini-Circuits [P.O. Box 166, Brooklyn, NY, 11235, USA: phone 714-934-4500]. In the U.K, contact Dale Electronics Ltd., Camberley, Surrey, 025 28 35094. In the Netherlands, contact "Colmex" B.V., 8050 AA Hattem, Holland, (0) 5206-41214/41217. I don't know the amount of their minimum order, but I've had Mini-Circuits in the USA respond to $25 orders on several occasions—which is certainly more reasonable than other companies.

The packages for the SRA/SBL devices are similar, being on the order of 20 mm long with 5-mm pin spacing. The principal difference between the packages for SRA and SBL devices is the height. In these packages, pin 1 is denoted by a blue bead insulator around the pin. Other pins are connected to the case or have a green (or other color) bead insulator. Also, the "MCL" logo on the top can be used to locate pin

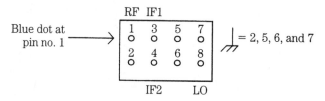

20-11 SBL/SBA package.

1: the "M" of the logo is directly over pin 1. Table 20-1 shows the characteristics of several DBMs in the SRA and SBL series, while Table 20-2 shows the pin assignments for the same devices.

In the standard series of devices, the RF input can accommodate signals up to +1 dBm (1.26 mW), although the LO input must see a +7 dBm (5 mW) signal level for proper operation. Given the 50-Ω input impedance of all ports of the SRA/SBL devices, the RF signal level must be kept below 700 mV p-p, while the LO wants to see 1400 mV p-p. It is essential that the LO level be maintained across the band of interest or else the mixing operation will suffer. Although the device will work down to +5 dBm, a great increase in spurious output and less port-to-port isolation are found. Spectrum analyzer plots of the output signal at low LO drive levels show considerable second- and third-order distortion products.

Notice the IF output of the SRA/SBL devices. Although some models in the series use a single IF output pin, most of these devices use two pins (3 and 4), and these must be connected together externally for the device to work.

As is true with most DBMs and all diode-ring DBM circuits, the SRA/SBL devices are sensitive to the load impedance at the IF output. Good mixing, and freedom from the LO/RF feedthrough problem, occurs when the mixer "looks" into a low VSWR load. For this reason, a good diplexer circuit is required at the output. In experiments, I've found that unterminated SBL-1-1 mixers produce nearly linear mixing when not properly terminated—and that's not desirable in a frequency converter.

Table 20-1. Mixer specifications

Type no.	LO/RF (MHz)	IF (MHz)	Mid-Band Loss (dB)
SRA-1	0.5–500	DC-500	5.5–7.0
SRA-1TX	0.5–500	DC-500	5.5–7.0
SRA-1W	1–750	DC-750	5.5–7.5
SRA-1-1	0.1–500	DC-500	5.5–7.5
SRA-2	1–1000	0.5–500	5.5–7.5
SBL-1	1–500	DC-500	5.5–7.0
SBL-1X	10–1000	5–500	6.0–7.5
SBL-1Z	10–1000	DC-500	6.5–7.5
SBL-1-1	0.1–400	DC-400	5.5–7.0
SBL-3	0.025–200	DC-200	5.5–7.5

Table 20-2. Mixer pin-outs

Type number	LO	RF	IF	GND	Case GND
			Pin/function		
SRA-1	8	1	3,4	2,5,6,7	2
SRA-1TX	8	1	3,4	2,5,6,7	2
SRA-1-1	8	1	3,4	2,5,6,7	2
SRA-1W	8	1	3,4	2,5,6,7	
2,5,6,7					
SRA-2	8	3,4	1	2,5,6,7	
2,5,6,7					
SRA-3	8	1	3,4	2,5,6,7	2
SBL-1	8	1	3,4	2,5,6,7	—
SBL-1-1	8	1	3,4	2,5,6,7	—
SBL-3	8	1	3,4	2,5,6,7	—

The effects of the termination impedance on the operation of the diode DBM circuit are shown in Fig. 20-12. The oscilloscope photo in Fig. 20-12A shows the mixer output directly at pins 3 and 4 of an SBL-1-1. Notice that it is a complex waveform of mixed signals and reflects the fact that the mixer is not properly terminated. The waveform at Fig. 20-12B was taken at the output of the low-pass filter when the high-pass filter and its associated dummy load was disconnected. At Fig. 20-12C is the resultant when properly matched high-pass (and dummy load) and low-pass filters were in operation. The waveform was a nearly clean sine wave of 455 kHz. Another waveform, not shown, was obtained when there was no high-pass filter, but the DBM was terminated in a resistance matched to its characteristic impedance and a low-pass filter. It was a distorted sine wave, and it reflected that other products were present in the output signal.

Figure 20-13 shows a typical SRA/SBL circuit: RF drive ($\leq +1$ dBm) is applied to pin 1, and the +7 dBm LO signal is applied to pin 8. The IF signal is output through pins 3 and 4, which are strapped together. All other pins (2, 5, 6, and 7) are grounded.

The diplexer circuit consists of a high-pass filter (C_1/L_1) that is terminated in a 50-Ω dummy load for the unwanted frequencies, and a low-pass filter ($L_2/L_3/C_4$) for the desired frequencies. All capacitors and inductors are selected to have a reactance of 50 Ω at the IF frequency.

Sometimes 1-dB resistor π-pad attenuators are used at the inputs and the IF output of the DBM. In some cases, the input attenuators are needed to prevent overload of the DBM (overload causes spurious product frequencies to be generated and might cause destruction of the device). In other cases, the circuit designer is attempting to "swamp out" the effects of source or load-impedance variations. Although this method works, it is better to design the circuit to be insensitive to such fluctuations rather than to use a swamping attenuator. The reason is that the resistive attenuator causes a signal loss and adds to the noise generated in the circuit (no resistor can be totally noise-free). A good alternative is to use a stable amplifier with

A

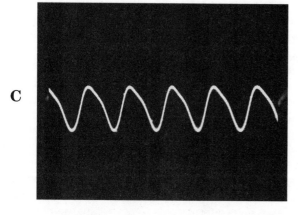

C

20-12
Waveforms of mixer:
(A) improper termination,
(B) with portion of diplexer
removed, and (C) properly
terminated DBM.

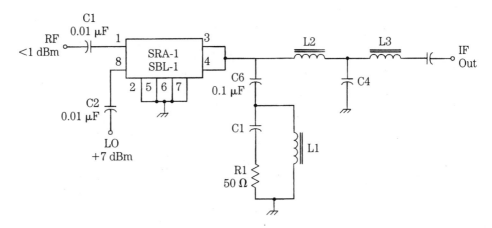

20-13 Output termination for SBL/SBA.

50-Ω input and output impedances, and that is not itself sensitive to impedance variation to isolate the DBM.

Mini-Circuits devices related to the SRA-1 and SBL-1 incorporate MAR-x series MMIC amplifiers internal to the DBM. One series of devices places the amplifier in the LO circuit so that much lower levels of LO signal will provide proper mixing. Another series places the amplifier in the IF output port. This amplifier accomplishes two things: it makes up for the inherent loss of the mixer and it provides greater freedom from load variations that can affect the regular SRA/SBL devices.

Bipolar transconductance cell DBMs

Active mixers made from bipolar silicon transistors formed into Gilbert transconductance cell circuits are also easily available. Perhaps the two most common devices are the Signetics NE-602 device and the LM-1496 device (Maplins catalog no. QH47B, p. 463 in 1993 edition) [Maplins, P.O. Box 3, Rayleigh, Essex SS6 8LR, England; phones (0702) 554161 (credit card orders) and (0702) 552911 (inquiries)].

The NE-602 device is shown in block diagram form in Fig. 20-14A and in partial circuit form in Fig. 20-14B. It offers 20 dB of conversion gain, good sensitivity, and good noise figure, but suffers a bit in the third-order intercept (which is only 15 dBm referenced to a matched input). Consequently, it is recommended that the LO signal be restricted to 200 mV, and the RF input be restricted to 70 mV, into the 1500-Ω input impedances of these terminals.

The LM-1496 device is shown in Figs. 20-15A through 20-15C. Figure 20-15A shows the internal circuitry, and Figs. 20-15B and 15C show the DIP and metal can packages, respectively. Pins 7 and 8 form the local oscillator (or "carrier" in communications terminology) input, and pins 1 and 4 form the RF input. These push–pull inputs are also sometimes labeled "high-level signal" (pins 7 and 8) and "low-level signal" (pins 1 and 4) inputs. Dc bias (pin 5) and gain adjustment (pins 2 and 3) are also provided.

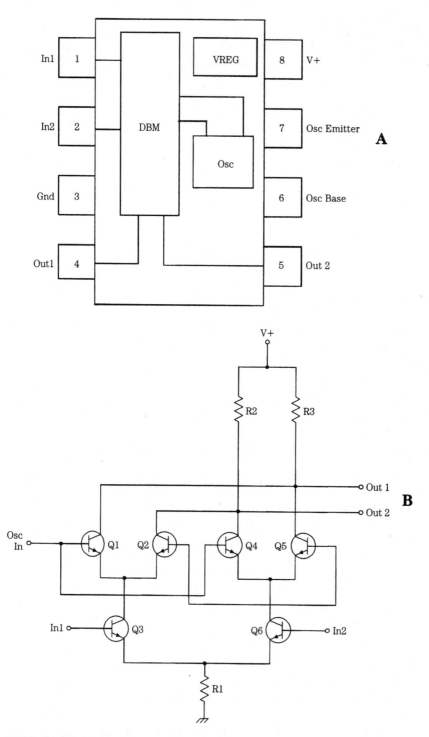

20-14 (A) NE-602 block diagram and (B) DBM transconductance cell.

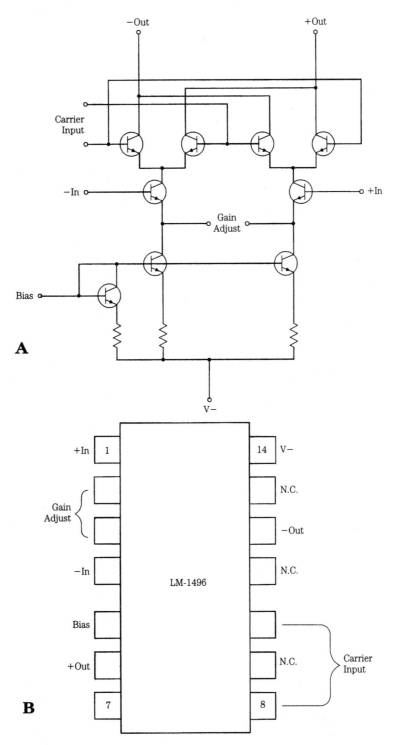

20-15 LM-1496 DBM: (A) internal circuit, (B) 14-pin DIP package, and (C) 10-pin metal can package.

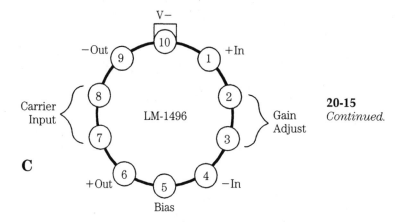

20-15
Continued.

Figure 20-16 shows the basic LM-1496 mixer circuit in which the RF and carrier inputs are connected in the single-ended configuration. The respective signals are applied to the input pins through dc-blocking capacitors C_1 and C_2; the alternate pin inputs in both cases are bypassed to ground through capacitors C_3 and C_4.

The output network consists of a 9:1 broadband RF transformer that combines the two outputs and reduces their impedance to 50 Ω. The primary of the transformer is resonated to the IF frequency by capacitor C_5.

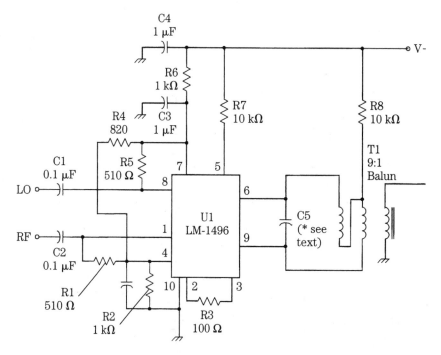

20-16 Mixer circuit based on LM-1496.

Figure 20-17 shows a circuit that uses the LM-1496 device to generate double-sideband suppressed-carrier (DSBSC) signals. When followed by a 2.5- to 3-kHz bandpass filter that is offset from the IF frequency, this circuit will also generate single-sideband (SSB) signals. Commonly, a crystal oscillator will generate the carrier signal (V_c), and the audio stages produce the modulating signal (V_m) from an audio oscillator or microphone input stage. I once saw a circuit that is very similar to this one in a signal generator/test set used to service both amateur radio and marine HF-SSB radio transceivers. It was the signal source to test the receiver sections of the transceivers. The carrier was set to 9 MHz, and both lower sideband (LSB) and upper sideband (USB) KVG crystal filters were used to select the desired sideband. An alternate scheme that is cheaper uses a single 9-MHz crystal filter, but two different crystals at frequencies either side of the crystal passband. One crystal generates the USB signal, and the other generates the LSB signal.

For single sideband to be useful, it has to be demodulated to recover the audio modulation. The circuit of Fig. 20-18 will do that job nicely. It uses a LM-1496 DBM as a product detector. This type of detector works on CW, SSB, and DSB signals (all three require a local oscillator injection signal) and produces the audio resultant from heterodyning the local carrier signal against the SSB IF signal in the receiver. All SSB receivers use some form of product detector at the end of the IF chain, and many of them use the LM-1496 device in a circuit similar to Fig. 20-18.

Preamplifiers and postamplifiers

There is often justification for using amplifiers with DBM circuits. The inputs can be made more sensitive with preamplifiers. When an amplifier is used following the output of a passive DBM ("postamplifier"), it will make up for the 5- to 8-dB loss typical of passive DBMs. In either case, the amplifier provides a certain amount of isolation of the input or output port of the DBM, which frees the DBM from the effects of source or load impedance fluctuations. In these cases, the amplifier is said to be acting as a *buffer amplifier*.

Figure 20-19 shows two popular amplifiers. They can be used for either preamplifier or postamplifier service because they each have 50-Ω input and output impedances. The circuit in Fig. 20-19A is based on the 2N5109 RF transistor and provides close to 20 dB of gain throughout the HF portion of the spectrum. A small amount of stabilizing degenerative feedback is provided by leaving part of the emitter resistance (R_3) unbypassed, while properly bypassing the remaining portion of the emitter resistance. Additional feedback occurs because of the 4:1 impedance transformer in the collector circuit of the transistor. This transformer can be home-brewed using an FT-44-43 or FT-50-43 toroidal ferrite core bifilar wound with 7 to 10 turns of no. 26 AWG enameled wire (or its equivalent in other countries).

The circuit in Fig. 20-19B is based on the Mini-Circuits MAR-x series of MMIC devices. These chips provide 13- to 26-dB gain, at good noise figures, for frequencies from near-dc to 1000 MHz (or more in some models, e.g., 1500 or 2000 MHz). The MAR-1 device shown in the circuit diagram is capable of 15-dB performance to 1000 MHz. The input and output capacitors can be disk ceramic types up to about 100 MHz, but above that frequency "chip" capacitors should be used. Values of 0.01 μF should be

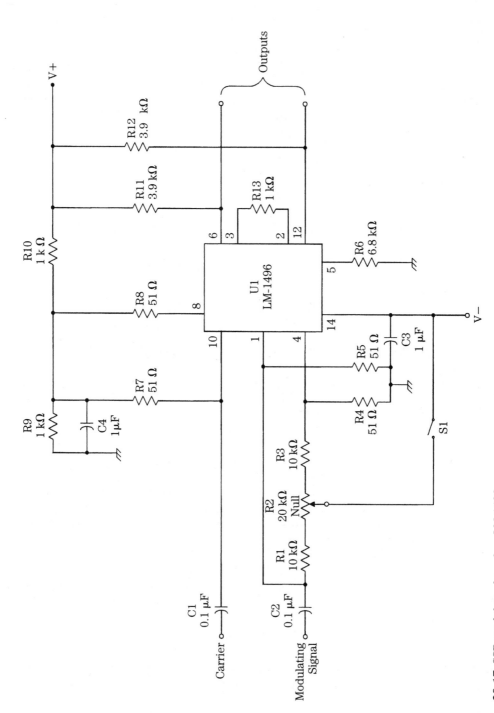

20-17 SSB modulator based on LM-1496.

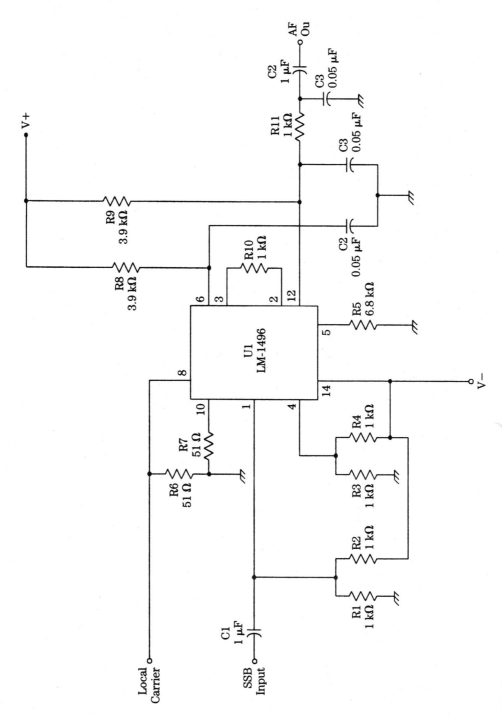

20-18 Product detector circuit.

used in the low-HF region (<10 MHz), 0.001 μF can be used up to 100 MHz, and 100 pF above 100 MHz. The RF choke (RFC1) should be 2.5 mH in the low-HF region, 1 mH from about 10 to 30 MHz, 100 μH from 30 to 100 MHz, and 10 μH above 100 MHz. These values are not critical and are given only as guidelines. Although it might be a bit tricky to get a 1-mH RFC to operate well at 100 MHz, there is no really hard boundary for these bands.

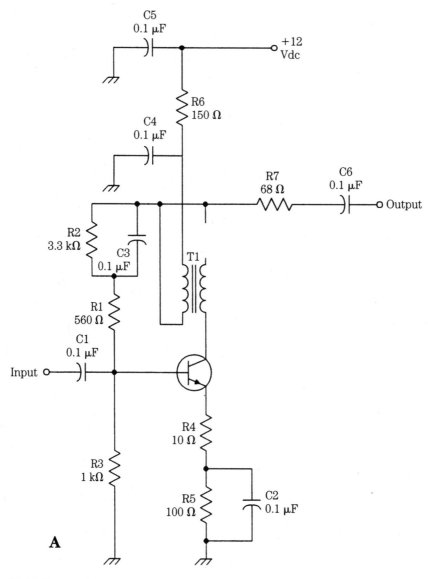

20-19 Postamplifiers: (A) NPN and (B) MAR-x.

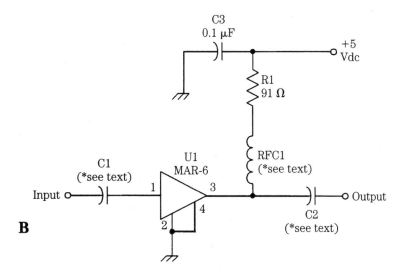

B

20-19 *Continued.*

Conclusion

Doubly balanced mixer circuits work better than most other mixer circuits in radio receivers, SSB transmitters/receivers, instrumentation, and measurement situations. They are easy to use in most cases, and, where properly designed into a circuit, they yield good results with minimum effort.

References

Carr, J. J. (1992). *NE-602 Primer,* Elektor Electronics, January, pp. 20–25.
Makhinson, J. (1993). "A High Dynamic Range MF/HF Receiver Front End," *QST,* February, pp. 23–28.

21
CHAPTER

PIN diodes and their uses

Most modern radio transceivers (i.e., *trans*mitter/re*ceiver*) use "relayless" switching to go back and forth between the receive and transmit states. In many cases, this switching is done with PIN diodes. Similarly, IF filters or front-end bandpass filters are selected with a front-panel switch that handles only direct current. How? Again, PIN diodes. These interesting little components allow us to do switching at RF, IF, and audio frequencies without routing the signals themselves all over the cabinet. This chapter shows how these circuits work.

The P-I-N or "PIN" diode is different from the PN junction diode (see Fig. 21-1A): it has an insulating region between the P- and N-type material (Fig. 21-1B). It is therefore a multiregion semiconductor device, despite having only two electrodes. The I-region is not really a true semiconductor insulator but rather is a very lightly doped N-type region. It is called an *intrinsic* region because it has very few charge carriers to support the flow of an electrical current.

When a forward-bias potential is applied to the PIN diode, charge carriers are injected into the I-region from both N and P regions. But the lightly doped design of the intrinsic region is such that the N- and P-type charge carriers don't immediately recombine (as in PN junction diodes). There is always a delay period for recombination. Because of this delay phenomena, a small, but finite number of carriers in the I-region are always uncombined. As a result, the resistivity of the I-region is very low.

One application that results from the delay of signals passing across the intrinsic region is that the PIN diode can be used as a RF phase shifter. In some microwave antennas, phase shifting is accomplished by the use of one or more PIN diodes in series with the signal line. Although other forms of RF phase shifters (e.g., phase shifter) are usable at those frequencies, the PIN diode remains popular.

Figure 21-1C shows some of the package styles used for PIN diodes at small signal power levels. All but one of these shapes is familiar to most readers, although the odd-shaped flat package is probably recognized only by people with some experience in UHF (and higher) switching circuits. The NTE-553 and ECG-553 PIN diode will dissipate 200 mW and uses the standard cylindrical package style. The NTE-555 and ECG-555 device, on the other hand, uses the UHF flat package style and can dis-

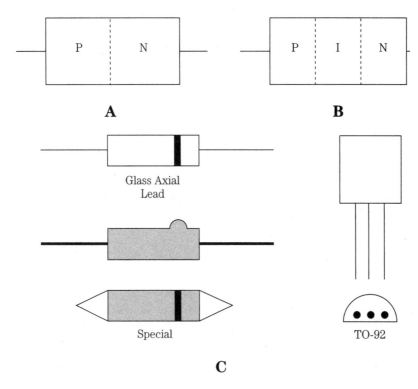

21-1 (A) PN diode, (B) PIN diode, and (C) PIN diode packages.

sipate 400 mW. I used these diodes for the experiments performed to write this chapter because they are service shop replacement lines, and both ECG and NTE are widely distributed in local parts stores. An alternative that might be harder to come by is the MPN3404, which uses the TO-92 plastic package (Fig. 21-1C).

Radio-frequency ac signals can pass through the PIN device and in fact see it under some circumstances as merely a parallel plate capacitor. We can use PIN diodes as electronic switches for RF signals and as a RF delay line or phase-shifter or as an amplitude modulator.

PIN diode switch circuits

PIN diodes can be used as switches in either series or parallel modes. Figure 21-2 shows two similar switch circuits. In the circuit of Fig. 21-2A, the diode (D_1) is placed in series with the signal line. When the diode is turned on, the signal path has a low resistance; when the diode is turned off, it has a very high resistance (thus providing the switching action). When switch S_1 is open, the diode is unbiased so the circuit is open by virtue of the very high series resistance. But when S_1 is closed, the diode is forward-biased and the signal path is now a low resistance. The ratio of off/on resistances provides a measure of the isolation provided by the circuit. A pair of radio-frequency chokes (RFC_1 and RFC_2) are used to provide a high impedance to RF signals, while offering low dc resistance.

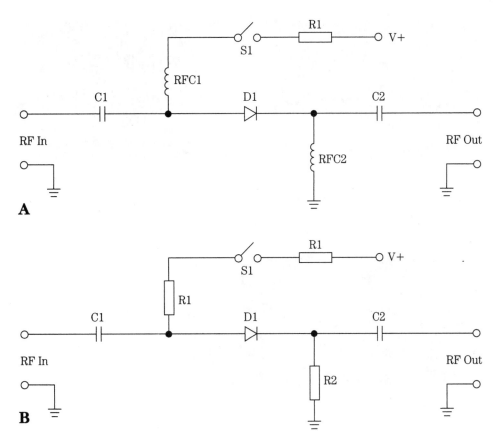

21-2 Series PIN diode switching circuits: (A) resistor loaded and (B) RF choke loaded.

Figure 21-2B is similar to Fig. 21-2A, except that the RF chokes are deleted and a resistor is added. Figure 21-3 shows a test that I performed on the circuit of Fig. 21-2B using a 455-kHz IF signal (the 'scope was set to show only a few cycles of the 455 kHz). The oscilloscope trace in Fig. 21-3A shows the ON position, where +12 Vdc was connected through switch S1 to the PIN diode's current-limiting resistor. This signal is 1200 mV peak-to-peak. The trace in Fig. 21-3B shows the same signal when the switch was OFF (i.e., +12 Vdc disconnected) but with the oscilloscope set to the same level. It appears to be a straight line. Increasing the sensitivity of the 'scope showed a level of 12 mV getting through. This means that this simplest circuit provides a 100:1 on/off ratio, which is 40 dB of isolation. [Note: for this experiment, ECG-555 and NTE-555 hot-carrier PIN diodes were used].

Figure 21-4 shows the same switch when a square wave is used to drive the PIN diode control voltage line rather than +12 Vdc. This situation is analogous to a CW keying waveform. The photo represents one on/off cycle. With the resistor and capacitor values shown, a pronounced switching transient is present.

Figure 21-5 shows the circuit for a shunt PIN diode switch. In this case, the diode is placed across the signal line, rather than in series with it. When the diode is

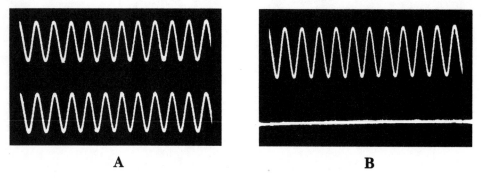

21-3 PIN diode switching action: (A) switch ON and (B) switch OFF (in both cases upper trace is input and lower trace is output).

21-4
PIN diode switching using square-wave chopper waveform.

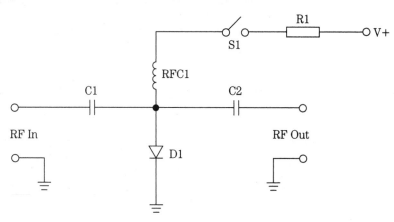

21-5 Shunt PIN diode switch.

turned off, the resistance across the signal path is high, so circuit operation is unimpeded. But when the diode is turned on (S_1 closed), a near-short-circuit is placed across the line. This type of circuit is turned off when the diode is forward-biased. This action is in contrast to the series switch, in which a forward-biased diode is used to turn the circuit on.

A combination series-shunt circuit is shown in Fig. 21-6. In this circuit, D_1 and D_2 are placed in series with the signal line and D_3 is in parallel with the line. D_1 and

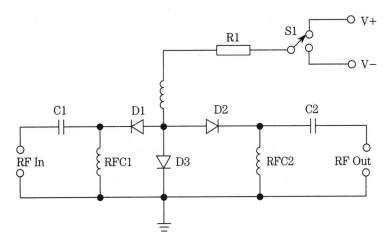

21-6 Series-shunt PIN diode switch.

D_2 will turn on with a positive potential applied and D_3 turns on when a negative potential is applied. When switch S_1 is in the ON position, a positive potential is applied to the junction of the three diodes. As a result, D_1 and D_2 are forward-biased and thus take on a low resistance. At the same time, D_3 is hard-reverse-biased, so it has a very high resistance. Signal is passed from input to output essentially unimpeded (most PIN diodes have a very low series resistance).

But when S_1 is in the OFF position, the opposite situation obtains. In this case, the applied potential is negative, so D_1/D_2 are reverse-biased (and take on a high series resistance) while D_3 is forward-biased (and takes on a low series resistance). This circuit action creates a tremendous attenuation of the signal between input and output.

PIN diode applications

PIN diodes can be used as either a variable resistor or as an electronic switch for RF signals. In the latter case, the diode is basically a two-valued resistor with one value being very high and the other being very low. These characteristics open several possible applications.

When used as a switch, PIN diodes can be used to switch devices, such as attenuators, filters, and amplifiers in and out of the circuit. It has become standard practice in modern radio equipment to switch dc voltages to bias PIN diodes rather than to directly switch RF/IF signals. In some cases, the PIN diode can be used to simply short out the transmission path to bypass the device.

The PIN diode will also work as an amplitude modulator. In this application, a PIN diode is connected across a transmission line or inserted into one end of a piece of microwave waveguide. The audio modulating voltage is applied through a RF choke to the PIN diode. When a CW signal is applied to the transmission line, the varying resistance of the PIN diode causes the signal to be amplitude modulated.

A popular police radar jamming technique among another generation of "bandit techies" was to place a PIN diode in a cavity, λ/4 from the rear that was resonant at the radar gun frequency and fed at its open end with a horn antenna and then modulate it with an audio sine wave equal to the doppler shift expected of either 125 or 0 MPH speeds. Don't try it, by the way, the cops have got better guns–even if they do cause an—ahem—"personal problem" when left on the seat between their legs for any length of time.

Another application is shown in Fig. 21-7. Here, a pair of PIN diodes are used as a transmit–receive (TR) switch in a radio transmitter; models from low-HF to microwave use this technique. Where you see a so-called "relayless TR switch," it is almost certain that a PIN diode network (such as Fig. 21-7) is in use. When switch S_1 is open, diodes D_1 and D_2 are unbiased, so they present a high impedance to the signal. Diode D_1 is in series with the transmitter signal so it blocks it from reaching the antenna; diode D_2, on the other hand, is across the receiver input so does not attenuate the receiver input signal at all. But when switch S_1 is closed, the opposite situation occurs: both D_1 and D_2 are now forward-biased. Diode D_1 is now a low resistance in series with the transmitter output signal so the transmitter is effectively connected to the antenna. Diode D_2 is also a low resistance and is across the receiver input and causes it to short out. The isolation network can be either a quarter-wavelength transmission line, microstrip line designed into the printed circuit board, or a LC pi-section filter.

Transmitters up to several kilowatts have been designed using this form of switching, and almost all current VHF/UHF portable "handi-talkies" use PIN diode switching. Higher power circuits require larger diodes than were covered in this chapter, but they are easily available from industrial distributors.

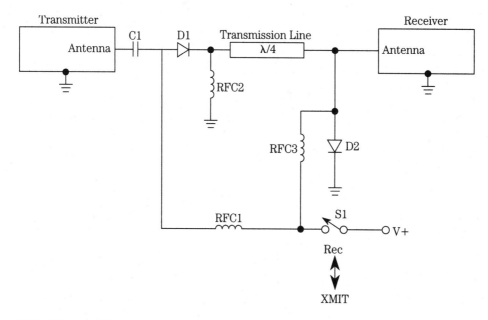

21-7 PIN diode T/R switching.

Figure 21-8 shows how multiple IF bandpass filters are selected, with only dc being routed around the cabinet between the circuitry and the front panel. A set of input and output PIN diode switches are connected as shown and fed with a switch that selects either −12 Vdc or +12 Vdc alternately. When the switch is in the position shown, the +12 V is to the filter 1 switches, so filter 1 is activated. When the switch is in the opposite position, the alternate filter is turned on. This same arrangement can be used in the front end of the receiver, or the local oscillator, to select LC components for different bands.

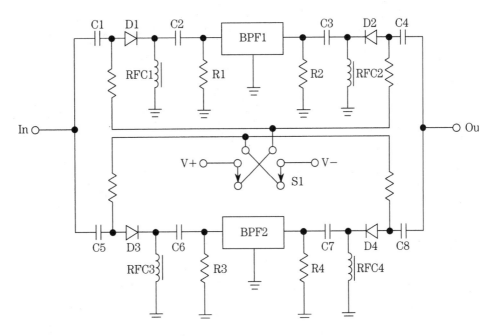

21-8 PIN diode bandpass filter switching.

Another filter selection method is shown in Fig. 21-9. This circuit is a partial representation of the front-end circuitry for the Heathkit SW-7800 general-coverage shortwave receiver. The entire circuit differs from this variant in that a total of six filter sections are used rather than just two (Fig. 21-9). In each bandpass filter (BPF$_1$ and BPF$_2$), a network of inductor and capacitor elements are used to set the center and both edge frequencies of the band to be covered.

The circuit of Fig. 21-9 uses a switch (S_1) to apply or remove the +12-Vdc bias potential on the diodes, but in the actual receiver, this potential is digitally controlled. The digital logic elements sense which of 30 bands are being used and selects the input RF filter accordingly.

Another application for PIN diodes is as a voltage-variable attenuator in RF circuits. Because of its variable resistance characteristic, the PIN diode can be used in a variety of attenuator circuits. One of the simplest is the shunt attenuator of Fig. 21-10. The front end of this circuit is a bank of selectable bandpass filters per

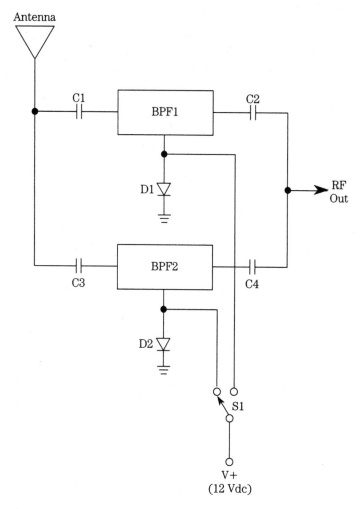

Antenna

21-9 RF input receiver bandpass filter switching.

Fig. 21-9, so we will not dwell on that topic now. But at the output of the filter bank is a shunted from the signal line to ground a series combination of a capacitor (C_1) and a PIN diode (D_1).

The PIN diode acts like an electronically variable resistor. The resistance across the diode's terminals is a function of the applied bias voltage. This voltage, hence the degree of attenuation of the RF signal, is proportional to the setting of potentiometer, R_1. The series resistor (R_2) is used to limit the current when the diode is forward-biased. This step is necessary because the diode becomes a very low resistance when a certain rather low potential is exceeded. This circuit is used as a RF gain control on some modern receivers. Another PIN diode attenuator circuit (the "bridge-tee" attenuator) is shown in Fig. 21-10. It is capable of providing a 10-dB attenuation range over a very wide bandwidth.

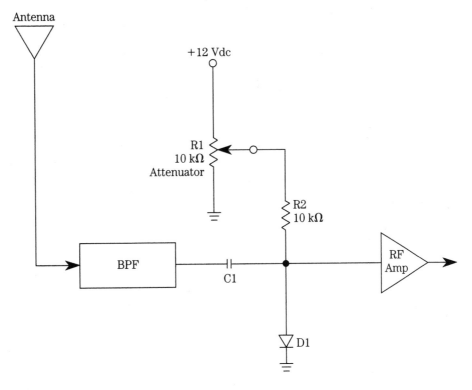

21-10 PIN diode RF attenuator.

Conclusion

PIN diodes are little known but are nonetheless very useful. If you build any RF projects that could involve traditional switching then you will probably find that the PIN diode makes a dandy substitute, which avoids the problems associated with routing signal around switching circuits. Try it; you'll like it.

22
CHAPTER

UHF and microwave diodes, transistors, and integrated circuits

Very few people needed to know anything about microwaves and microwave devices—until now. For many decades, all service technicians and others only needed to know about devices that worked up to the highest UHF TV channel, which was just short of the microwave region. Even then, many local servicers would consider the entire UHF TV tuner as a black box component and replace it as a unit when it went bad. But today, with satellite TVRO units increasingly popular, two-way communications heading into the low-microwave region, and a host of other applications on those frequencies, it becomes necessary for the service technician to learn the ins and outs of the solid-state devices used on microwave frequencies. This chapter takes a close look at solid-state microwave devices.

Diode devices

Di-ode. *Di* means "two", and *-ode* is derived from "electrode." Thus, the word "diode" refers to a two-electrode device. Although in conventional low-frequency solid-state terminology, the word *diode* indicates only PN junction diodes; in the microwave world, we see both PN diodes and others as well. Some diodes, for example, are multiregion devices, and still others have but a single semiconductor region. The latter depend on bulk-effect phenomena to operate. This section examines some of the various microwave devices that are available on the market.

World War II created a huge leap in microwave capabilities because of the needs of the radar community. Even though solid-state electronics was in its infancy, developers were able to produce crude microwave diodes for use primarily as mixers in superheterodyne radar receivers. Theoretical work done in the late 1930s, 1940s, and early 1950s predicted microwave diodes that would be capable of not only nonlinear mixing in superheterodyne receivers, but also of amplification and oscillation as well. These predictions were carried forward by various researchers. By 1970, a variety of commercial devices were on the market.

Early microwave diodes

The development of solid-state microwave devices suffered the same sort of problems as microwave vacuum tube development. Just as electron transit time limits the operating frequency of vacuum tubes, a similar phenomena affects semiconductor devices. The transit time across a region of semiconductor material is determined not only by the thickness of the region, but also by the electron saturation velocity, about 10^7 cm/s, which is the maximum velocity attainable in a given material. In addition, minority carriers are in a "storage" state that prevents them from being easily moved, causing a delay that limits the switching time of the device. Also, like the vacuum tube, interelectrode capacitance limits the operating frequency of solid-state devices. Although turned to an advantage in variable capacitance diodes, capacitance between the N and P regions of a PN junction diode seriously limits operating frequency and/or switching time.

During World War II, microwave diode (such as the 1N21 and 1N23) devices were developed. The capacitance problem was partially overcome by using the point contact method of construction (shown in Fig. 22-1). The point contact diode uses a

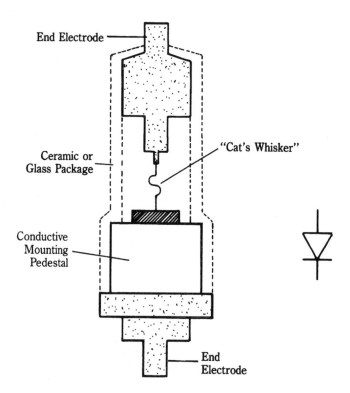

22-1 PN junction diode.

"cat's whisker" electrode to effect a connection between the conductive electrode and the semiconductor material.

Point contact construction is a throwback to the very earliest days of radio when a commonly used detector in radio receivers was the naturally occurring galena crystal (a lead compound). Those early "crystal set" radios used a cat's whisker electrode to probe the crystal surface for the spot where rectification of radio signals could occur.

The point contact diode of World War II was used mainly as a mixer element to produce an intermediate frequency (IF) signal that was the difference between a radar signal and a local oscillator (LO) signal. Those early superheterodyne radar receivers did not use RF amplifier stages ahead of the mixer, as was common in lower frequency receivers. Although operating frequencies to 200 GHz have been achieved in point contact diodes, the devices are fragile and are limited to very low power levels.

Schottky barrier diodes

The Schottky diode (also called the *hot carrier diode*) is similar in mechanical structure to the point contact diode, but is more rugged and is capable of dissipating more power. The basic structure of the Schottky diode is shown in Fig. 22-2A, and the equivalent circuit is shown in Fig. 22-2B. This form of microwave diode was predicted in 1938 by William Schottky during research into majority carrier rectification phenomena.

A "Schottky barrier" is a quantum mechanical potential barrier set up by virtue of a metal to semiconductor interface. The barrier is increased by increasing reverse bias potential and reduced (or nearly eliminated) by forward-bias potentials. Under reverse-bias conditions, the electrons in the semiconductor material lack sufficient energy to breach the barrier, so no current flow occurs. Under forward-bias conditions, however, the barrier is lowered sufficiently for electrons to breach it and flow from the semiconductor region into the metallic region.

The key to junction diode microwave performance is its switching time, or switching speed as it is sometimes called. In general, the faster the switching speed (i.e., shorter the switching time), the higher the operating frequency. In ordinary PN junction diodes, switching time is dominated by the minority carriers resident in the semiconductor materials. The minority carriers are ordinarily in a rest or "storage" state so are neither easily nor quickly swept along by changes in the electric field potential. But majority carriers are more mobile, so respond rapidly to potential changes. Thus, Schottky diodes, by depending on majority carriers for switching operation, operate with very rapid speeds—well into the microwave region for many models.

Operating speed is also a function of the distributed circuit RLC parameters (see Fig. 22-2B). In Schottky diodes that operate into the microwave region, series resistance is about 3 to 10 Ω, capacitance is around 1 pF and series lead inductance tends to be 1 nH or less.

The Schottky diode is basically a rectifying element, so it finds a lot of use in mixer and detector circuits. The double balanced mixer (DBM) circuit, which is used for a variety of applications from heterodyning to the generation of double-sideband suppressed-carrier signals to phase sensitive detectors, is usually made

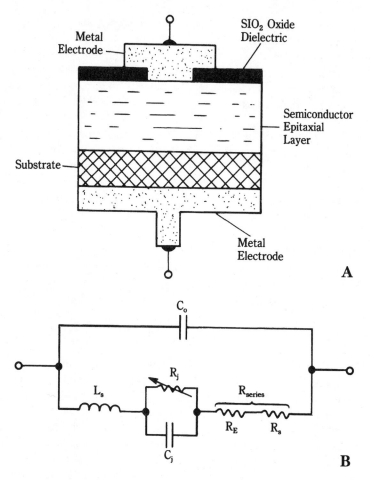

22-2 (A) Schottky barrier diode and (B) equivalent circuit.

from matched sets of Schottky diodes. Many DBM components work into the low end of the microwave spectrum.

Tunnel diodes

The tunnel diode was predicted by L. Esaki of Japan in 1958 and is often referred to as the "Esaki diode." The tunnel diode is a special case of a PN junction diode in which carrier doping of the semiconductor material is on the order of 1019 or 1020 atoms per cubic centimeter, or about 10 to 1000 times higher than the doping used in ordinary PN diodes. The potential barrier formed by the depletion zone is very thin; on the order of 3 to 100 angstroms, or about 10^{-6} cm. The tunnel diode operates on the quantum mechanical principle of tunneling (hence its names), which is a majority carrier event. Ordinarily, electrons in a PN junction device lack the energy required to climb over the potential barrier created by the depletion zone under reverse-bias conditions. The electrons are thus said to be trapped in a "potential well." The current, therefore, is theoretically zero in that case. But if the carriers are numerous enough, and the barrier zone is thin enough, then tunneling will occur. That is, electrons will disappear on one side of the barrier (i.e., inside the potential well) and then reappear on the other side of the barrier to form an electrical current. In other words, they appear to have tunneled under or through the potential barrier in a manner that is said to be analogous to a prisoner tunneling through the wall of the jailhouse.

Do not waste a lot of energy attempting to explain tunneling phenomena, for indeed even theoretical physicists do not understand the phenomena very well. The tunneling explanation is only a metaphor for a quantum event that has no corresponding action in the nonatomic world. The electron does not actually go under, over, or around the potential barrier but rather it disappears here and then reappears elsewhere. Although the reasonable person is not comfortable with such seeming magic, until more is known about the quantum world, we must be content with accepting tunneling as merely a construct that works for practical purposes. All that we can state for sure about tunneling is that if the right conditions are satisfied then there is a certain probability per unit of time that tunneling will occur.

There are three basic conditions that must be met to ensure a high probability of tunneling occurring:

1. Heavy doping of the semiconductor material to ensure large numbers of majority carriers;
2. Very thin depletion zone (3 to 100 A); and
3. For each tunneling carrier, there must be a filled energy state from which it can be drawn on the source side of the barrier, and an available empty state of the same energy level to which it can go on the other side of the barrier.

Figure 22-3 shows an *I*-vs-*V* curve for a tunnel diode. Notice the unusual shape of this curve. Normally, in familiar materials that obey Ohm's law, the current (I) will increase with increasing potential (V) and decrease with decreasing potential. Such behavior is said to represent positive resistance ($+R$) and is the behavior normally found in common conductors. But in the tunnel diode, there is a negative resistance zone (NRZ in Fig. 22-3) in which increasing potential forces a decreasing current.

Points A, B, and C in Fig. 22-3 show us that tunnel diodes can be used in monostable (A and C), bistable (flip-flop from A-to-C or C-to-A), or astable (oscillate about B) modes of operation, depending on applied potential. These modes give us some insight into typical applications for the tunnel diode: switching, oscillation, and amplification. A tunnel diode has a negative resistance ($-R$) property, which indicates the ability to oscillate in relaxation oscillator circuits. Unfortunately, the power output levels of the tunnel diode are restricted to only a few milliwatts because the applied dc potentials must be less than the bandgap potential for the specific diode.

Figure 22-4A shows the usual circuit symbol for the tunnel diode, and Fig. 22-4B shows a simplified, but adequate, equivalent circuit. The tunnel diode can be modeled as a negative resistance element shunted by the junction capacitance of the diode. The series resistance is the sum of contact resistance, lead resistance, and bulk material resistances in the semiconductor material, and the series inductance is primarily lead inductance.

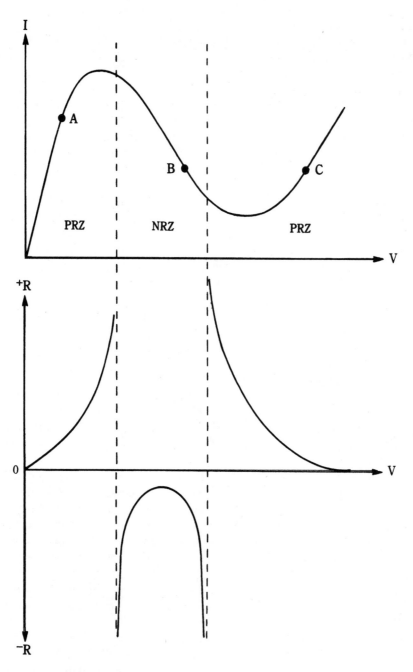

22-3 Tunnel diode characteristic curves

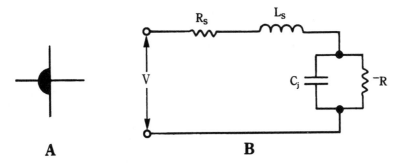

22-4 (A) Tunnel diode symbol and (B) equivalent circuit.

The tunnel diode will oscillate if the total reactance is zero ($X_L - X_c = 0$), and if the negative resistance balances out the series resistance. We can calculate two distinct frequencies with regard to normal tunnel diodes: resistive frequency (F_r) and the self-resonant frequency (F_s). These frequencies are given by the following equations.

Resistive frequency:

$$F_r = \frac{1}{2\pi RC_j} \sqrt{\frac{R}{R_S} - 1} \qquad (22\text{-}1)$$

Self-resonant frequency:

$$F_s = \frac{1}{2\pi} \sqrt{\frac{1}{L_S C_j} - \frac{1}{(RC_j)^2}}, \qquad (22\text{-}2)$$

where

F_r = the resistive frequency in hertz (Hz)
F_s = the self-resonant frequency in hertz (Hz)
R = the absolute value of the negative resistance in ohms
R_S = the series resistance in ohms
C_j = the junction capacitance in farads
L_S = the series inductance in henrys.

The negative resistance phenomena, also called *negative differential resistance* (NDR), is a seemingly strange phenomena in which materials behave electrically contrary to Ohm's law. In ohmic materials (i.e., those that obey Ohm's law), we normally expect current to increase as electric potential increases. In these "positive resistance" materials, $I = V/R$. In negative resistance (NDR) devices, however, there might be certain ranges of applied potential in which current decreases with increasing potential.

In addition to the *I*-vs-*V* characteristic, certain other properties of negative resistance materials distinguish them from ohmic materials. First, in ohmic materials, we normally expect to find the voltage and current in phase with each other unless an inductive or capacitive reactance is also present. Negative resistance, however, causes current and voltage to be 180° out of phase with each other. This relationship is a key to recognizing negative resistance situations. Another significant difference is that a positive resistance dissipates power and negative resistance generates power (i.e., converts power from a dc source to another form). In other words, a +R

device absorbs power from external sources and a $-R$ device supplies power (converted from dc) to the external circuit.

Two forms of oscillatory circuit are possible with NDR devices: resonant and unresonant. In the resonant type, depicted in Fig. 22-5, current pulses from the NDR device shock excite a high-Q LC tank circuit or resonant cavity into self-excitation. The oscillations are sustained by repetitive reexcitation. Unresonant oscillator circuits use no tuning. They depend on the device dimensions and the average charge carrier velocity through the bulk material to determine operating frequency. Tunnel diodes generally work in the resonant mode. Some NDR devices, such as the Gunn, will operate in either resonant or unresonant oscillatory modes.

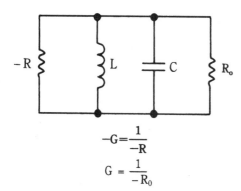

$$-G = \frac{1}{-R}$$

$$G = \frac{1}{-R_0}$$

Criteria for Oscillator: $|-G_1| \geq G_0$

22-5

Negative resistance oscillator circuit.

Introduction to negative resistance $(-R)$ devices

Earlier, you were introduced to the concept of negative resistance $(-R)$ and its inverse, negative conductance $(-G = 1/-R)$. To review the facts, the negative resistance phenomena, also called negative differential resistance (NDR), is a seemingly strange phenomena in which materials behave electrically contrary to Ohm's law. In ohmic materials (i.e., those that obey Ohm's law), we normally expect current to increase as electric potential increases. In these "positive resistance" materials, $I = V/R$. In negative resistance (NDR) devices, however, there might be certain ranges of applied potential in which current decreases with increasing potential.

In addition to the strange I-vs-V characteristic, certain other properties of negative resistance materials distinguish them from ohmic materials. First, in ohmic materials, we normally expect to find the voltage and current in-phase with each other, unless an inductive or capacitive reactance is also present. Negative resistance, however, causes current and voltage to be 180° out of phase with each other. This relationship is a key to recognizing negative-resistance situations.

In the resonant type of negative resistance circuit, depicted in Fig. 22-5, current pulses from the NDR device shock excite a high-Q LC tank circuit or resonant cavity into self-excitation. The oscillations are sustained by repetitive reexcitation. Unresonant oscillator circuits use no tuning. They depend on the device dimensions and the average charge carrier velocity through the bulk material to determine operating frequency. Some NDR devices, such as the Gunn diode (later in this chapter), will operate in either resonant or unresonant oscillatory modes.

Transferred electron devices (e.g., Gunn diodes)

Gunn diodes are named after John B. Gunn of IBM, who, in 1963, discovered a phenomenon since then called the *Gunn Effect*. Experimenting with compound semiconductors such as gallium arsenide (GaAs), Gunn noted that the current pulse became unstable when the bias voltage increased above a certain crucial threshold potential. Gunn suspected that a negative resistance effect was responsible for the unusual diode behavior. Gunn diodes are representative of a class of materials called *transferred electron devices* (TED).

Ordinary two-terminal small-signal PN diodes are made from elemental semiconductor materials, such as silicon (Si) and germanium (Ge). In the pure form before charge carrier doping is added, these materials contain no other elements. TED devices, on the other hand, are made from compound semiconductors; that is, materials that are chemical compounds of at least two chemical elements. Examples are gallium arsenide (GaAs), indium phosphide (InP), and cadmium telluride (CdTe). Of these, GaAs is the most commonly used.

N-type GaAs used in TED devices is doped to a level of 1014 to 1017 carriers per centimeter at room temperature (373°K). A typical sample used in TED/Gunn devices will be about 150×150 μM in cross-sectional area and 30 μM long.

Two-valley TED model

The operation of TED devices depends on a semiconductor phenomenon found in compound semiconductors called the *two-valley model* (for InP material, there is also a *three-valley model*). The two-valley model is also called the *Ridley/ Watkins/ Hilsum theory*. Figure 22-6 shows how this model works in TEDs. In ordinary elemental semiconductors, the energy diagram shows three possible bands: valence band, forbidden band, and conduction band. The forbidden band contains no allowable energy states so it contains no charge carriers. The difference in potential between valence and conduction bands defines the forbidden band, and this potential is called the *bandgap voltage*.

In the two-valley model, however, there are two regions in the conduction band in which charge carriers (e.g., electrons) can exist. These regions are called *valleys* and are designated the upper valley and the lower valley. According to the RWH theory, electrons in the lower valley have low effective mass (0.068) and consequently a high mobility (8000 cm^2/V-s). In the upper valley, which is separated from the lower valley by a potential of 0.036 electron volts (eV), electrons have a much higher effective mass (1.2) and lower mobility (180 cm^2/V-s) than in the lower valley.

At low electric field intensities (0 to 3.4 kV/cm), electrons remain in the lower valley and the material behaves ohmically. At these potentials, the material exhibits positive differential resistance (PDR). At a certain critical threshold potential (V_{th}), electrons are swept from the lower valley to the upper valley (hence the name *transferred electron devices*). For GaAs the electric field must be about 3.4 kV/cm, so V_{th}

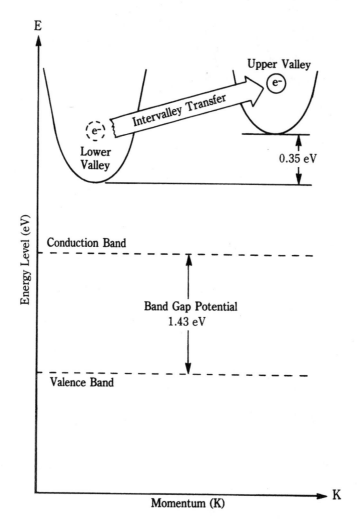

22-6 Transferred electron diode operation.

is the potential that produces this strength of field. Because V_{th} is the product of the electric field potential and device length, a 10-μM sample of GaAs will have a threshold potential of about 3.4 V. Most GaAs TED devices operate at maximum dc potentials in the 7- to 10-V range.

The average velocity of carriers in a two-valley semiconductor (such as GaAs) is a function of charge mobility in each valley and the relative numbers of electrons in each valley. If all electrons are in the lower valley then the material is in the highest average velocity state. Conversely, if all electrons were in the upper valley, then the average velocity is in its lowest state. Figure 22-7A shows drift velocity as a function of electric field, or dc bias.

In the PDR region, the GaAs material is ohmic, and the drift velocity increases linearly with increasing potential. It would continue to increase until the saturation

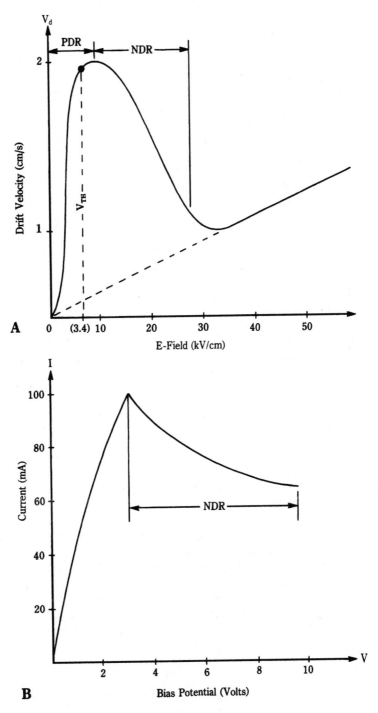

22-7 (A) Drift velocity vs electric field, (B) *I*-vs-*I*_{bias}, and (C) diode current vs time.

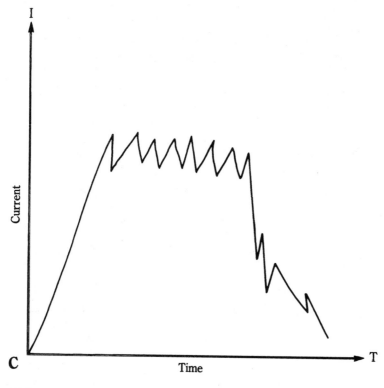

22-7 *Continued.*

velocity is reached (about 107 cm/s). As the voltage increases above threshold potential, which creates fields greater than 3.4 kV/cm, more and more electrons are transferred to the upper valley so that the average drift velocity drops. This phenomena gives rise to the negative resistance (NDR) effect. Figure 22-7B shows the NDR effect in the *I*-vs-*V* characteristic.

Figure 22-7C shows the *I*-vs-*T* characteristic of the Gunn diode operating in the NDR region. Ordinarily, you would expect a smooth current pulse to propagate through the material. But notice the oscillations (i.e., Gunn's instabilities) superimposed on the pulse. This oscillating current makes the Gunn diode useful as a microwave generator.

For the two-valley model to work, several criteria must be satisfied. First, the energy difference between lower and upper valleys must be greater than the thermal energy (K_T) of the material; K_T is about 0.026 eV, so GaAs with 0.036 eV differential energy satisfies the requirement. Second, in order to prevent hole-electron pair formation the differential energy between valleys must be less than the forbidden band energy (i.e., the bandgap potential). Third, electrons in the lower valley must have high mobility, low density of state, and low effective mass. Finally, electrons in the upper valley must be just the opposite: low mobility, high effective mass, and a high density of state. It is sometimes claimed that ordinary devices use so-called "warm" electrons (i.e., 0.026 eV) and TED devices use "hot" electrons greater than 0.026 eV.

Gunn diodes

The Gunn diode is a transferred electron device that is capable of oscillating in several modes. In the unresonant transit-time (TT) mode, frequencies between 1 and 18 GHz are achieved, with output powers up to 2 W (most are on the order of a few hundred milliwatts). In the resonant limited space-charge (LSA) mode, operating frequencies to 100 GHz and pulsed power levels to several hundred watts (1% duty cycle) have been achieved.

Figure 22-8A shows a diagram for a Gunn diode, and an equivalent circuit is shown in Fig. 22-8B. The active region of the diode is usually 6 to 18 μm long. The N⁺ end regions are ohmic materials of very low resistivity (0.001 Ω-cm), and are 1 to 2 μm

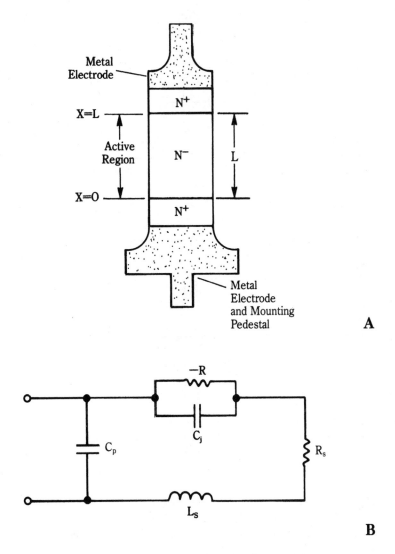

22-8 (A) Gunn diode and (B) equivalent circuit.

thick. The function of the N^+ regions is to form a transition zone between the metallic end electrodes and the active region. In addition to improving the contact, the N^+ regions prevent migration of metallic ions from the electrode into the active region.

Domain growth

The mechanism underlying the oscillations of a Gunn diode is the growth of Ridley domains in the active region of the device. An electron domain is created by a bunching effect (Fig. 22-9) that moves from the cathode end to the anode end of

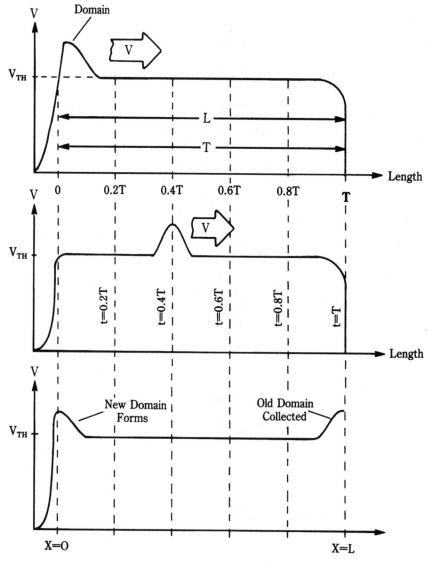

22-9 Domain formation and propagation.

the active region. When an old domain is collected at the anode, a new domain forms at the cathode and begins propagating. The propagation velocity is close to the saturation velocity (10^7 cm/s). The time required for a domain to travel the length (L) of the material is called the transit time (T_t), which is:

$$T_t = \frac{1}{V_s} \qquad (22\text{-}3)$$

where

T_t = the transit time in seconds (s)
L = the length in centimeters (cm)
V_s = the saturation velocity (10^7 cm/s)

Figure 22-10 graphically depicts domain formation. You might recognize that the "domain" shown here is actually a dipole, or double domain. The area ahead of the domain forms a minor depletion zone, and in the area of the domain, there is a bunching of electrons. Thus, there is a difference in conductivity between the two regions, and this conductivity is different still from the conductivity of the rest of the active region. There is also a difference between the electric fields in the two domain poles and also in the rest of the material. The two fields equilibrate outside of the domain. The current density is proportional to the velocity of the domain, so a current pulse forms.

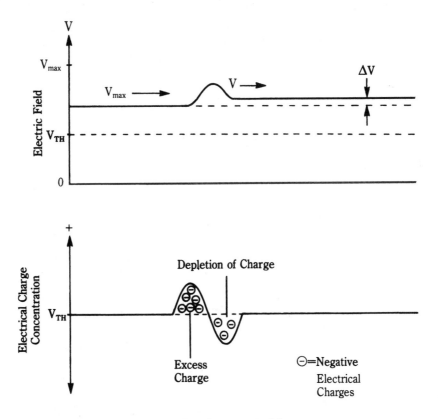

22-10 Domain formation in semiconductor material.

Gunn operating modes

Transferred electron devices (i.e., Gunn diodes) operate in several modes and submodes. These modes depend in part on device characteristics and in part on external circuitry. For the Gunn diode, the operating modes are stable amplification (SA) mode, transit time (TT) mode, limited space-charge (LSA) mode, and bias circuit oscillation (BCO) mode.

Stable amplification (SA) modes In this mode, the Gunn diode will behave as an amplifier. The requirement for SA-mode operation is that the product of doping concentration (N_o) and effective length of the active region (L) must be less than $1012/cm^2$. Amplification is limited to frequencies in the vicinity of v/L, where v is the domain velocity and L is the effective length.

Transit time (Gunn) mode The transit time (TT), or Gunn, mode is unresonant and depends on device length and applied dc bias voltage. The dc potential must be greater than the critical threshold potential (V_{th}). Because of the Gunn effect, current oscillations in the microwave region are superimposed on the current pulse. Operation in this mode requires that the $N_o L$ product be $1012/cm^2$ to $1014/cm^2$. Operating frequency F_o is determined by device length, or rather the transit time of the pulse through the length of the material. Because domain velocity is nearly constant, and is often close to the electron saturation velocity ($V_s = 10^7$ cm/s), length and transit time are proportional to each other. The operating frequency is inversely proportional to both the length of the device and the transit time. The length of the active region in the Gunn diode determines operating frequency. The frequency varies from about 6 GHz for an 18-μm sample (counting 1.5 μm for each N+ electrode region) to 18 GHz for a 6-μm sample. The operating frequency in the TT mode is approximately:

$$F_o = \frac{V_{dom}}{L_{eff}} \tag{22-4}$$

where

F_o = the operating frequency in hertz (Hz)
V_{dom} = the domain velocity in centimeters per second (cm/s)
L_{eff} = the effective length in centimeters (cm)

Operation in the TT mode provides efficiencies of 10% or less, with 4 to 6% being most common. Output powers are usually less than 1000 mW, although 2000 mW has been achieved.

Limited space charge (LSA) mode The LSA mode depends on shock exciting a high-Q resonant tank circuit or tuned cavity with current pulses from the Gunn diode. The LSA mode and its submodes are also called *accumulation mode, delayed domain mode,* and *quenched domain mode.* These various names reflect variations on the LSA theme. For the LSA mode, the $N_o L$ product must be $1012/cm^2$ or higher and the N_o/F quotient must be between 2 × 105 and 2 × 104 s/cm^3.

Figure 22-11A shows a simplified circuit for a LSA-mode Gunn oscillator, and Fig. 22-11B shows the waveforms. The circuit consists of the Gunn diode shunted by either a LC tank circuit (as shown) or a tuned cavity that behaves like a tank circuit. The criterion for resonant oscillation in a negative resistance circuit is sim-

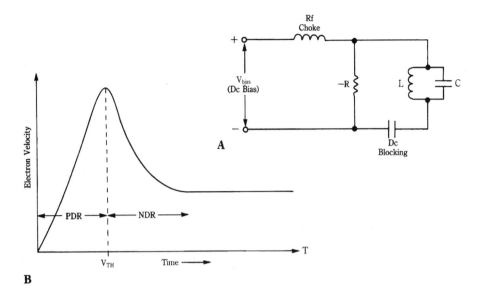

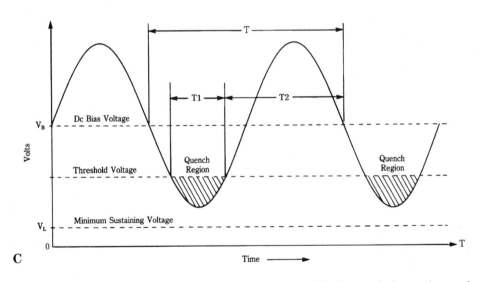

22-11 (A) Oscillator circuit for negative resistance mode, (B) characteristic waveforms of circuit, and (C) output voltage.

ple: the negative conductance $(-G = 1/-R)$ must be greater than or equal to the conductance represented by circuit losses:

$$-G = G_o. \qquad (22\text{-}5)$$

At turn-on, transit time current pulses (Fig. 22-12) hit the resonant circuit and shock excite it into oscillations at the resonant frequency. These oscillations set up

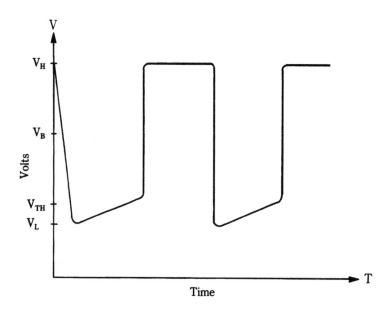

22-12 LSA output pulses.

a sine-wave voltage across the diode (Fig. 22-12) that adds to the bias potential. The total voltage across the diode is the algebraic sum of the dc bias voltage and the RF sine-wave voltage. The dc bias is set so that the negative swing of the sine wave forces the total voltage below the critical threshold potential, V_{th}. Under this condition, the domain does not have time to build up and the space charge dissipates. The domain quenches during the period when the algebraic sum of the dc bias and the sine-wave voltage is below V_{th}. The LSA oscillation period ($T = 1/F$) is set by the external tank circuit and the diode adapts to it. The period of oscillation is found from:

$$T = 2\pi\sqrt{LC} + \frac{1}{\left(R\dfrac{V_b}{V_{th}}\right)}$$

(22-6)

where

T = the period in seconds (s)
L = the inductance in henrys (H)
C = the capacitance in farads (F)
R = the low-field resistance in ohms (Ω)
V_{th} = the threshold potential
V_b = the bias voltage.

The LSA mode is considerably more efficient than the transit-time mode. The LSA mode is capable of 20% efficiency so that you can produce at least twice as much RF output power from a given level of dc power drawn from the power supply as transit time operation. At duty factors of 0.01, the LSA mode is capable of delivering hundreds of watts of pulsed output power. The output power of any oscillator or amplifier is the product of three factors: dc input voltage (V), dc input current (I), and the conversion efficiency (n, a decimal fraction between 0 and 1):

$$P_o = nVI, \tag{22-7}$$

where

P_o = the output power
n = the conversion efficiency factor (0-1)
V = the applied dc voltage
I = the applied dc current

For the Gunn diode case, a slightly modified version of this expression is used:

$$P_o = nMV_{th}LN_oeVA, \tag{22-8}$$

where

n = the conversion efficiency factor (0-1)
v = the average drift velocity
V_{th} = the threshold potential (kV/cm)
M = the multiple V_{dc}/V_{th}
L = the length in centimeters
n_o = the donor concentration
e = the electric charge (1.6×10^{-19} coulomb)
A = the area of the device in cm^2

Bias circuit oscillation (BCO) Gunn mode This mode is quasiparasitic in nature and occurs only during one of the normal Gunn oscillating modes (TT or LSA). If product F_1 is very, very small then BCO oscillations can occur at frequencies from 0.01 to 100 MHz.

Gunn diode applications

Gunn diodes are often used to generate microwave RF power in diverse applications ranging from receiver local oscillators, to police speed radars, to microwave communications links. Figure 22-13 shows two possible methods for connecting a Gunn diode in a resonant cavity. Figure 22-13A shows a cavity that uses a loop-coupled output circuit. The output impedance of this circuit is a function of loop size and position, with the latter factor dominating. The loop positioning is a trade-off between maximum output power and oscillator frequency stability.

Figure 22-13B shows a Gunn diode mounted in a section of flanged waveguide. RF signal passes through an iris to be propagated through the waveguide to the load. In both Figs. 22-13A and 22-13B, the exact resonant frequency of the cavity is set by a tuning screw inserted into the cavity space.

IMPATT diodes

The avalanche phenomenon is well-known in PN junction diodes. If a reverse bias potential exceeds a certain critical threshold, then the diode breaks down and the reverse current increases abruptly from low "leakage" values to a very high value. The principal cause of this phenomena is secondary electron emission (i.e., charge carriers become so energetic as to be able to knock additional valence elec-

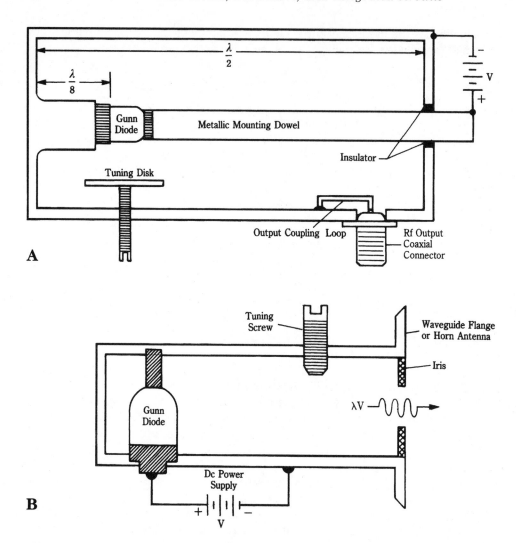

22-13 (A) Gunn diode operated in a cavity and (B) Gunn diode in a cavity antenna.

trons out of the crystal lattice to form excess hole-electron pairs). The common zener diode works on this principle.

The onset of an avalanche current in a PN junction is not instantaneous, but rather, there is a short phase-delay period between the application of a sufficient breakdown potential and the creation of the avalanche current. In 1959, W. T. Read of Bell Telephone Laboratories postulated that this phase delay could create a negative resistance. It took until 1965 for others (Lee and Johnson) at Bell Labs to create the first of these "Read diodes." Johnson generated about 80 mW at 12 GHz in a silicon PN junction diode. Today, the class of diodes of which the Read device is a member are referred to collectively as Impact Avalanche Transit Time (IMPATT) diodes. The name IMPATT reflects the two different mechanisms at work: Avalanche (impact ionization) and Transit time (drift).

Figure 22-14 shows a typical IMPATT device based on a N+-P-I-P+ structure. Other structures are also known, but the NPIP of Fig. 22-14A is representative. The "+" indicates a higher than normal doping concentration, as indicated by the profile in Fig. 22-14C. The doping profile ensures an electric field distribution (Fig. 22-14B) that is higher in the P region in order to confine avalanching to a small zone. The I region is an intrinsic semiconductor that is lightly doped to have a low charge carrier density. Thus, the I region is a near insulator, except when charge carriers are injected into it from other regions.

The IMPATT diode is typically connected in a high-Q resonant circuit (LC tank or cavity). Because avalanching is a very noisy process, noise at turn-on "rings" the tuned circuit and creates a sine-wave oscillation at its natural resonant frequency (Fig. 22-14D). The total voltage across the NPIP structure is the algebraic sum of the dc bias and tuned-circuit sine wave.

The avalanche current (I_o) is injected into the I region and begins propagating over its length. Notice in Fig. 22-14E that the injected current builds up exponentially until the sine wave crosses zero then drops exponentially until the sine wave

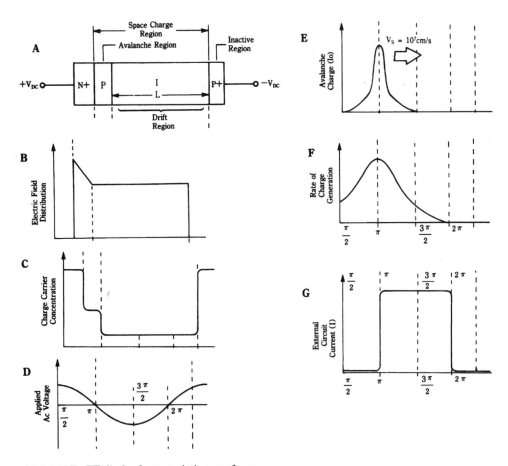

22-14 IMPATT diode characteristic waveforms.

reaches the negative peak. This current pulse is thus delayed 90° with respect to the applied voltage.

Compare now the external circuit current pulse in Fig. 22-14G with the tuned circuit sine wave (Fig. 22-14D). Notice that transit time in the device has added additional phase delay, so the current is 180° out of phase with the applied voltage. This phase delay is the cause of the negative resistance characteristic. Oscillation is sustained in the external resonant circuit by successive current pulses reringing it. The resonant frequency of the external resonant circuit should be:

$$F = \frac{V_d}{2L},$$
(22-9)

where

F = the frequency in hertz (Hz)
V_d = the drift velocity in centimeters per second (cm/s)
L = the length of the active region in centimeters (cm)

IMPATT diodes typically operate in the 3- to 6-GHz region, although 100-GHz operation has been achieved. These devices typically operate at potentials in the 75- to 150-Vdc range. Because an avalanche process is used, the RF output signal of the IMPATT diode is very noisy. Efficiencies of single drift-region devices (such as Fig. 22-14A) are about 6 to 15%. By using double-drift construction (Fig. 22-15), efficiencies can be improved to the 20 to 30% range. The double-drift device uses electron conduction in one region and hole conduction in the other. This operation is possible because the two forms of charge carrier drift approximately in phase with each other.

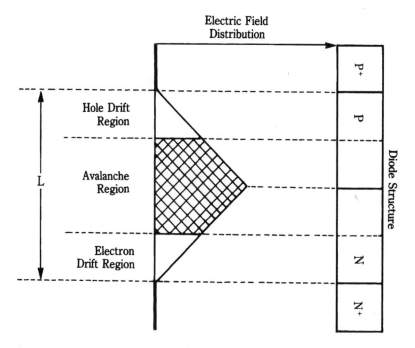

22-15 PPNN double-drift IMPATT diode.

TRAPATT diodes

Gunn and IMPATT diodes operate at frequencies of 3 GHz or above. Operation at lower frequencies (e.g., 500 MHz to 3 GHz) was left to transistor, which also limited available RF output power to a great extent. Gunn and IMPATT devices cannot operate at lower frequencies because it was difficult to increase transit time in those devices. You might assume that it is only necessary to lengthen the active region of the device to increase transit time. But certain problems arose in long structures, domains were found to collapse, and sufficient fields were hard to maintain.

A solution to the transit time problem is found in a modified IMPATT structure that uses P$^+$-N-N$^-$ regions (Fig. 22-16). The P+ region is typically 3 to 8 μM, and the N-region is typically 3 to 13 μm. The diameter of the device might be 50 to 750 μm, depending on the power level required. The first of these diodes was produced by RCA in 1967. It produced more than 400 W at 1000 MHz, at an efficiency of 25%. Since then, frequencies as low as 500 MHz, and powers to 600 W, have

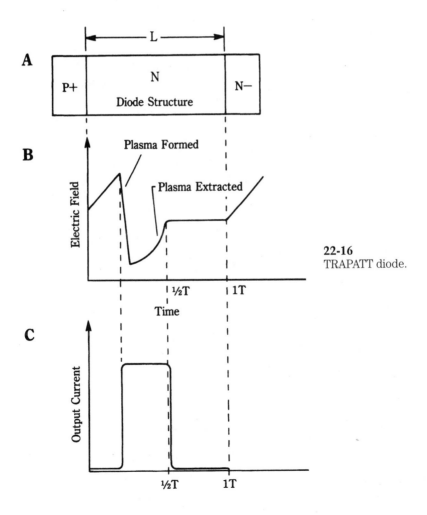

22-16
TRAPATT diode.

been achieved. Today, efficiencies in the 60 to 75% range are common. One device was able to provide continuous tuning over a range of 500 to 1500 MHz.

The name of the P+-N-N- device reflects its operating principle: trapped plasma avalanche triggered transit (TRAPATT)—the method by which transit time is increased by formation of a plasma in the active region. A plasma is a region of a large number of disassociated holes and electrons that do not easily recombine. If the electric field is low, then the plasma is trapped. That is, the charge carriers are swept out of the N region slowly (see Fig. 22-16B).

The output is a harmonic-rich, sharp rise-time current pulse (Fig. 22-16C). To become self-oscillatory, this pulse must be applied to a low-pass filter at the input of the transmission line or waveguide that is connected to the TRAPATT. Harmonics are not passed by the filter, so they are reflected back to the TRAPATT diode to trigger the next current pulse.

BARITT diodes

The BARITT diode (Fig. 22-17) consists of three regions of semiconductor material forming a pair of abrupt PN junctions, one each P+-N and N-P+. The name of this device comes from a description of its operation: barrier injection transit time. The BARITT structure is designed so that the electric field applied across the end electrodes causes a condition at or near punch-through. That is, the depletion zone is formed throughout the entire N region of the device. Current is formed by sweeping holes into the N region of the device. Under ordinary circumstances, the P+-N junction is forward-biased and the N-P+ is reverse-biased. The depleted N region forms a potential barrier into which holes are injected into the N region from the forward-biased junction. These charge carriers then drift across the N region at the saturation velocity of 10^7 cm/sec, forming a current pulse. There are three conditions for proper BARITT operation.

The electrical field across the device must be great enough to force charge carriers in motion to drift at the saturation velocity (10^7 cm/s); The electrical field must

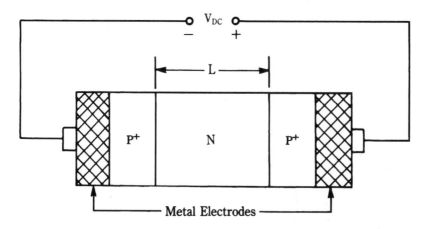

22-17 BARITT diode.

be great enough to create the punch-through condition, and the electrical field must not be great enough to cause avalanching to occur.

The normal circuit configuration for BARITT devices is in a resonant LC tank or cavity. At turn-on, noise pulses will initially shock-excite the resonant circuit into self-oscillation, and successive current pulses supply the energy to continue the oscillations. Figure 22-18 shows the relationship between the oscillatory sine-wave voltage, the internal current, and the external current. As in the other diodes, the output energy comes from tapping the energy in the resonant circuit.

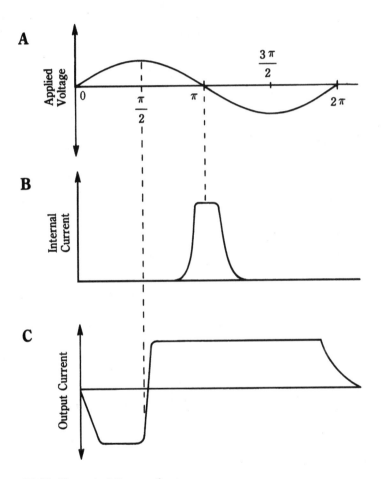

22-18 Characteristic waveforms.

UHF and microwave RF transistors

Transistors were developed right after World War II and by 1955 were being used in consumer products. Those early devices were limited to audio and low-RF frequencies, however. As a result, only solid-state audio products and AM-band radios were widely available in the 1950s. Development continued, however, and by

1963, solid-state FM broadcast and VHF communications receivers were on the market. Microwave applications, however, remained elusive.

Early transistors were severely frequency-limited by a number of factors, including electron saturation velocity, base structure thickness (which affects transit time), base resistance, and device capacitances. In the latter category are junction capacitances and stray capacitances resulting from packaging. When combined with stray circuit inductances and circuit resistances, the device capacitance significantly rolled-off upper operating frequencies.

The solution to the operating frequency limitation was in developing new semiconductor materials (e.g., gallium arsenide, different device internal geometries, and new device construction and packaging methods). Today, transistor devices operate well into the microwave region, and 40-GHz devices are commercially obtainable; 90-GHz devices have been demonstrated in Japanese laboratories. Transistors have replaced other microwave amplifiers in many applications—especially in low-noise receiver amplifiers.

Semiconductor overview

Semiconductors fall into a gray area between good conductors and poor conductors (i.e., insulators). The conductivity of these materials can be varied by "doping" with impurities, changes in temperature and by light (as in the case of phototransistors). Prior to the development of microwave transistors, most devices were made of *Group IV-A* materials, such as silicon (Si) and germanium (Ge). The term *Group IV-A* refers to the group occupied by these materials on the Periodic Table of chemical elements. Some modern microwave transistors are made of Group III-A semiconductors, such as gallium and indium (see Table 22-1).

Figure 22-19 shows an energy level diagram for semiconductor materials. Two permissible bands represent states that are allowed to exist: the conduction band and the valence band. The region between these permitted bands is a forbidden band. This band represents energy states that are not allowed to exist. The width of the forbidden band is the difference between the conduction band energy (E_c) and valence band energy (E_v). Called the bandgap energy (E_{BG}), this parameter is unique for each type of material. For silicon at 25°C the bandgap energy is 1.12 electron volts (eV); for germanium, it is 0.803 eV. A Group III-A material called *gallium arsenide* (GaAs) is used in microwave transistors and has a bandgap energy of 1.43 eV.

Table 22-1. Mobility (cm²/V-s)

Material	Group	Electron	Hole
Ge	IV-A	3900	1900
Si	IA-A	1350	600
GaAs	III-A	8500	400
InP	III-A	4600	150

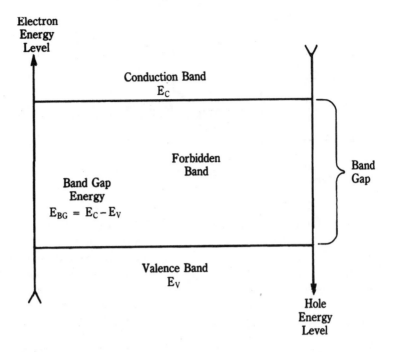

22-19 Energy bands diagram.

Charge carrier mobility is a gauge of semiconductor material activity and is measured in units of cm²/V-s. Electron mobility is typically more vigorous than hole mobility as Table 22-1 demonstrates. Notice particularly the spread for Group III-A semiconductors.

Bipolar transistors

A bipolar transistor is one that uses both types of charge carriers in its operation. In other words, both electrons and holes are used for conduction, meaning that both N-type and P-type materials are needed. Figure 22-20 shows the basic structure of a bipolar transistor. In this case, the device uses two N-type regions and a single P-type region, thereby forming an NPN device. A PNP device has exactly the opposite arrangement. Because NPN devices predominate in the microwave world, however, we will consider only that type of device. Silicon bipolar NPN devices have been used in the microwave region up to about 4 GHz. At higher frequencies, designers typically use field-effect transistors.

The first transistors were of the point-contact type of construction in which "cat's whisker" metallic electrodes were placed in rather delicate contact with the semiconductor material. A marked improvement soon became available in the form of the diffused junction device. In those devices, N- and P-type impurities are diffused into the raw semiconductor material. Metallization methods are used to deposit electrode contact pads onto the surface of the device. Mesa devices use a growth technique to build up a substrate of semiconductor material into a tablelike structure.

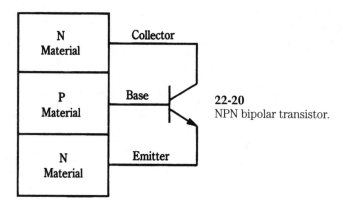

22-20
NPN bipolar transistor.

Modern silicon bipolar transistors tend to be NPN devices of either planar (Fig. 22-21A) or epitaxial (Fig. 22-21B) design. The planar transistor is a diffusion device on which surface passivation is provided by a protective layer of silicon dioxide (SiO_2). The layer provides protection of surfaces and edges against contaminants migrating into the structure.

The epitaxial transistor (Fig. 22-21B) uses a thin, low-conductivity "epitaxial layer" for part of the collector region (the remainder of the collector region is high-

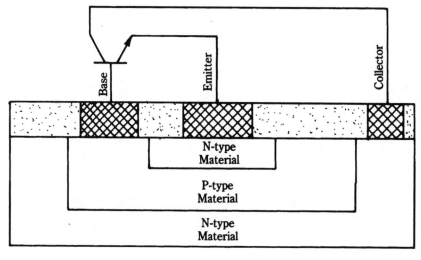

A

 Ohmic Contact Electrode

 Protective SiO_2 layer

22-21 (A) Details of planar construction, (B) epitaxial construction, (C) doping profile, and (D) SICOS construction.

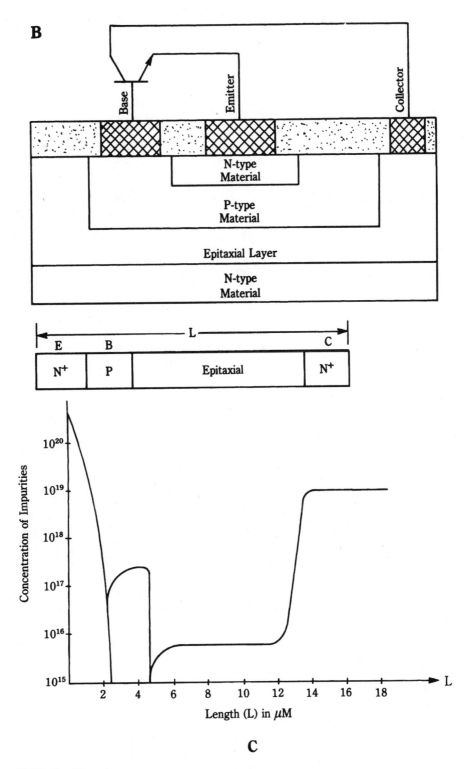

22-21 *Continued.*

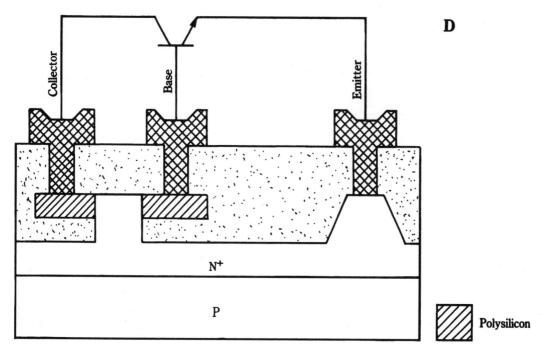

22-21 *Continued.*

conductivity material). The epitaxial layer is laid down as a condensed film on a substrate made of the same material. Figure 22-21C shows the doping profile versus length for an epitaxial device. Notice that the concentration of impurities, hence region conductivity, is extremely low for the epitaxial region.

The sidewall contact structure (SICOS) transistor is shown in Fig. 22-21D. Regular NPN transistors exhibit a high junction capacitance and a relatively large electron flow across the subemitter junction to the substrate. These factors reduce the maximum cutoff frequency and available current gain. A SICOS device with a 0.5-μm base width offers a current gain (H_{fe}) of 100 and a cutoff frequency of 3 GHz.

Heterojunction bipolar transistors (HBT) have been designed with maximum frequencies of 67 GHz on base widths of 1.2 μm; current gains of 10 to 20 are achieved on very small devices and up to 55 for larger base widths. The operation of HBTs depends in part on the use of a thin coating of $NA_2S \cdot 9H_2O$ material, which reduces surface charge pair recombination velocity. At very low collector currents, a device with a 0.15-μm base width and 40-μm $\times$ 100-μm emitter produced current gains in excess of 3800, with 1500 being also achieved at larger collector currents (on the order of 1 mA/cm^2). Compare with non-HBT devices which typically have gain figures under 100.

The Johnson relationships

Microwave bipolar transistors typically obey a set of equations called the *Johnson relationships*. A set of six equations are useful for determining device limitations.

NOTE: In the following equations, terms are defined when first used; subsequent uses are not redefined.

Equation I: Voltage-frequency limit:

$$\frac{V_{max}}{2\pi\left(\dfrac{l}{v}\right)} = \frac{E_{max}V_s}{2\pi},$$

(22-10)

where

V_{max} = the maximum allowable voltage ($E_{max}L_{min}$)
V_s = the material saturation velocity
E_{max} = the maximum electric field
(l/v) = the average charge carrier time at average charge velocity through the length of the material.

Equation II: Current-frequency limit:

$$\frac{I_{max}X_{C_0}}{2\pi\left(\dfrac{l}{v}\right)} = \frac{E_{max}V_s}{2\pi},$$

(22-11)

where

I_{max} = the maximum device current
X_{C_0} = the reactance of the output capacitance (i.e., $1/(2\pi(l/v)C_0)$).

Equation III: Power-frequency limit:

$$\frac{\sqrt{P_{max}X_{C_0}}}{2\pi\left(\dfrac{l}{v}\right)} = \frac{E_{max}V_s}{2\pi},$$

(22-12)

where
P_{max} is the maximum power.

Equation IV: Power gain-frequency limit:

$$\frac{\sqrt{G_{max}KTV_{max}}}{e} = \frac{E_{max}V_s}{2\pi},$$

(22-13)

where

G_{max} = the maximum power gain
K = Boltzmann's constant (1.38×10^{-23})
T = the temperature in degrees Kelvin
e = the electronic charge (1.6×10^{-19}).

Equation V: Maximum gain:

$$G_{max} = \frac{Z_0}{Z_\epsilon}\sqrt{\frac{F_t}{F}},$$

(22-14)

where

F = the operating frequency
Z_o = the real component of output impedance
Z_{in} = the real component of the input impedance.

Equation VI: Impedance ratio:

$$\frac{Z_o}{Z_\epsilon} = \frac{C_\epsilon}{C_o} = \frac{\dfrac{I_{max} T_b}{\left(\dfrac{KT}{e}\right)}}{\dfrac{I_{max} T_b}{V_{max}}}, \tag{22-15}$$

where

C_{in} = the input capacitance
C_o = the output capacitance
T_b = charge transit time.

Cutoff frequency (F_t)

The cutoff frequency (F_t) is the frequency at which current gain drops to unity. Several factors affect cutoff frequency. First, the saturation velocity for charge carriers in the semiconductor material; second is the time required to charge the emitter-base junction capacitance (T_{eb}); third, the time required to charge the base-collector junction capacitance (T_{cb}); fourth, the base region transit time (T_{bt}); and fifth, base-collector depletion zone transit time (T_{bc}). These times add together to give us the emitter-collector transit time (T). The expression for cutoff frequency is:

$$F_t = \frac{1}{2\pi T} \tag{22-16}$$

or

$$F_T = \frac{1}{2\pi (T_{eb} + T_{bt} + T_{bc} + T_{cb})}, \tag{22-17}$$

where

F_t is the frequency in hertz (Hz)
Times are in seconds.

Bipolar transistor geometry

Two of the problems that limited the high-frequency performance of early transistors were emitter and base resistances and base transit time. Unfortunately, simply reducing the base width in order to improve transit time, and reduce overall resistance, also reduces current and voltage-handling capability. Although some improvements were made, older device internal geometries were clearly limited. The solutions to these problems lay in the geometry of the base-emitter junction. Figure 22-22 shows three geometries that yielded success in the microwave region: interdigital (Fig. 22-22A), matrix (Fig. 22-22B), and overlay (Fig. 22-22C). These construction geometries yielded the thin, wide-area, low-resistance base regions that are needed to increase the operating frequency.

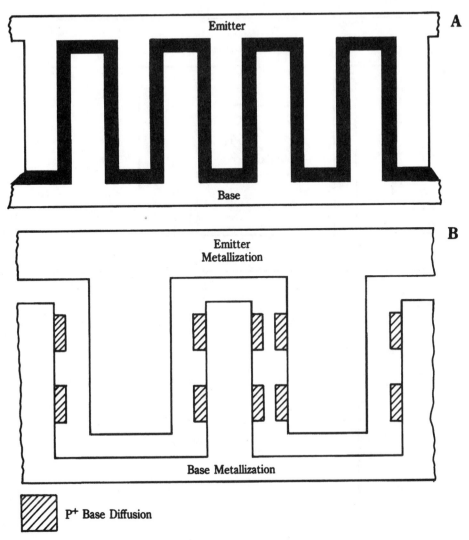

22-22 (A) Interdigital emitter construction, (B) matrix construction, and (C) overlay construction.

Low-noise bipolar transistors

Geometries that provide the low resistance needed to increase operating frequency also serve to reduce the noise generated by the device. There are three main contributors to noise in a bipolar transistor: thermal noise, shot noise in the e-b circuit, and shot noise in the b-c circuit. The thermal noise is a function of temperature and base region resistance. By reducing the resistance (a function of internal geometry), we also reduce the noise. The shot noise produced by the P-N junction is a function of the junction current. There is an optimum collector current (I_{co}) at which noise figure is best. Figure 22-23 shows noise figure in decibels plotted against the collector current for a particular device. The production of optimum noise figure at practical collector currents is a function of junction efficiency.

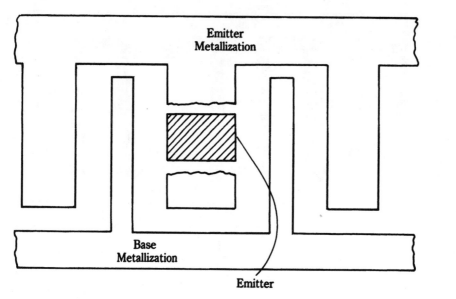

C

22-22 *Continued.*

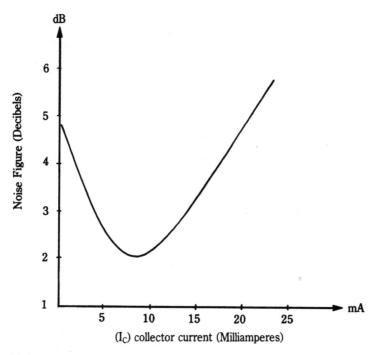

22-23 Noise figure vs collector current.

Field-effect transistors

Now look at the field-effect transistor (FET), as used in microwave systems. The FET operates by changing the conductivity of a semiconductor "channel" by varying the electric field in the channel. Two elementary types are found: junction field-effect transistors (JFET) and metal-oxide semiconductor field-effect transistors (MOSFET), also sometimes called by the name *insulated-gate field-effect transistor* (IGFET). In the microwave world, a number of devices are either adaptations or modifications of these basic types.

Figure 22-24A shows the structure of a "generic" JFET device that works on the depletion principle. The channel in this case in N-type semiconductor material and the gate is made of P-type material. A gate-contact metallized electrode is deposited over the gate material. In normal operation, the PN junction is reverse-biased and the applied electric field extends to the channel material.

The electric field repels charge carriers (in this example, electrons) in the channel, creating a depletion zone in the channel material. The wider the depletion zone, the higher the channel resistance. Because the depletion zone is a function of applied gate potential, the gate potential tends to modulate channel resistance. For a constant drain-source voltage, varying the channel resistance also varies the channel current. Because an input voltage varies an output current (I_o/V_{in}), the FET is called a *transconductance amplifier*. The MOSFET (Fig. 22-24B) replaces the semiconductor material in the gate with a layer of oxide insulating material. Gate metallization is overlaid on the insulator. The electric field is applied across the insulator in the manner of a capacitor. The operation of the MOSFET is similar to the JFET in broad terms and in a depletion-mode device the operation is very similar. In an enhancement-mode device channel resistance drops with increased signal voltage.

Microwave FETs

The microwave field-effect transistor represented a truly giant stride in the performance of semiconductor amplifiers in the microwave region. Using Group III-A materials, notably gallium arsenide (GaAs), the FET presses cutoff frequency performance way beyond the 3- or 4-GHz limits achieved by silicon bipolar devices. The gallium arsenide field-effect transistor (GaAsFET) offers superior noise performance (noise figure less than 1 dB is achieved!) over silicon bipolar devices, improved temperature stability, and higher power levels. The GaAsFET can be used as a low-noise amplifier (LNA), class-C amplifier, or oscillator. GaAsFETs are also used in monolithic microwave integrated circuits (MIMIC), extremely high-speed analog-to-digital (A/D) converters, analog/RF applications, and in high-speed logic devices. In addition to GaAs, AlGaAs and InGaAsP materials are also used.

Figure 22-25 shows the principal player among microwave field-effect transistors: the metal semiconductor field-effect transistor (MESFET), also called the *Schottky barrier transistor* (SBT) or *Schottky barrier field-effect transistor* (SBFET). The epitaxial "active layer" is formed of N-type GaAs doped with either sulfur or tin ions, with a gate electrode formed of evaporated aluminum. The source and drain electrodes are formed of gold germanium (AuGe) or gold telluride (AuTe) or AuGeTe alloy. Another form of microwave JFET is the high-electron mobility tran-

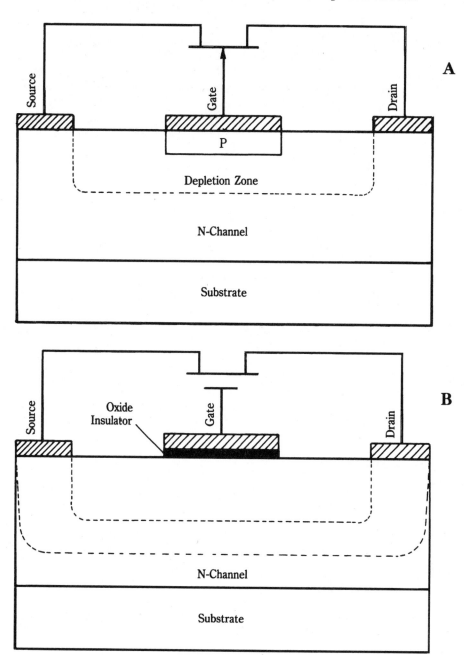

22-24 (A) JFET transistor and (B) MOSFET transistor.

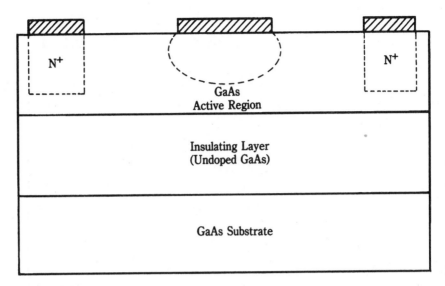

22-25 MESFET transistor.

sistor (HEMT), shown in Figs. 22-26A and 22-26B. The HEMT is also known as the *two-dimensional electron GaAsFET* (TEGFET) and *heterojunction FET* (HFET). Devices in this category produce power gains up to 11 dB at 60 GHz and 6.2 dB at 90 GHz. Typical noise figures are 1.8 dB at 40 GHz and 2.6 dB at 62 GHz. Power levels at 10 GHz have approached 2 W (CW)/mm of emitter periphery dimension ("emitter" is the gate structure junction in JFET-like devices).

HEMT devices are necessarily built with very thin structures in order to reduce transit times (necessary to increasing frequency because of the $1/T$ relationship). These devices are built using ion implantation, molecular beam epitaxy (MBE), or metal organic chemical vapor deposition (MOCVD). The power output capability of field-effect transistors has climbed rapidly over the past few years. Devices in the 10-GHz region have produced 2-W/mm power densities and output levels in the 4- to 5-W region. Figure 22-27 shows a typical power FET internal geometry used to achieve such levels.

An advantage of FETs over bipolar devices is that input impedance tends to be high and relatively frequency-stable. It is thus easier to design wideband power amplifiers and either fixed or variable-frequency (tunable) power amplifiers.

Noise performance

Only a few years ago, users of transistors in the high-UHF and microwave regions had to contend with noise figures in the 8- to 10-dB range. A fundamental truism states that no signal below the noise level can be detected without sophisticated computer signals processing (which takes time). Thus, reducing the noise floor of an amplifier helps greatly in building more sensitive real-time microwave systems.

Figure 22-27 shows the typical range of noise figures in decibels expected from various classes of transistors. The silicon bipolar transistor shows a flat noise figure

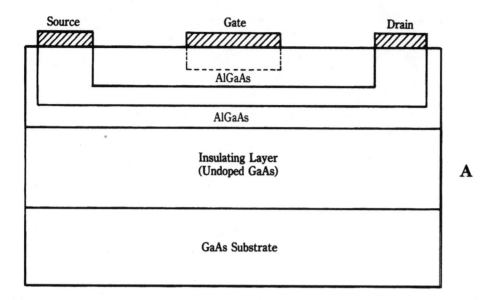

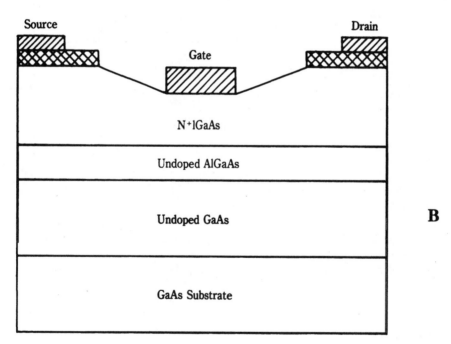

22-26 (A) HEMT transistor and (B) HFET transistor.

(curve A) up to a certain frequency and a sharp increase in noise thereafter. The microwave FET (curve B), on the other hand, shows increases in both high- and low-frequency regions. The same is also true for the HEMT (curve C), but to a lesser degree. Only the supercooled (−260°C!) HEMT performs better.

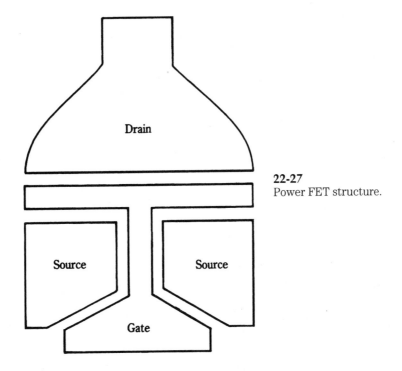

22-27
Power FET structure.

Most MESFET and HEMT devices show increases in noise at low frequencies. Some such devices also tend to become unstable at those frequencies. Thus, microwave and UHF devices might, contrary to what your instincts might suggest, fail to work at frequencies considerably below the optimum design frequency range.

Selecting transistors

The criteria for selecting bipolar or field-effect transistors in the microwave range depends on application. The importance of noise figure, for example, becomes apparent in the "front end" of a microwave satellite receiver system. In other applications, power gain and output might be more important. Obviously, gain and noise figure are both important—especially in the front ends of receiver systems. For example, earth communications or receiver terminals typically use a parabolic dish antenna with a low-noise amplifier ("LNA") at the feedpoint. When selecting devices for such applications, however, beware of seemingly self-serving transistor specification data sheets. For example, consider the noise figure specification. As you saw in Fig. 22-28, there is a strong frequency dependence regarding noise figure. Yet device data sheets often list the maximum frequency and minimum noise figure—even though the two rarely coincide with one another! When selecting a device, consult the N_F-vs-F curve.

Device gain can also be specified in different ways, so caution is in order. There are at least three ways to specify gain: maximum allowable gain (G_{max}), gain at optimum noise figure (GNF), and the insertion gain. The maximum obtainable gain oc-

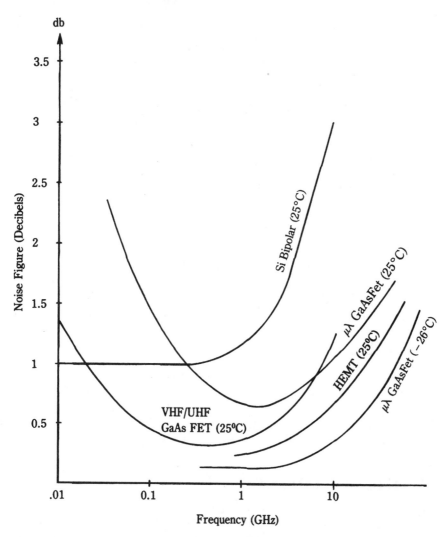

22-28 Noise figure vs frequency for several devices.

curs usually at a single frequency where the input and output impedances are conjugately matched (i.e., reactive part of impedance canceled out and resistive part transformed for maximum power transfer). The noise figure gain is an impedance-matching situation in which noise figure is optimized, but not necessarily power gain. Rarely, perhaps never, are the two gains the same.

Another specification to examine is the 1-dB compression point. This critical point is that at which a 10-dB increase in the input signal level results in a 9-dB increase in the output signal level. Operation too near this point can cause distortion in a linear amplifier, so it must be avoided.

UHF/microwave RF integrated circuits

Very wideband amplifiers (i.e., those operating from near-dc to UHF or into the microwave regions) have traditionally been very difficult to design and build with consistent performance across the entire passband. Many such amplifiers have either gain irregularities, such as "suck-outs" or peaks. Others suffer large variations of input and output impedance over the frequency range. Still others suffer spurious oscillation at certain frequencies within the passband. Barkhausen's criteria for oscillation requires loop gain of unity or more and 360° (in phase) feedback at the frequency of oscillation. At some frequency, the second of these criteria might be met by adding the normal 180° phase shift inherent in the amplifier to phase shift because of stray RLC components. The result will be oscillation at the frequency where the RLC phase shift was an additional 180°. In addition, in the past, only a few applications required such amplifiers. Consequently, such amplifiers were either very expensive or didn't work nearly as well as claimed. Hybrid Microwave Integrated Circuit (HMIC) and Monolithic Microwave Integrated Circuit (MMIC) devices were low-cost solutions to the problem.

What are HMICs and MMICs?

MMICs are tiny "gain block" monolithic integrated circuits that operate from dc or near dc to a frequency in the microwave region. HMICs, on the other hand, are hybrid devices that combine discrete and monolithic technology. One product (Signetics NE-5205) offers up to +20 dB of gain from dc to 0.6 GHz, and another low-cost device (Minicircuits Laboratories, Inc. MAR-x) offers +20 dB of gain over the range from dc to 2 GHz, depending on the model. Other devices from other manufacturers are also offered, and some produce gains to +30 dB and frequencies to 18 GHz. Such devices are unique in that they present input and output impedances that are a good match to the 50 or 75 Ω normally used as system impedances in RF circuits.

Monolithic integrated circuit devices are formed through photoetching and diffusion processes on a substrate of silicon or some other semiconductor material. Both active devices (such as transistors and diodes) and some passive devices can be formed in this manner. Passive components, such as on-chip capacitors and resistors, can be formed using various thin- and thick-film technologies. In the MMIC device, interconnections are made on the chip via built-in planar transmission lines.

Hybrids are a level closer to regular discrete circuit construction than ICs. Passive components and planar transmission lines are laid down on a glass, ceramic, or other insulating substrate by vacuum deposition or other methods. Transistors and unpackaged monolithic "chip dies" are cemented to the substrate and then connected to the substrate circuitry via mil-sized gold or aluminum bonding wires. Because the material in this series could sometimes apply to either HMIC or MMIC devices, the convention herein shall be to refer to all devices in either subfamily as *Microwave Integrated Circuits* (MIC), unless otherwise specified.

Three things specifically characterize the MIC device. First is simplicity. As you will see in the circuits that follow, the MIC device usually has only input, output, ground, and power supply connections. Other wideband IC devices often have up to 16 pins, most of which must be either biased or capacitor-bypassed. The second feature of the MIC is the very wide frequency range (dc-GHz) of the devices. And the third is the constant input and output impedance over several octaves of frequency.

Although not universally the case, MICs tend to be unconditionally stable because of a combination of series and shunt negative feedback internal to the device. The input and output impedances of the typical MIC device are a close match to either 50 or 75 Ω, so it is possible to make a MIC amplifier without any impedance-matching schemes—a factor that makes it easier to broadband than if tuning was used. A typical MIC device generally produces a standing wave ratio (SWR) of less than 2:1 at all frequencies within the passband, provided that it is connected to the design system impedance (e.g., 50 Ω). The MIC is not usually regarded as a low-noise amplifier (LNA), but it can produce noise figures (NF) in the 3- to 8-dB range. Some MICs are LNAs, however, and the number available should increase in the near future.

Narrowband and passband amplifiers can be built using wideband MICs. A narrowband amplifier is a special case of a passband amplifier and is typically tuned to a single frequency. An example is the 70 MHz IF amplifier used in microwave receivers. Because of input and/or output tuning, such an amplifier will respond only to signals in the 70 MHz frequency band.

Very wideband amplifiers

Engineering wideband amplifiers, such as those used in MIC devices, seems simple but has traditionally caused a lot of difficulty for designers. Figure 22-29A shows the most fundamental form of MIC amplifier; it is a common-emitter NPN bipolar transistor amplifier. Because of the high-frequency operation of these devices, the amplifier in MICs are usually made of a material such as gallium arsenide (GaAs). In Fig. 22-29A, the emitter resistor (R_e) is unbypassed so introduces a small amount of negative feedback into the circuit. Resistor R_e forms series feedback for transistor Q_1. The parallel feedback in this circuit is provided by collector-base bias resistor R_f. Typical values for R_f are in the 500-Ω range and for R_e are in the 4- to 6-Ω range. In general, the designer tries to keep the ratio R_f/R_e high in order to obtain higher gain, higher output power compression points, and lower noise figures. The input and output impedances (R_o) are equal and are defined by the patented equation:

$$R_f + \sqrt{R_f R_e}, \qquad (22\text{-}18)$$

where
R_o = the output impedance in ohms
R_f = the shunt feedback resistance in ohms
R_e = the series feedback resistance in ohms.

A more common form of MIC amplifier circuit is shown in Fig. 22-29B. Although based on the Darlington amplifier circuit, this amplifier still has the same sort of series and shunt feedback resistors (R_e and R_f) as the previous circuit. All resistors except R_{bias} are internal to the MIC device. A Darlington amplifier, also called a *Darlington pair* or *superbeta transistor,* consists of a pair of bipolar transistors (Q_1 and Q_2) connected so that Q_1 is an emitter follower driving the base of Q_2, and with both collectors connected in parallel with each other. The Darlington connection permits both transistors to be treated as if they were a single transistor with higher-

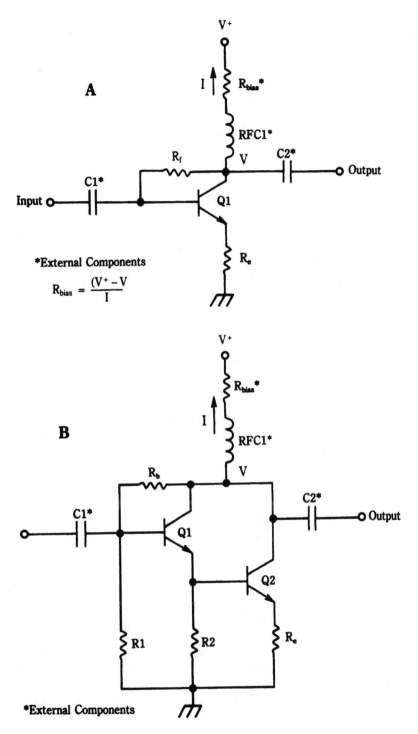

22-29 (A) Simple wideband RF amplifier, (B) Darlington cascade circuit, and (C) generic MIC amplifier.

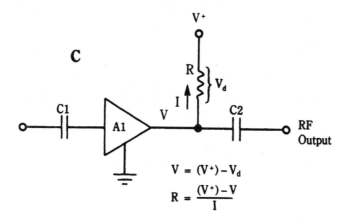

22-29 *Continued.*

than-normal input impedance and a beta gain (β) equal to the product of the individual beta gains. For the Darlington amplifier, therefore, the beta (β or H_{fe}) is:

$$\beta_o = (\beta_{Q_1})(\beta_{Q_2}), \tag{22-19}$$

where

β_o = the beta gain of the Q_1/Q_2 pair
β_{Q_1} = the beta gain of Q_1
β_{Q_2} = the beta gain of Q_2.

You should be able to see two facts. First, the beta gain is very high for a Darlington amplifier that is made with relatively modest transistors. Second, the beta of a Darlington amplifier made with identical transistors is the square of the common beta rating.

External components

Figures 22-29A and 22-29B show several components that are usually external to the MIC device. The bias resistor (R_{bias}) is sometimes internal, although on most MIC devices it is external. RF choke RFC_1 is in series with the bias resistor and is used to enhance operation at the higher frequencies; RFC_1 is considered optional by some MIC manufacturers. The reactance of the RF choke is in series with the bias resistance and increases with frequency according to the $2pFL$ rule. Thus, the transistor sees a higher impedance load at the upper end of the passband than at the lower end. Use of RFC_1 as a "peaking coil" thus helps overcome the adverse effect of stray circuit capacitance that ordinarily causes a similar decreasing frequency-dependent characteristic. A general rule is to make the combination of R_{bias} and X_{RFC_1} form an impedance of at least 500 W at the lowest frequency of operation. The gain of the amplifier might drop about 1 dB if RFC_1 is deleted. This effect is caused by the bias resistance shunting the output impedance of the amplifier. The capacitors are used to block dc potentials in the circuit. They prevent intracircuit potentials from affecting other circuits as well as potentials in other circuits from affecting MIC operation. More is said about these capacitors later, but for now understand that practical capacitors are not ideal; real capacitors are complex RLC circuits. Although the L and R components are negligible at low frequencies, they are substantial in the microwave region. In addition, the LC characteristic forms a self-resonance that can either "suck out" or enhance gain at specific frequencies. The result is an uneven frequency-response characteristic at best and spurious oscillations at worst.

General MMIC amplifier

Figure 22-29C shows a "generic" circuit representing MIC amplifiers in general. As you will see when you look at an actual product, this circuit is nearly complete. The MIC device usually has only input, output, ground, and power connections; some models don't have a separate dc power input. There is no dc biasing, no bypassing (except at the dc power line), and no seemingly "useless" pins on the package. MICs use either microstrip packages, like UHF/microwave small-signal transistor packages or small versions of the miniDIP or metallic IC packages. Some HMICs are packaged in larger transistorlike cases and others are packaged in special hybrid packages. The dc bias resistor (R_{bias}) connected to either the power supply terminal (if any) or the output terminal must be set to a value that limits the current to the advice and drops the supply voltage to a safe value. MIC devices typically require a low dc voltage (4 to 7 Vdc) and a maximum current of about 15 to 25 mA depending on the type. There might also be an optimum current of operation for a specific device. For example, one device advertises that it will operate over a range of 2 to 22 mA, but that the optimum design current is 15 mA. The value of resistor needed for R_{bias} is found from Ohm's law:

$$R_{bias} = \frac{(V_f) - V}{I_{bias}},\qquad (22\text{-}20)$$

where

R_{bias} = in ohms
$V+$ = dc power supply potential in volts
V = rated MIC device operating potential in volts
I_{bias} = operating current in amperes.

The construction of amplifiers based on MIC devices must follow microwave practices. This requirement means short, wide, low-inductance leads made of printed circuit foil and stripline construction. Interconnection conductors behave like transmission lines at microwave frequencies so they must be treated as such. In addition, capacitors should be capable of passing the frequencies involved, yet have as little inductance as possible. In some cases, the series inductance of common capacitors forms a resonance at some frequency within the passband of the MMIC device. These "resonant" circuits can sometimes be detuned by placing a small ferrite bead on the capacitor lead. Microwave "chip" capacitors are used for ordinary bypassing.

MIC technology is currently able to provide very low-cost microwave amplifiers with moderate gain and noise figure specifications and better performance is available at higher cost; the future holds promise of even greater advances. Manufacturers have extended MIC operation to 18 GHz and have dropped noise figures substantially. In addition, it is possible to build the entire front end of a microwave receiver into a single HMIC or MMIC, including the RF amplifier, mixer, and local oscillator stages.

Although MIC devices are available in a variety of package styles, those shown in Fig. 22-30 are typical of many. Because of the very high-frequency operation of these devices, MICs are packaged in stripline transistorlike cases. The low-inductance leads for these packages are essential in UHF and microwave applications.

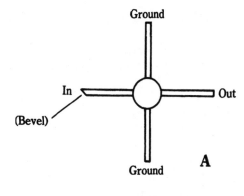

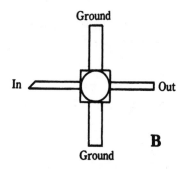

22-30
RF transistor and MIC packages.

Cascade MMIC amplifiers

MMIC devices can be connected in cascade (Fig. 22-31) to provide greater gain than is available from only a single device, although a few precautions must be observed. It must be recognized, for example, that MICs possess a substantial amount of gain from frequencies near dc to well into the microwave region. With all cascade amplifiers, you must prevent feedback from stage to stage. Two factors must be addressed. First, as always, is component layout. The output and input circuitry external to the MIC must be physically separated to prevent coupling feedback. Second, it is necessary to decouple the dc power supply lines that feed two or more stages. Signals carried on the dc power line can easily couple into one or more stages, resulting in unwanted feedback.

Figure 22-31 shows a method for decoupling the dc power line of a two-stage MIC amplifier. In lower-frequency cascade amplifiers, the $V+$ ends of resistors R_1 and R_2 would normally be joined together and connected to the dc power supply. Only a single capacitor would be needed at that junction to ensure adequate decoupling between stages. But as the operating frequency increases, the situation becomes more complex, in part caused by the nonideal nature of practical

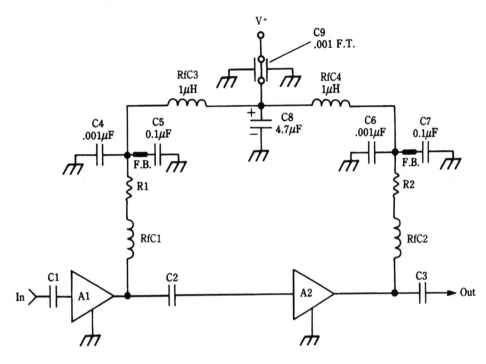

22-31 Cascade MIC amplifiers.

components. For example, in an audio amplifier the electrolytic capacitor used in the power supply ripple filter might provide sufficient decoupling. At RF frequencies, however, electrolytic capacitors are essentially useless as capacitors (they act more like resistors at those frequencies).

The decoupling system in Fig. 22-31 consists of RF chokes RFC_3 and RFC_4 and capacitors C_4 through C_9. The RF chokes help block high frequency ac signals from traveling along the power line. These chokes are selected for a high reactance at VHF through microwave frequencies, while having a low dc resistance. For example, a 1-μH RF choke might have only a few milliohms of dc resistance, but (by $2\pi FL$) has a reactance of more than 3000 Ω at 500 MHz. It is important that RFC_3 and RFC_4 be mounted so as to minimize mutual inductance because of the interaction among their respective magnetic fields.

Capacitors C_4 through C_9 are used for bypassing signals to ground. Notice that a wide range of values, and several types of capacitors, are used in this circuit. Each has its own purpose. Capacitor C_8, for example, is an electrolytic and is used to decouple very low-frequency signals (i.e., those up to several hundred kilohertz). Because C_8 is ineffective at higher frequencies, it is shunted by capacitor C_9, shown in Fig. 22-31 as a feedthrough capacitor. Such a capacitor is usually mounted on the shielded enclosure housing the amplifier. Capacitors C_5 and C_7 are used to bypass signals in the HF region. Because these capacitors are likely to exhibit substantial series inductance, they will form undesirable resonances within the amplifier passband. Ferrite beads are sometimes installed on each capacitor lead in order to detune capacitor self-resonances. Like C_9, capacitors C_4 and C_6 are used to decouple signals in the VHF-and-up region. These capacitors must be of microwave "chip" construction or they might prove ineffective above 200 MHz or so.

Gain in cascade RF amplifiers

In low-frequency amplifiers, you might reasonably expect the composite gain of a cascade amplifier to be the product of the individual stage gains:

$$G_{\text{total}} = G_1 + G_2 + G_3 + \cdots + G_n, \qquad (22\text{-}21)$$

where

G_1–G_n = the gains in decibels.

Although that reasoning is valid for low-frequency voltage amplifiers, it fails for RF amplifiers (especially in the microwave region), where input–output standing-wave ratio (SWR) becomes significant. In fact, the gain of any RF amplifier cannot be accurately measured if the SWR is greater than about 1.15:1.

There are several ways in which SWR can be greater than 1:1, and all of them involve an impedance mismatch. For example, the amplifier might have an input or output resistance other than the specified value. This situation can arise because of design errors or manufacturing tolerances. Another source of mismatch is the source and load impedances. If these impedances are not exactly the same as the amplifier input or output impedance, respectively, a mismatch will occur.

An impedance mismatch at either input or output of the single-stage amplifier will result in a gain mismatch loss (M.L.) of:

$$\text{M.L.} = -10\log\left(1 - \left(\frac{\text{SWR} - 1}{\text{SWR} + 1}\right)^2\right), \qquad (22\text{-}22)$$

where

M.L. = mismatch loss in decibels (dB)
SWR = standing wave ratio (dimensionless).

In a cascade amplifier we have the distinct possibility of an impedance mismatch, hence an SWR, at more than one point in the circuit. An example (Fig. 22-32) might be where neither the output impedance (R_o) of the driving amplifier (A_1) nor the input impedance (R_i) of the driven amplifier (A_2) are matched to the 50-Ω (Z_o) microstrip line that interconnects the two stages. Thus, R_o/Z_o or its inverse forms one SWR, although R_i/Z_o or its inverse forms the other. For a two-stage cascade amplifier the mismatch loss is:

$$\text{M.L.} = 20\log\left(1 \pm \left(\frac{\text{SWR}_1 - 1}{\text{SWR}_1 + 1}\right)\left(\frac{\text{SWR}_2 - 1}{\text{SWR}_2 + 1}\right)\right). \qquad (22\text{-}23)$$

The mismatch loss can very from a negative loss resulting in less system gain (G_b) to a "positive loss" (which is actually a gain in its own right), resulting in greater system gain (G_a). The reason for this apparent paradox is that it is possible for a mismatched impedance to be connected to its complex conjugate impedance.

Attenuators in amplifier circuits?

It is common practice to place attenuator pads in series with the input and output signal paths of microwave circuits to "swamp-out" impedance variations that adversely affect circuits which ordinarily require either impedance matching or a

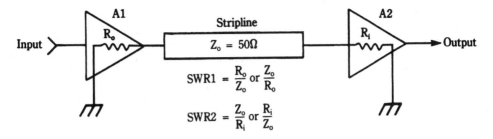

22-32 Strip-line impedance matching between to MICs.

constant impedance. Especially when dealing with devices such as LC filters (low-pass, high-pass, and bandpass), VHF/UHF amplifiers, matching networks, and MIC devices, it is useful to insert 1-dB, 2-dB, or 3-dB resistor attenuator pads in the input and output lines. The characteristics of many RF circuits depend on seeing the design impedance at input and output terminals. With the attenuator pad (see Fig. 22-33) in the line, source and load impedance changes don't affect the circuit nearly as much.

The attenuator tactic is also sometimes useful when confronted with seemingly unstable very wideband amplifiers. Insert a 1-dB pad in series with both input and output lines of the unstable amplifier. This tactic will cost about 2 dB of voltage gain, but it often cures instabilities that arise out of frequency-dependent load or source-impedance changes.

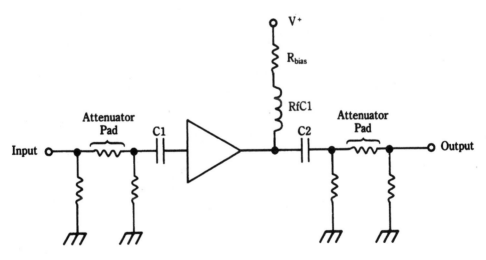

22-33 Input and output attenuators provide additional stability.

Noise figure in cascade amplifiers

It is common in microwave systems—especially in communications and radar receivers—to place a low-noise amplifier (LNA) at the input of the system. The amplifiers that follow the LNA need not be of LNA design, so they are less costly. The question is sometimes asked: "why not use a LNA in each stage?" The answer can be deduced from Friis' equation:

$$NF_{\text{total}} = N_{F_1} + \frac{N_{F_2} - 1}{G_1} + \frac{N_{F_3} - 1}{G_1 G_2} + \cdots + \frac{N_{F_n} - 1}{G_1 G_2 \ldots G_n}, \quad (22\text{-}24)$$

where

(Note: All quantities are dimensionless ratios rather than decibels)
NF_{total} = system noise figure
N_{F_1} = noise figure of stage-1
N_{F_2} = noise figure of stage-2
N_{F_n} = noise figure of stage-nth
G_1 = gain of stage-1
G_2 = gain of stage-2
G_n = gain of nth stage.

A lesson to be learned from this equation is that the noise figure of the first stage in the cascade chain dominates the noise figure of the combination of stages. Notice that the overall noise figure increased only a small amount when a second amplifier was used.

MAR-x series devices (a MIC example)

The Mini-Circuits Laboratories MAR-x series of MIC devices (Fig. 22-34A) offers gains from +13 dB to +20 dB and top-end frequency response of either 1 or 2 GHz depending on the type. The package used for the MAR-x device (Fig. 22-34B) is similar to the case used for modern UHF and microwave transistors. Pin 1 (RF input) is marked by a color dot and a bevel. The usual circuit for the MAR-x series devices is shown in Fig. 22-34A. The MAR-x device requires a voltage of +5 Vdc on the output terminal and must derive this potential from a dc supply of greater than +7 Vdc.

The RF choke (RFC$_1$) is called *optional* in the engineering literature on the MAR-x, but it is recommended for applications where a substantial portion of the total bandpass capability of the device is used. The choke tends to preemphasize the higher frequencies and thereby overcomes the deemphasis normally caused by circuit capacitance; in traditional video amplifier terminology that coil is called a "peaking coil" because of this action (i.e., it peaks up the higher frequencies).

It is necessary to select a resistor for the dc power supply connection. The MAR-x device wants to see +5 Vdc at a current not to exceed 20 mA. In addition, $V+$ must be greater than +7 V. Thus, you need to calculate a dropping resistor (R_d) of:

$$R_d = \frac{(V+) - 5\,(\text{Vdc})}{I}, \quad (22\text{-}25)$$

where

R_d is in ohms.
I is in amperes.

In an amplifier designed for +12-Vdc operation, you might select a trial bias current of 15 mA (i.e., 0.015 A). The resistor value calculated is 467 Ω (a 470-Ω, 5%, film resistor should be satisfactory). As recommended, a 1-dB attenuator pad is inserted in the input and output lines. The $V+$ is supplied to this chip through the output terminal.

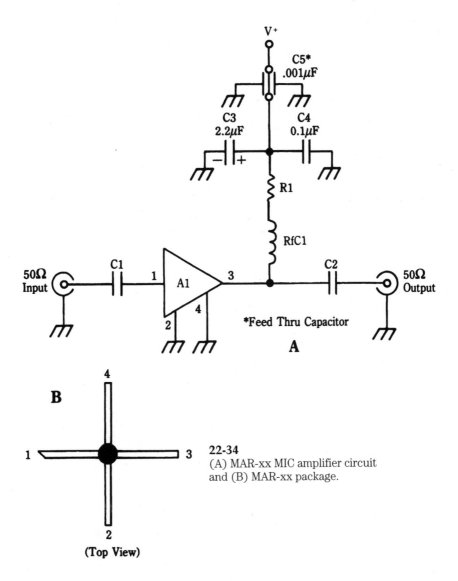

22-34
(A) MAR-xx MIC amplifier circuit
and (B) MAR-xx package.

Combining MMIC amplifiers in parallel

Figures 22-35 through 22-37 show several MIC applications involving parallel combinations of MIC devices. Perhaps the simplest of these is the configuration of Fig. 22-35. Although each input must have its own dc-blocking capacitor to protect the MIC device's internal bias network, the outputs of two or more MICs might be connected in parallel and share a common power-supply connection and output coupling capacitor. Several advantages are realized with the circuit of Fig. 22-35.

First, the power output increases even though total system gain (P_o/P_m) remains the same. As a consequence, however, drive power requirements also increase. The output power increases 3 dB when two MICs are connected in parallel and 6 dB when four are connected (as shown). The 1-dB output-power compression point

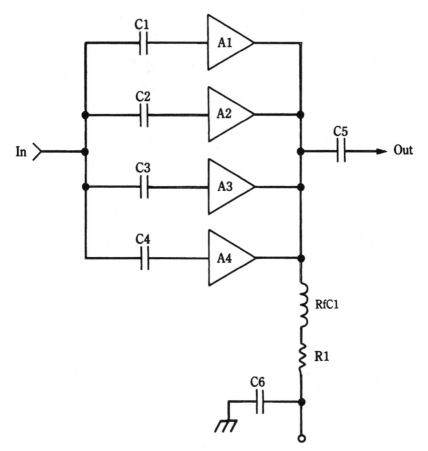

22-35 Parallel MAR-xx devices.

also increases in parallel amplifiers in the same manner: 3 dB for two amplifiers and 6 dB for four amplifiers in parallel.

The input impedance of a parallel combination of MIC devices reduces to R_i/N, where N is the number of MIC devices in parallel. In the circuit shown in Fig. 22-35, the input impedance would be $R_i/4 = 12.5\ \Omega$ if the MICs are designed for 50-Ω service. Because 50/12.5 represents a 4:1 SWR, some form of input impedance matching must be used. Such a matching network can be either broadbanded or frequency-specific as the need dictates. Figures 22-36 through 22-38 show methods for accomplishing impedance matching.

The method shown in Fig. 22-36 is used at operating frequencies up to about 100 MHz and is based on broadband ferrite toroidal RF transformers. These transformers dominate the frequency response of the system because they are less broadbanded than the usual MIC device. This type of circuit can be used as a gain block in microwave receiver IF amplifiers (which are frequently in the 70-MHz region) or in the exciter section of Master Oscillator Power Amplifier (MOPA) transmitters.

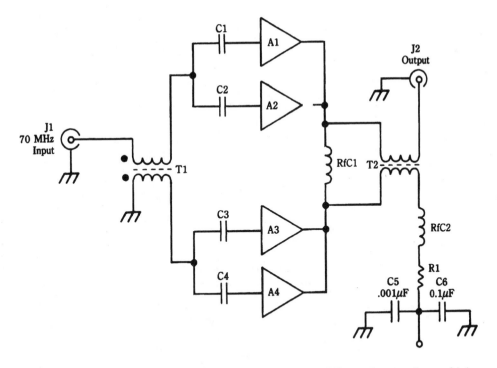

22-36 HF/low-band VHF amplifier using broadband toroidal RF transformers for combining and matching.

Another method is more useful in the UHF and microwave regions. Figure 22-37 shows several forms of the Wilkinson power divider circuit. A LC network version is shown here for comparison, although coaxial or stripline transmission line versions are used more often in microwave applications. The LC version (Fig. 22-37A) is used to frequencies of 150 MHz. This circuit is bidirectional so it can be used as either a power splitter or power divider. RF power applied to port C is divided equally between port A and port B. Alternatively, power applied to ports A/B are summed together and appear at port C. The component values are found from the following relationships:

$$R = 2Z_o \tag{22-26}$$

$$L = \frac{70.7}{2\pi F_o} \tag{22-27}$$

$$C = \frac{1}{141.4\pi F_o}, \tag{22-28}$$

where

    R is in ohms
    L is in henrys (H)
    C is in farads (F)
    F_o is in hertz (Hz).

Figure 22-37B shows a coaxial-cable version of the Wilkinson divider that can be used at frequencies up to 2 GHz. The lower frequency limit is set by practicality

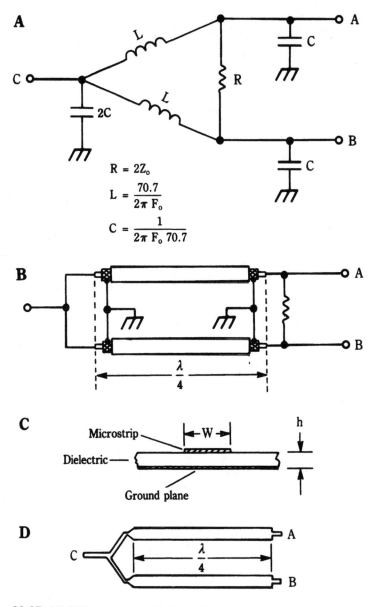

$$R = 2Z_0$$

$$L = \frac{70.7}{2\pi F_0}$$

$$C = \frac{1}{2\pi F_0\, 70.7}$$

22-37 (A) Wilkinson power divider, (B) coaxial cable version, (C) microstrip version (end view), and (D) plan view of strip-line version.

because the transmission-line segments become too long to be handled easily. The upper frequency limit is set by the practicality of handling very short lines and by the dielectric losses, which are frequency dependent. The transmission line segments are each a quarter wavelength; their length is found from:

$$L = \frac{2952\,V}{F}, \tag{22-29}$$

where
> L = the physical length of the line in inches
> F = the frequency in megahertz (MHz)
> V = the velocity factor of the transmission line (0 to 1)

An impedance transformation can take place across a quarter-wavelength transmission line if the line has a different impedance than the source or load impedances being matched. Such an impedance-matching system is often called a *Q-section.* The required characteristic impedance for the transmission line is found from:

$$(Z_o)' = \sqrt{Z_L Z_o}, \tag{22-30}$$

where
> $(Z_o)'$ = the characteristic impedance of the quarter-wavelength section
> Z_L = the load impedance
> Z_o = the system impedance (e.g., 50 Ω).

In the case of parallel MIC devices, the nominal impedance at port C of the Wilkinson divider is one-half of the reflected impedance of the two transmission lines. For example, if the two lines are each 50-Ω transmission lines, then the impedance at port C is 50/2 Ω = 25 Ω. Similarly, if the impedance of the load (i.e., the reflected impedance) is transformed to some other value, then port C sees the parallel combination of the two transformed impedances. In the case of a parallel MIC amplifier, you might have two devices with 50-Ω input impedance each. Placing these devices in parallel halves the impedance to 25 Ω, which forms a 2:1 SWR with a 50-Ω system impedance. But if the quarter-wavelength transmission line transforms the 50-Ω input impedance of each device to 100 Ω, then the port C impedance is 100/2 = 50 Ω . . . which is correct. At the upper end of the UHF spectrum, and in the microwave spectrum, it might be better to use a stripline transmission line instead of coaxial cable.

A stripline (see Fig. 22-37C) is formed on a printed circuit board. The board must be double-sided so that one side can be used as a ground plane and the stripline is etched into the other side. The length of the stripline depends on the frequency of operation; either half-wave or quarter-wave lines are usually used. The impedance of the stripline is a function of three factors: (1) stripline width (w), (2) height of the stripline above the groundplane (h), and (3) dielectric constant (ϵ) of the printed circuit material:

$$Z_o = 377 \frac{h}{w\sqrt{\epsilon}}, \tag{22-31}$$

where
> h = the height of the stripline above the groundplane
> w = the width of the stripline (h and w in same units)
> Z_o = the characteristic impedance in ohms.

The stripline transmission line is etched into the printed circuit board, as in Fig. 22-37D. Strip-line methods use the printed wiring board to form conductors, tuned circuits, and so on. In general, for microwave operation, the conductors must be very wide (relative to their simple dc and RF power carrying size requirements) and very short in order to reduce lead inductance. Certain elements, the actual strip-lines, are

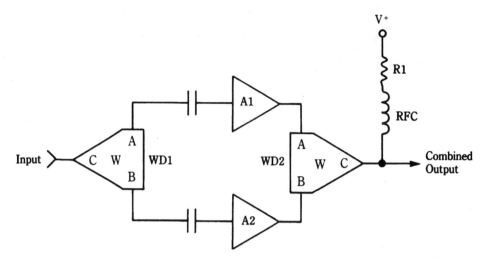

22-38 General circuit for a multi-MIC amplifier using Wilkinson dividers.

transmission-line segments and follow transmission-line rules. The printed circuit material must have a large permitivity and a low loss tangent. For frequencies up to about 3 GHz, it is permissible to use ordinary glass-epoxy double-sided board ($\epsilon = 5$), but for higher frequencies, a low-loss material, such as Rogers Duroid ($\epsilon = 2.17$), must be used.

When soldering connections in these amplifiers, it is important to use as little solder as possible and keep the soldered surface smooth and flat as possible. Otherwise, the surface wave on the strip-line will be interrupted—and operation suffers.

Figure 22-38 shows a general circuit for a multi-MIC amplifier based on one of the above discussed Wilkinson power dividers. Because the divider can be used as either splitter or combiner, the same type can be used as input (WD_1) and output (WD_2) terminations. In the case of the input circuit, port C is connected to the amplifier main input and ports A/B are connected to the individual inputs of the MIC devices. At the output circuit, another divider (WD_2) is used to combine power output from the two MIC amplifiers and direct it to a common output connection. Bias is supplied to the MIC amplifiers through a common dc path at port C of the Wilkinson dividers.

23
CHAPTER

LC RF filter circuits

Low-pass, high-pass, bandpass, and notch

Filters are frequency-selective circuits that pass some frequencies and reject others. Filters are available in several different flavors: low-pass, high-pass, bandpass, and notch. All of these filters are classified according to the frequencies that they pass (or reject). The breakpoint between the accept band and the reject band is usually taken to be the frequency at which the passband response falls off -3 dB. The four different types are characterized below.

Low-pass filters pass all frequencies below the cutoff frequency defined by the -3-dB point (Fig. 23-1A). These filters are useful for removing the harmonic content of signals or eliminating interfering signals above the cutoff frequency.

High-pass filters pass all frequencies above the -3-dB cutoff point (Fig. 23-1B). These filters are useful for eliminating interference from strong signals below the cutoff point. For example, a person using a shortwave receiver might wish to install a 1800-kHz high-pass filter to eliminate signals from strong AM broadcast stations.

Bandpass filters pass all frequencies between lower (F_L) and upper (F_H) -3-dB points, while rejecting those outside the F_H–F_L range. Bandpass filters are either wideband (low Q) or narrow band (high Q), as shown in Fig. 23-1C and 23-1D, respectively.

Q is the quality factor of the bandpass filter and is defined as the ratio of the center frequency to the bandwidth. For example, if a filter is centered on 10,000 kHz, and it has a 25-kHz bandwidth between the lower and upper -3-dB points, then the Q is 10,000/25 = 400.

Notch filters (band reject filters), pass all frequencies except those between lower and upper cutoff frequencies (Fig. 23-1E). Some notch filters are made broad, but many are very narrow. The latter are designed to suppress a single frequency. For example, a local FM broadcast signal will often interfere with television or two-way radio services. A notch filter tuned to that frequency will wipe it out. Similarly, on an AM

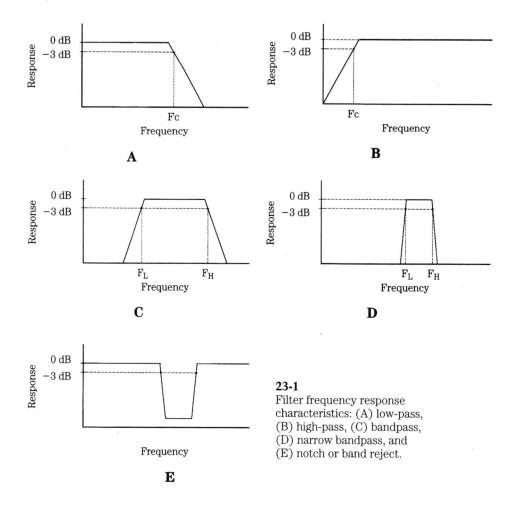

23-1
Filter frequency response
characteristics: (A) low-pass,
(B) high-pass, (C) bandpass,
(D) narrow bandpass, and
(E) notch or band reject.

band receiver, if it is being desensitized by a strong local station, reducing reception of all other frequencies on the AM band, a single-frequency notch filter can be used to suppress that one station's signal, leaving the other frequencies untouched.

Filter applications

Figure 23-2 shows several different types of filter applications. In Fig. 23-2A, the filter is placed between the antenna and the antenna input of a receiver. Its function is to remove unwanted signals before they reach the front end. Whether a high-pass, low-pass, or bandpass filter is used depends on the local situation (i.e., the frequencies that you wish to eliminate).

There are several good reasons for using a filter ahead of a receiver, even when the strong local signal is not within the normal passband of the receiver. The narrow passband selectivity of the receiver is set in the IF amplifiers and the RF ("front-end") selectivity is a lot broader. As a result, strong signals often reach the input

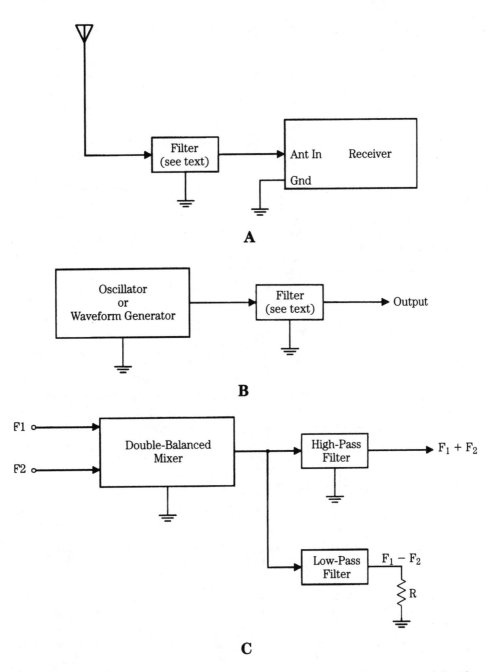

23-2 Filter applications: (A) ahead of a receiver antenna input to reduce unwanted signals, (B) at the output of oscillators and other waveform generators, (C) at the output of a double balanced mixer in a diplexer impedance-matching circuit.

stage, which will be either an RF amplifier or mixer. In either case, the unwanted signal can drive the input of the receiver into a nonlinear region of operating, creating either harmonics or intermodulation distortion products. These spurious signals not only are audible in the receiver (in some combinations), but also take up part of the receiver's dynamic range.

Figure 23-2B shows a filter placed at the output of an oscillator circuit. If the output signal is a pure sine wave, it will contain only one frequency (i.e., the desired oscillation frequency). But if there is even a little distortion, then harmonics will be present. These harmonics will adversely affect some circuits that the oscillator drives, so they must be eliminated.

The most usual filter used on oscillator outputs is the low-pass filter. The filter is designed with a cutoff frequency that is between the desired fundamental frequency and its second harmonic.

Another type of filter is a very high-Q bandpass filter. These filters are very narrow. The purpose of using such a filter is to reduce the phase noise on the oscillator signal. It takes a very narrow-band filter to do this trick, and both the oscillator frequency and the passband of the filter must be stable or the poly will be unsuccessful.

The circuit in Fig. 23-2C shows the use of two filters at the output of a double balanced mixer (DBM). The DBM receives two input frequencies, F_1 and F_2, and will output a spectrum of $mF_1 \pm nF_2$, where m and n are integers greater than 1 (on non-DBM mixers, m and n might be 1, indicating that F_1, F_2, or both may appear in the output). It is not sufficient to pick off the signal in the desired band and reject the signals in the unwanted band. Those unwanted signals will be reflected back into the mixer and might adversely affect its operation. The proper strategy is to use a circuit that passes the undesired signals to a dummy load that has a resistance equal to the mixer output impedance. This arrangement allows the dummy load to absorb the unwanted signals rather than permits them to reflect back into the mixer circuit.

Filter construction

A filter must be constructed properly if it is to work correctly. The two main factors to consider are layout and shielding.

Proper layout involves two main issues. First, keep the output and input ends of the circuit physically separated so as to prevent coupling of signal energy between them. Second, make sure that all inductors in the filter are shielded, arranged at right angles to each other, or wound on toroidal cores. All three of these approaches are used because they reduce coupling between inductors. Shielding keeps the magnetic field of the coil contained within the metallic shield. When cylindrical, solenoid-wound coils (i.e., those with a length greater than the diameter) are placed at right angles with respect to each other then magnetic coupling is minimized. A toroidal coil form is doughnut shaped and it naturally contains the magnetic field because of its geometry.

Shielding of the entire filter circuit is necessary to keep outside signal energy from getting into the filter and to ensure that the only signal reaching the load (i.e., the circuit being driven by the filter) has passed through the filter circuit. Figure 23-3 shows a sample filter (such as those covered in this chapter) enclosed within a shielded box. The signal input and output jacks. (J_1 and J_2) are coaxial connectors.

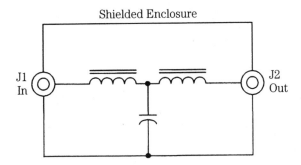

Shielded Enclosure

23-3
Filter success often depends on layout and construction. Shielded enclosures are a must!

The box should be either a die-cast aluminum box with a tight-fitting lid (some brands are pretty sloppy, so be careful), a sheet-metal aluminum box with overlapping lips for a RF seal, or a box specifically intended for RF work (these can be identified by the RF "finger" gaskets around the edge of the cover).

Filter design approach

The design of inductor-capacitor (LC) filters for radio frequency (RF) circuits is often presented in its most arcane mathematical form. Use of that design approach allows you to optimize designs. However, this form is also above the capabilities of many people who could put filters to good use. A different approach is needed, and that approach is the normalized 1-MHz model. In this design approach, a model is built by calculating the component values for 1 MHz. The component values can be scaled for any frequency by dividing the 1-MHz value by the desired frequency, expressed in megahertz.

A limitation on this approach is that it assumes that the input and output impedances are both equal to 50 Ω (i.e., the design impedance of the 1-MHz model filter). Because RF systems tend to have 50-Ω impedances; however, this restriction does not pose a big problem in most cases.

Low-pass filters

Low-pass LC filters are recognized by having the inductor (or inductors) in series with the signal path and the capacitors shunted across the signal path. The low-pass filter (LPF) attenuates all signal frequencies above its cutoff and passes all frequencies below the cutoff.

Figure 23-4 shows two basic single-section LPFs. The t-filter configuration is shown in Fig. 23-4A, and the pi-filter configuration is shown in Fig. 23-4B. The values of the capacitors and inductors are calculated using the equations:

$$L_{\mu H} = \frac{K_1}{F_{MHz}},$$ (23-1)

and

$$C_{pF} = \frac{K_1}{F_{MHz}}.$$ (23-2)

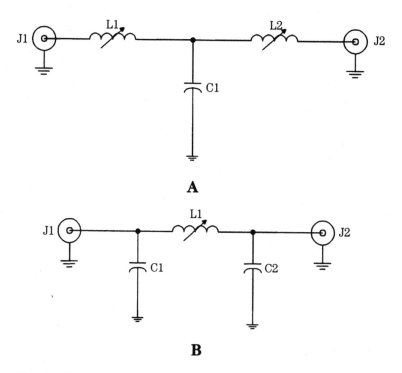

23-4 (A) Three-element, T-section low-pass filter and (B) three-element, pi-section low-pass filter.

These equations will also be used to calculate the values of the components in the other filters as well, although the numbering of the constants (K) will be different. The values of constants K_1 and K_2 are found from Table 23-1.

Table 23-1. Filter design constants for Fig. 23-4

	K_1	K_2
Fig. 23-4A (t filter)	7.94	6360
Fig. 23-4B (π filter)	15.88	3180

Example: Calculate the component values for both t-filter and pi-filter configurations for a low-pass filter with a cut-off frequency of 35 MHz.

t filter:

$$L_{\mu H} = K_1/F_{MHz}$$

$$L_{\mu H} = 7.94/35 = 0.27\,\mu H$$

$$C_{pF} = K_2/F_{MHz}$$

$$C_{pF} = 6360/35 = 182\,pF$$

π filter:
$$L_{\mu H} = k_1/F_{MHz}$$

$$L_{\mu H} = 15.88/35 = 0.454 \, \mu H$$
$$C_{pF} = K_2/F_{MHz}$$
$$C_{pF} = 3180/35 = 91 \, pF.$$

The inductors can be either homewound or purchased. Although it's possible to use adjustable inductances and capacitances in these filter circuits, it is not recommended that they be adjusted in the circuit. The adjustable components can allow one to obtain the specific values called for in the equations, but they should be preset to the value prior to being connected into the circuit. This job can be done using either a LC bridge or a digital LC meter, such as are found on some digital multimeters.

The inductors should be single components, wound, or selected and set for the specific inductance required. The capacitors, on the other hand, can be made up from several capacitors in series and parallel in order to obtain the correct value. Remember when doing this, however, that tolerances can make the whole thing less than useful. Most capacitors have tolerances of 5 or 10%, unless otherwise noted. It is best to use as close a value capacitor as possible, and that could involve hand-selecting capacitors, according to actual capacitance using a meter or bridge.

Each section of the filter provides a certain degree of attenuation, as indicated by the steepness of the roll-off slope beyond the cutoff frequency. Cascading sections will increase the roll-off slope, so they will also increase the attenuation obtained at any given frequency in the stopband. Figure 23-5 shows t-filter and pi-filter circuit in the two-section version. The calculation constants are shown in Table 23-2.

High-pass filters

A high-pass filter (HPF) attenuates signal frequencies below the −3-dB cutoff frequency and passes signal frequencies above the cutoff frequency.

Single-section high-pass filters are shown in Figs. 23-6A and 23-6B. These are the inverse of the LPF single-section filters in that the capacitors are in series with the signal path, and the inductors are across the signal path. The version in Fig. 23-6A is the t-filter configuration, and that in Fig. 23-6B is the pi-filter. The values for the components are found in Table 23-3.

Example: Find the component values for a single-section high-pass filter with a cutoff frequency of 40 MHz.
t filter (Fig. 23-6A):
$$L_1 = KL_1/F_{MHz}$$
$$L_1 = 3.97/40 = 0.1 \, \mu H$$

$$C_1 = KC_1/F_{\text{MHz}}$$
$$C_1 = 3180/40 = 79.5\,\text{pF}$$

π filter (Fig. 23-6B):

$$L_1 = KL_1/F_{\text{MHz}}$$
$$L_1 = 7.94/40 = 0.2\,\mu\text{H}$$
$$C_1 = KC_1/F_{\text{MHz}}$$
$$C_1 = 1590/40 = 40\,\text{pF}.$$

Two section high-pass filters are shown in Figs. 23-7A (t-filter) and 7B (pi-filter), and the 1-MHz model constants are shown in Table 23-4.

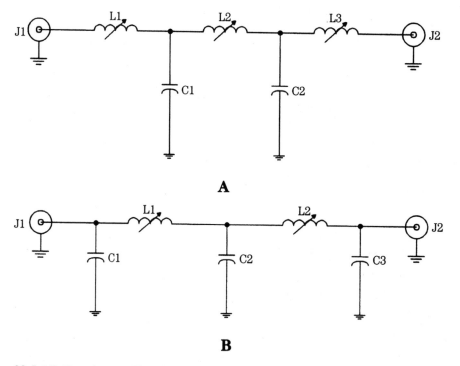

A

B

23-5 (A) Five-element, T-section low-pass filter and (B) five-element, pi-section low-pass filter.

Table 23-2. Filter design constants for Fig. 23-5

	K_{L_1}	K_{L_2}	K_{L_3}	K_{C_1}	K_{C_2}	K_{C_3}
Fig. 23-5A	9.126	15.72	9.126	4365	4365	—
Fig. 23-5B	10.91	10.91	—	3650	6287	3650

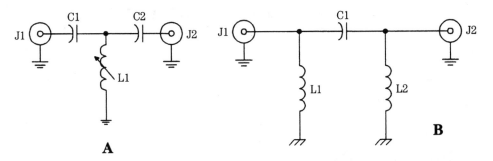

23-6 (A) Three-element, T-section high-pass filter and (B) three-element, pi-section high-pass filter.

Table 23-3. Filter design constants for Fig. 23-6

	K_{L_1}	K_{L_2}	K_{C_1}	K_{C_2}
Fig. 23-6A	3.97	—	3180	3180
Fig. 23-6B	7.94	7.94	1590	—

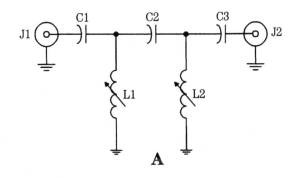

A

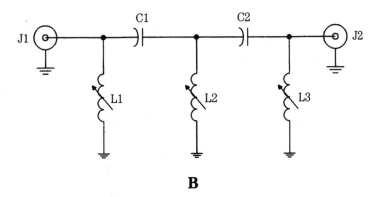

B

23-7 (A) Five-element, T-section high-pass filter and (B) five-element, pi-section high-pass filter.

Table 23-4. Filter design constants for Fig. 23-7

	K_{L_1}	K_{L_2}	K_{L_3}	K_{C_1}	K_{C_2}	K_{C_3}
Fig. 23-7A	5.8	5.8	—	2776	1662	2776
Fig. 23-7B	6.94	4.03	6.94	2321	2321	—

Bandpass filters

A bandpass filter passes only those frequencies between lower and upper -3-dB cutoff frequencies and attenuates all others. One simple way to obtain a bandpass frequency response characteristic is to cascade low-pass and high-pass sections (Fig. 23-8). The LPF is designed with a cutoff frequency equal to the high cutoff frequency in the passband, and the HPF is designed to cutoff at the low point in the desired passband. The values are as calculated for the individual sections.

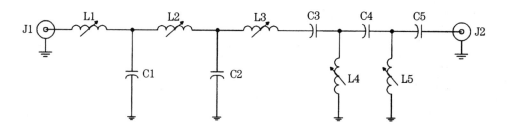

23-8 Bandpass filter made by cascading low-pass and high-pass sections.

Enhancing AM band reception

Figure 23-9 shows a bandpass filter for the AM broadcast band. This type of filter can be used between the antenna and antenna input of an AM broadcast band receiver in order to reduce any out-of-band signals that might interfere with reception or reduce the performance of the receiver. This filter consists of a 2000-kHz low-pass filter cascaded with a 500-kHz high-pass filter.

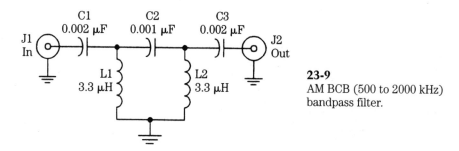

23-9
AM BCB (500 to 2000 kHz)
bandpass filter.

Removing the AM band

Many shortwave receivers are afflicted with a marginal front end—even though the rest of the receiver is pretty decent. For example, sensitivity and IF selectivity might be real good, but if the third-order intercept point, dynamic range, and intermodulation performance of the receiver aren't up to snuff then the receiver is likely to overload, desensitize, or hear signals that aren't really there.

One of the principal sources of front-end overload is local AM band signals. The broadcasters tend to be very local (i.e., down the block or less than a mile away for many listeners) and very strong. Although a 1000-W station might not impress anyone in the 25-m shortwave band, it can have a very large local field strength two blocks away. And the 50,000-W clear-channel "blowtorches" really raise havoc. The signal need not be in the same band, or near the same frequency, as you are trying to tune in to. The front ends of a lot of receivers, even in relatively expensive receivers, can be marginal enough that strong signals will get past the tuning or filtering to affect the operation of the input amplifier or mixer stage.

The only cure for the problem is to eliminate the offending signals. In other words, place a high-pass filter ahead of the receiver that will attenuate the AM signals, but pass all higher frequencies. All frequencies below 2000 kHz are severely attenuated in these filters, so the offending AM signals will be reduced considerably.

Figure 23-10A is a high-pass filter with a cutoff frequency of about 1800 kHz. This is a classic five-element, T-section high-pass filter. It is identified as a high-pass filter because the capacitors are in series with the signal line and the inductors are shunted across the signal line (a low-pass filter reverses these). This filter can be built using ordinary NPO disk ceramic capacitors for C_1, C_2, and C_3. Alternatively, silver-mica units can be used. To obtain the 2000-pF value in silver mica, it might be necessary to parallel two 1000-pF units. The 3.3-μH coils can be "store-bought" or wound on toroid core forms. Two popular forms for this particular application are the T-50-2 (red) and T-50-15 (red/white) cores. The A_L value for the T-50-2 (red) is 49 and for the T-50-15 (red/white) is 135. These values translate to 26 turns of wire for the T-50-2 (red) and 17 turns for the T-50-15 (red/white).

Another variant is shown in Fig. 23-10B. This circuit uses a 977-kHz mid-band suppression feature, consisting of two series-resonant circuits (C_4/L_1 and C_5/L_2), along with the normal configuration of series capacitors (C_1, C_2, and C_3). Still another variation is shown in Fig. 23-10C. This filter has a slightly lower -3-dB cutoff frequency (i.e., 1600 kHz, but it is capable of a higher degree of attenuation than the previous filters, especially close to the cutoff frequency; e.g., at the second harmonic of the cutoff frequency). This filter provides up to -40 dB of suppression. Notice, however, that maintaining the theoretical suppression depends in part on good layout and construction practices. Each section of the filter should be enclosed in its own shielded compartment of a well-made shielded enclosure; coaxial input and output connectors are used at J_1 and J_2.

Notice in Fig. 23-10C that the capacitance values are nonstandard. These capacitances can be approximated by standard capacitance values if two or more capacitors are placed in either series or parallel, as needed. Try various combinations until the closest fit is obtained. A much closer fit can then be achieved by measuring the actual capacitance of capacitors proposed for use in the circuit. All capacitors have a tolerance on the nominal value, so the actual value is different from the marked value by a small amount.

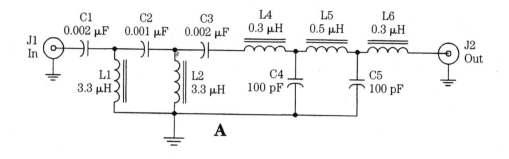

A

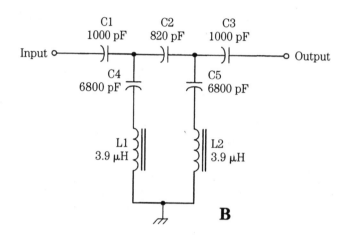

B

High-Pass Filter 1600 kHz (−40 dB)

Three-Section Shielded Enclosure

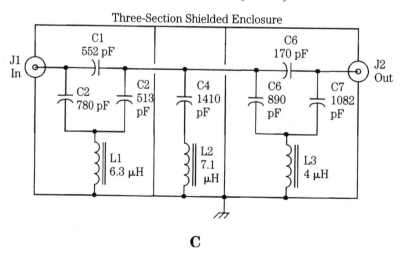

C

23-10 High-pass AM band suppression filters (A) using the five-element, T-section design, (B) with mid-band suppression, and (C) AM band "Thunder Lizard" filter.

Notch filters

A notch filter is a breed of bandstop filter with a high Q (i.e., a narrow bandwidth). They are designed to attenuate a single frequency or, actually, a narrow band of frequencies around a single center frequency.

Notch filters can be used to attenuate a single offending signal. For example, if you live near a FM broadcaster, you might notice that the signal is spread over a wide area of the band or that it appears at several spots on the band. Although this could be the result of a FM broadcaster's transmitter in need of repair, it's much more likely caused by spurious response of your receiver from front-end overload.

Two basic ways are used to notch out a single frequency, although they are often combined for improved effectiveness. A parallel-resonant trap (Fig. 23-11A) has an impedance that is very high at the resonant frequency and very low at frequencies removed from resonance. A parallel-resonant trap placed in series with the signal path will, therefore, severely attenuate frequencies at or near its resonant frequency and pass all others. A series-resonant trap (Fig. 23-11B) is just the opposite: it has a very low impedance at its resonant frequency and a high impedance at all other frequencies. A series-resonant trap shunted across should attenuate signals at or close to the resonant frequency.

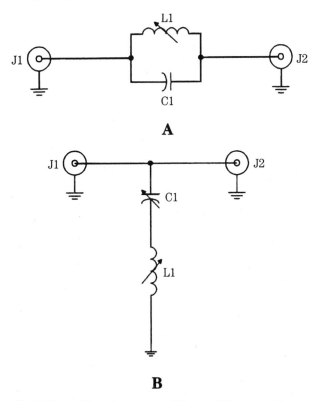

A

B

23-11 Notch filters/wavetraps: (A) parallel resonant in series with the signal path and (B) series resonant in parallel with the signal path.

The resonant frequency is found from:

$$F_{\text{MHz}} = \frac{159}{\sqrt{LC}}, \tag{23-3}$$

where

F is in megahertz (MHz)
L is in microhenrys (μH)
C is in picofarads (pF).

In the practical sense, this equation is not too terribly useful because the better approach is to select a trial value for either inductance or capacitance (depending on which is more convenient) and then calculate the required value for the other component. These equations are:

$$C_{\text{pF}} = \frac{25{,}280}{F_{\text{MHz}}^2 L_{\mu\text{H}}} \tag{23-4}$$

or

$$L_{\mu} = \frac{25{,}280}{F_{\text{MHz}}^2 C_{\text{pF}}}. \tag{23-5}$$

The parallel- and series-resonant circuits can be connected together to form a multisection notch filter (Fig. 23-12A). Indeed, this is the approach normally taken in commercial wavetrap circuits. The capacitances shunted across L_1 and L_3 are shown as two units in parallel to indicate that it will probably be necessary to make special, nonstandard values from two or more capacitors.

A multisection notch filter for the FM broadcast band is shown in Fig. 23-12B. This type of circuit is often sold in video stores as a "FM wave trap," and it is used to eliminate FM "herringbone" interference patterns on VHF television receivers. This filter consists of three filter sections in cascade: two parallel-resonant filters sandwiching a series-resonant filter between them.

One other form of notch filter can be used especially at VHF frequencies, although in theory it would work from dc to daylight. A transmission line will reflect its terminating impedance every half wavelength back down the line. Thus, if the end of a transmission line is short-circuited then the 0-Ω impedance will appear at points every half wavelength back toward the source from the short. This fact makes it possible to use a half-wavelength shorted transmission line stub as a notch filter; both coaxial cable and twin-lead can be used in this capacity (Fig. 23-13).

The length of the transmission line is the physical half-wavelength for the design frequency multiplied by the velocity factor of the transmission line. The velocity factor is a decimal number between 0 and 1 that indicates the speed of signal propagation in the line as a fraction of the speed of light:

$$L_{\text{meters}} = \frac{150\,V}{F_{\text{MHz}}}. \tag{23-6}$$

For example, suppose you need a half-wave shorted stub to remove a 100-MHz FM broadcast signal. The available coaxial cable is polyfoam cable with a velocity factor of 0.82; the length is:

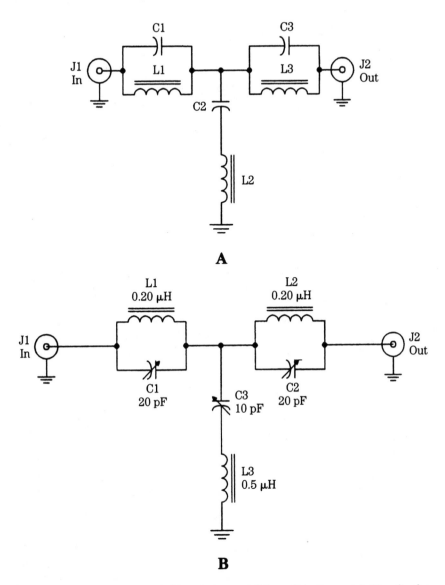

23-12 (A) Multisection notch filter design and (B) multisection notch filter for the FM BCB.

$$L_{\text{meters}} = \frac{(150)(0.82)}{100_{\text{MHz}}} \tag{23-7}$$

$$L_{\text{meters}} = \frac{123}{100}\,\text{meters} = 1.23\,\text{meters}. \tag{23-8}$$

In actual operation, the required length might be a little different, and the correct length must be found from experimentation. The reason is that the actual ve-

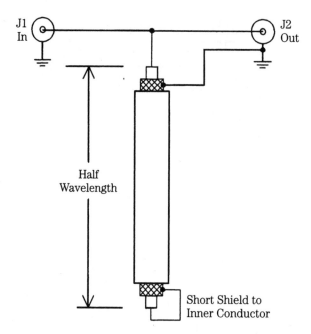

23-13 Use of a half-wavelength shorted stub across the signal line to attenuate a single-frequency signal.

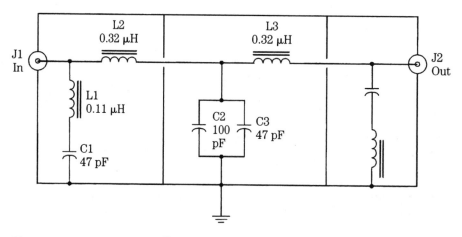

23-14 A 2- to 33-MHz bandpass filter.

locity factor might not be exactly as published and the length measurement will have a certain error. As a result, it's usually good practice to cut the trial length a few centimeters too long and then use trial and error to find the actual correct length. I've used a pin through the coax as a shorting bar to do the experimental trials, and it seemed to work well.

More on bandpass filters

A bandpass filter is designed to pass all frequencies that are above a lower cutoff frequency and all frequencies that are below an upper cutoff frequency. Figures 23-14 and 23-15 show two approaches to making a bandpass filter for the high-frequency (HF) shortwave bands. The circuit in Fig. 23-14 is designed for a frequency band of 2 to 33 MHz, so it will encompass the entire HF shortwave band while eliminating interference from LF and MW AM band broadcasters as well as VHF stations. Notice that each section is built inside its own shielded compartment. This is good layout practice for any multisection filter and it prevents interaction between the elements (especially coils).

The bandpass filter in Fig. 23-15 is not so elegant as the previous example but it serves the purpose. This approach to bandpass filter design cascades low-pass and high-pass sections. The input side of Fig. 23-15 is the AM BCB filter used in the last paragraph and the output section is a 33-MHz low-pass filter of similar design.

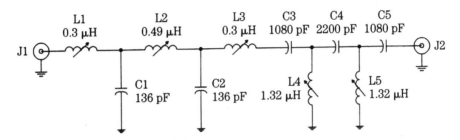

23-15 A 2- to 33-MHz bandpass filter based on cascading low-pass and high-pass sections.

Conclusion

Although RF filters involving LC elements are often presented as a difficult subject (which it is!), simplifications are possible using normalized 1-MHz models and a little arithmetic. The filters in this chapter are not necessarily optimized, but they will turn in more than merely adequate performance if you are diligent when constructing them.

Time-domain
reflectometry on a budget

Doping out transmission line difficulties can be a tedious and difficult chore—especially when the load end is not easily accessible. Although there are a number of different methods available, I want to discuss time-domain reflectometry. Although the official, bottled-in-bond, 100-proof time-domain reflectometer (TDR) is a terribly expensive piece of equipment, some TDR methods can be used by any amateur who has access to an oscilloscope. Will the results be as good as professional TDR equipment? In a word: no. Will the results be useful in troubleshooting a transmission line? In a single word: yes.

The basis of TDR

Time-domain reflectometry works on the principle that waves on a nonmatched transmission line reflect. The waveform seen at any given point along the line is the algebraic sum of the forward and reflected waveforms. In TDR measurements, we look at the waveform at the input end of the transmission line system.

Figure 24-1 shows the basic set-up for our impromptu TDR. A pulse generator, or other source of 1-MHz square waves, is applied simultaneously to the vertical input of an oscilloscope and the input end of the transmission line. This neat little trick is accomplished with an ordinary coaxial tee connector, either BNC or UHF, depending on your own situation. In my case, the pulse source and oscilloscope use BNC connectors, so I used a BNC tee connector. The PL-259 UHF connector on the end of the 200-foot length of RG-58 transmission line was converted using a SO-239 to BNC male adapter. In years gone by, that adapter was a $20 bench accessory, but today, the advent of VCRs and other such products has spawned a line of low-cost coax adapters that will darn near allow us to adapt anything to anything.

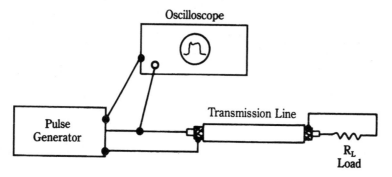

24-1 Set-up for time-domain reflectometry measurements.

The pulse source

Almost any source of square waves in the vicinity of 1 MHz can be used for the pulse generator. If you have one of those function generators that output pulses to 1 MHz or more, then use it. Be careful, however, if the output impedance is 600 Ω (as some are). For those, you might want to fashion a 600- to 50-Ω transformer or a simple resistor pad. Alternatively, you can build your own pulse source.

Many different forms of TTL oscillators can be used as a pulse source. Figure 24-2A shows the circuit for a pulse generator based on the Motorola MC-4024P device. This chip is not from the CMOS line, despite its 4000-series type number. It is a dual, TTL-compatible, voltage-controlled oscillator (VCO). It has a TTL part number 74424. The circuit is built inside of a shielded box. The operating frequency is given very approximately by the equation $F = 300/C$, where F is in megahertz (MHz) and C is in picofarads (pF). The actual operating frequency is not terribly critical, as long as it is somewhere near 1 MHz. For very short transmission lines, the operating frequency may have to be increased—experiment with it.

The output waveform is a square wave with a period of about 1.1 (μs) when $C_1 = 330$ pF. The half-cycle used in the experiment (see Fig. 24-2B) has a period of one-half that, or 550 (0.55 μs). If you want to clean up the rise and fall times of the output waveform then pass the output through either an inverter made from H-series TTL or a Schmitt trigger (7414). The inset to Fig. 24-2A shows an inverter made by strapping the inputs of a 74H00 NAND gate.

Figure 24-2B is our comparison for considering the waveforms to follow. Figure 24-2B was photographed from my oscilloscope with the coaxial cable disconnected. The photos to follow show what this pulse looks like when a reflected pulse hits it after returning down the transmission line.

Another alternate for our pulse generator is to use a crystal oscillator on 1 MHz, 2 MHz, and so on. These oscillators can then also serve as a marker generator for other purposes. The tuning calibration is a bear. I am contemplating a 20-MHz TTL crystal oscillator that will have TTL frequency dividers in cascade to obtain 10-MHz, 5-MHz, 2-MHz, 1-MHz, 500-kHz, 100-kHz, 50-kHz, and 10-kHz outputs. There is no reason why that marker generator can't be used as the pulse source in TDR measurements.

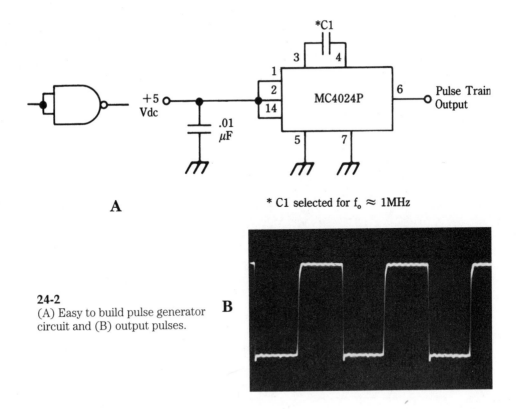

* C1 selected for $f_o \approx 1MHz$

24-2
(A) Easy to build pulse generator circuit and (B) output pulses.

Test set-up

The test set-up of Fig. 24-1 was built for these measurements. The load box at the antenna end of the transmission line was a little multi-Z dummy load that I use for various workshop applications and is shown in Fig. 24-3. I can select an external dummy load, a short circuit, and a total of 10 discrete impedances. When the external dummy load is disconnected, the load box sees an open transmission line in that switch position. Why have a load box? It is not part of the TDR, but helped to calibrate the system and generate the pictures that follow. You don't need one for your own TDR, even though I recommend load boxes as a workshop adjunct.

Why have 12 different loads, you say? Because I had a single-pole, 12-position rotary switch in my junk box, that's why. The impedance values shown were selected to (hopefully) represent a wide range of actual impedances typically encountered in amateur antennas.

Some actual measurements

Figure 24-4 shows two conditions that often show up on transmission lines: open and short-circuited. It is rarely of practical difference which occurs because both need to be corrected at the antenna end. However, should it be important, you can-

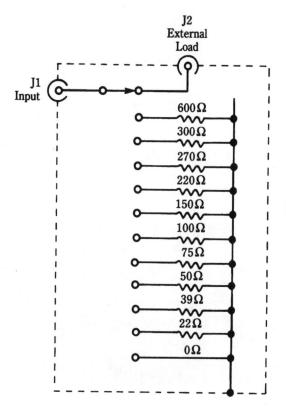

J2
External
Load

J1
Input

600Ω
300Ω
270Ω
220Ω
150Ω
100Ω
75Ω
50Ω
39Ω
22Ω
0Ω

24-3
Dummy load box.

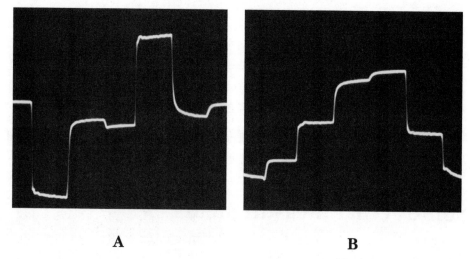

A

B

24-4 (A) Shorted line and (B) open line.

not differentiate which is which by the VSWR reading. In both cases, the entire incident wave is reflected. The only difference is the location of the nodes and antinodes, which are out of phase with each other. If the precise electrical length of the line is known, then you can infer from that whether the line is open or shorted. Otherwise, you make a TDR measurement.

Figure 24-4A shows the waveform when the load end of the line is short-circuited (in other words, $Z_L = 0$). The opposite case, an open-circuited line (impedance infinite) is shown in Fig. 24-4B. You can infer from the wave shape whether the line is open or shorted. In both cases, I experimented with several different lengths of RG-58/U coax up to 200 ft long and found similar waveforms in each case.

As you might suspect, impedances between zero and infinite are represented by various combinations of the two waveforms shown in Fig. 24-4. Figure 24-5 shows some of these waveforms. Figure 24-5A shows the supposedly matched 50-Ω case. If the system were perfect, the top edge of the pulse would be totally flat. The actual resistor used in the load box was a 51-Ω, 5-percenter. In addition, there is bound to be some reactance in the load and possible anomalies in the coax itself. I have per-

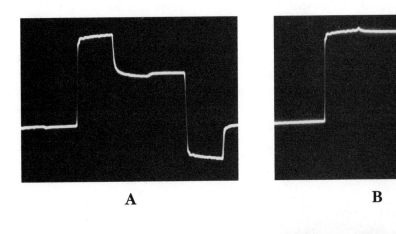

A

B

24-5
(A) Matched load and transmission line,
(B) $Z_L < Z_o$, and (C) $Z_L > Z_o$.

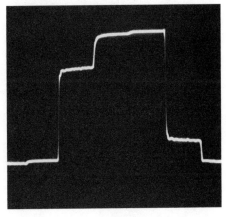

C

formed this little experiment before using a noninductive 200-Ω potentiometer as the load and was able to all but totally adjust out the lack of flatness. As you will see in a moment, the waveform in Fig. 24-5A represents a real load impedance greater than 50 Ω.

The waveform shown in Fig. 24-5B is for a 22-Ω load. This impedance is common on vertical antennas. Nominal impedance for one-quarter wavelength vertical is 37 Ω, but actual impedances are likely to be lower in practical installations (that's why Palomar Engineers sell their fantastic impedance transformer).

As the load impedance values slip above the surge impedance of the cable (50 Ω for RG-58/U), the waveform takes on a different shape. Instead of the reflected impedance causing a little rise in the flat edge, it causes a droop. Figure 24-5C, you can see that the amount of droop is related to how far above the surge impedance the load is.

25
CHAPTER

Solving frequency drift problems

Few circuit problems are less welcomed than frequency drift. This problem is characterized by the radio transmitter or receiver changing frequency not under the influence of the operator. Actually, two different problems are seemingly related: drift and shift. Drift is a gradual change of frequency, usually as a function of temperature. Shift, on the other hand, is an abrupt change of frequency. The causes of these two related phenomena are different, but they are often confused with each other.

There is also a difference between problems with new projects and drift on old equipment that once worked well. On a newly constructed project, or new equipment from less reputable sources, the problem might well be an inherent error in the design (some easily corrected, others not). On equipment that once worked well, however, you have a problem of a failed component.

Frequency shift problems

Resonant circuits in modern electronic equipment might be LC-tuned by a combination of inductance and capacitance or tuned by a piezoelectric resonator element (crystal). In either event, the cause of a sudden unwanted shift of operating frequency is usually some form of mechanical trauma somewhere in the circuit. In other words, some component is either broken or has an intermittent connection.

Figure 25-1 shows a partial circuit of an oscillator. The resonance of this circuit is determined by the combination of $C_1/C_2/C_3/L_1$. If any of these components fail, change value, or become disconnected then the resonant frequency of the circuit will shift. If coil L_1 fails, then the circuit will probably cease oscillating, so the fault becomes obvious. But what if one of the capacitors fails? In that case, the circuit might well continue oscillating, but at a different frequency than before.

The trimmer capacitor was selected for our example in Fig. 25-1 because those components seem especially at fault. After many years of experience, I can attest

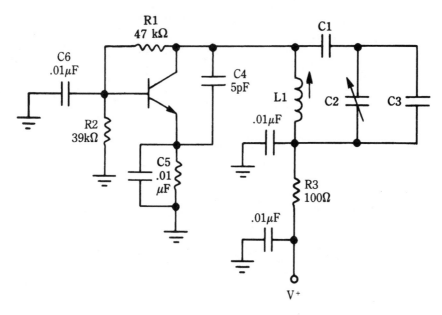

25-1 VHF VFO circuit.

that trimmers seem to have a high casualty rate. Perhaps the worst offenders are the half-turn type that use silver deposited on a pair of ceramic surfaces. The mica-compression types also fail, but seem less prone than the other types. The mechanism of failure seems to be looseness in the adjustment screw. Tap the capacitor gently with an insulated probe and note whether the shift occurs.

Don't make the mistake of assuming that the other forms are fault-free; far from it. Every form of fixed capacitor has at least a small failure rate, usually caused by disconnected leads inside the body of the capacitor.

Also, don't overlook the possibility that the problem is caused by the solder connections on the capacitor—especially where the capacitor is mounted on a printed wiring board that is easy to flex. Although some bad joints escape the factory QA inspector and subsequently last for years in service before failing, others die an early death because of trauma or flexing of the board. In addition to solder joint breaks, it is also possible that the printed wiring track is cracked. In both cases, a little solder and a hot iron will solve the problem.

One of my professional colleagues is a semiactive amateur operator with a receiver–transmitter pair that was once top of the line in amateur circles. He noticed that the formerly very good receiver dial calibration was now about 25 kHz off. On inspecting the LO VFO circuit, he found that it contained a number of small fixed capacitors in addition to the frequency setting trimmer and the main tuning variable capacitor. He made a few quick calculations of resonance to see which, if any, would result in about 25-kHz shift if it were open. He quickly identified a 27-pF unit that was part of the temperature-compensation circuit. A new 27-pF N1500 unit solved the problem. When you find a disk ceramic capacitor with a specified temperature coefficient, please don't make the mistake of thinking a low-temperature coefficient type is a better replacement. Also, don't use another disk with a different temperature coefficient.

On older receivers, be sure to examine the main tuning capacitor for problems. I once had a terrible problem with a piece of equipment that exhibited both shift and drift in magnificent proportions. Although employed in a communications shop at the time, I did not get around to fixing the receiver for several months ("cobbler's kid's shoes" syndrome). When I looked into the problem it turned out to be crud under the rotor grounding spring on the main tuning capacitor. That capacitor rotor is normally grounded to the chassis through its own frame but because the rotor must move, a brass or steel grounding leaf spring or spider clip is usually placed around the rotor at the front bearing. The spring or clip grounds the rotor to the frame. In this case, a lot of dirty grease built up under the spring. Rather than cause operation to cease, however, the result was massive frequency shifts, drift when it wasn't shifting, and a sad tendency toward microphonics.

Another problem on older receivers is poor grounding. I have seen frequency-shift problems on many car radios, two-way radios, and other RF equipment caused by poor grounding or cracked ground tracks on the printed circuit board. In one infamous FM model, the FM front end was grounded at only two points on the printed circuit. If that wasn't bad enough, the ground ran around the edge of the card and was stressed at one point, so cracks tended to develop. Although the circuit would not cease oscillating, the several extra inches of ground line on a flexing board caused operating frequency shifts. I suspect that the cause of the frequency shift was the added inductance of the printed circuit ground path. At VHF frequencies, the effect of distributed inductance or capacitance is more profound than at low frequencies. Again, judicious use of a soldering iron not only repaired the break but also added strength to the weak point.

Drift problems

The other frequency changing problem is old-fashioned drift. Unfortunately, most electronic components exhibit some temperature sensitivity. This sensitivity is usually measured in terms of a temperature coefficient, which specifies a certain shift of value in parts per million (ppm) per degree Celsius of temperature shift. The temperature coefficient (T_c) can be either positive or negative. A positive temperature coefficient (PT_c) indicates the value increases with increases in temperature. A negative temperature coefficient (NT_c) indicates that the value tends to decrease with increases in temperature.

Most inductors used in oscillator circuits have a PTC problem. The value of inductance (microhenrys) is determined by the coil dimensions and the number of turns of wire. Although the turn-count remains constant, the diameter, length, and the size of the wire used in the inductor is a function of temperature. Temperature coefficient of inductors can be minimized by design. If the inductor uses Litz wire (or some other low-T_c wire) and a low-T_c coil form then the temperature coefficient is reduced tremendously. Old-fashioned cardboard forms are terrible sources of drift. According to the old standard wisdom, the best forms are ceramic, but that is not true any longer. There are fiberglass and other synthetic materials that now provide better T_c characteristics.

Capacitors are also sources of T_c problems. Ceramic capacitors are available in either PT_c, NT_c, or no-T_c versions. Markings of "Nxxx" and "Pxxx" indicate the direction and amount of temperature coefficient. For example, an N750 is an NT_c device with a T_c of 750 ppm. The capacitor marked NPO has a low T_c but not precisely zero—as some people find out to their chagrin when subbing really low-T_c caps with NPO ceramic units. In general, silver mica and polystyrene capacitors are the lowest-T_c units available.

So, why not make all capacitors with low or nearly zero T_c? The reason is that they are sometimes used in temperature compensation circuits. Figure 25-2 shows two forms of oscillator: one LC-tuned and the other crystal-tuned. Both of these oscillators seem to have extra capacitors in the circuit. For example, C_2 in Fig. 25-2A and C_1/C_2 in Fig. 25-2B. In these cases, the extra capacitors are for temperature compensation. They are typically small-value compared with other circuit capacitors, but have pre-calculated T_cS that will cause the capacitance to change a predictable amount with changes in temperature. Both circuits, even the crystal oscillator, will change frequency with changes in circuit capacitance, so the T_c of the compensation capacitors forces the frequency to change in a predictable manner. The idea is to counter the PT_c of the inductor with an equal and opposite-direction T_c of the capacitors.

Unfortunately, even some manufacturers don't understand drift. Several years ago, I wrote a column for *World Radio*. In one column, I passed along a request for information sent in by an amateur working in the jungles of Central America. He had a low-cost transceiver that made a brief splash on the amateur market. His problem was that it drifted badly and in fact had always drifted (even new). My request was answered by an engineer at Stoner Communications who passed along the information that the rig had been designed by a consulting design engineer whose reputation is spotless. His name is well known to technically oriented amateurs, but I suspect he prefers privacy.

Contacting the designer, I found that the original prototypes and first production units had a drift spec of 100 Hz in the first 15 min and 50 Hz/hr thereafter; not terrific, but good for a cheap rig. Furthermore, those early rigs actually met the specification. So, what happened? The designer told me that the original design used Litz wire in the VFO inductor and a special low-T_c fiberglass form. The VFO also used DM-25 silver mica capacitors, except for a couple of ceramics used for temperature compensation. The inexperienced manufacturer used a "kid technician" to redesign the rig to make it cheaper to produce. The technician replaced the Litz wire with enameled wire, the coil form with an off-the-shelf ceramic type, and the DM-25 SM capacitors with NPO disk ceramics. The result was a disaster.

VHF problems

Figure 25-3A shows the block diagram of a FM receiver. The LO (example in Fig. 25-3B) is kept on channel by a dc control voltage from the automatic frequency control (AFC) output of the FM detector stage. The actual mechanism of frequency control is the variable capacitance (varactor) diode in the oscillator circuit (D_1 in Fig. 25-3B). This type of diode is used in both LC tuned circuits, as used in broadcast

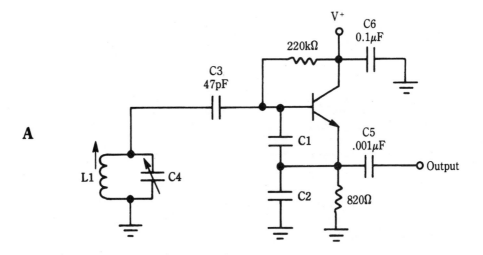

L1, C4: Determined by F$_0$

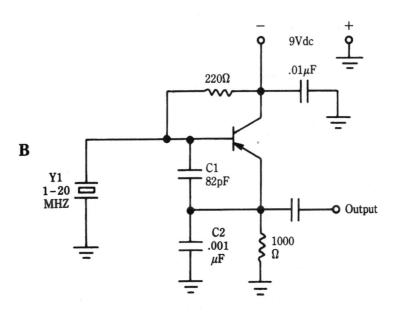

25-2 (A) Colpitts oscillator and (B) crystal Colpitts oscillator.

receivers, as well as crystal-controlled units used in communications equipment. If the varactor becomes defective then it will cause a frequency shift. Alternatively, if the diode is intermittent, then the circuit will jump on and off channel as the diode opens and closes the circuit. I have, however, seen quite a number of sets over the years in which a newly developed (as opposed to inherent) drift problem was caused by a defective diode. For some reason, the diode capacitance was a function of temperature. Although not a scientific experiment, I once measured about five diodes

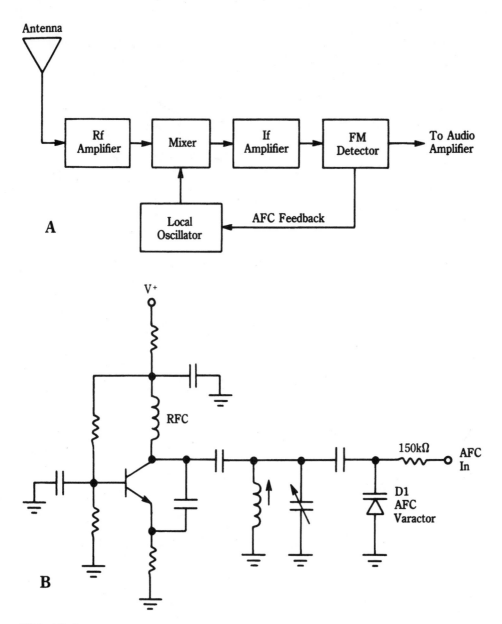

25-3 (A) Superheterodyne FM radio receiver and (B) local oscillator with AFC varactor diode.

that show this phenomenon and found that all had excessive leakage resistance in the reverse bias direction.

Don't overlook the dc power supply as a potential source of drift problems. Oscillators typically require fixed, well-regulated dc operating potentials for best stability. That dictum applies equally well to both LC-tuned and crystal oscillators. If the dc voltage or bias voltages change then expect an oscillator frequency shift.

The problem with dc voltage is especially acute in mobile equipment, whether communications or broadcast equipment. In the early 1960s, I worked in a car radio shop that dealt with Blaupunkt receivers (the German-made set in Porsche cars). One customer came in with an odd problem: the radio changed frequency with changes of engine speed. That set, which used vacuum tubes, had a germanium diode across the resonant circuit to limit oscillation amplitude. A gas-regulator tube supposedly kept the potentials applied to the circuit, but it was defective. Because the car power supply varies markedly with engine speed, the now unregulated power-supply voltage applied to the oscillator also varied. One lesson: although I frequently chuckled at a customer's diagnosis, I never again doubted the customer's description of the symptom ("... changes station when I pull away from a light ...").

Modern mobile electronic equipment is probably more prone than some of the older equipment to this type of fault. In communications equipment and FM broadcast receivers, the oscillator voltage is typically regulated. Figure 25-4A shows the zener diode regulator typically used in many car radio circuits, and Figure 25-4B shows the three-terminal IC voltage regulator that keeps the oscillator voltage to 10 V in a certain mobile transmitter master oscillator. If any of these components faults, then shift or drift will result. The nominal 12-V vehicle power supply is actually quite variable. With the engine off, my own car measured 11.8-Vdc on one meter and 12.05-Vdc on my DMM. When the engine is started, however, the voltage varies from 12.3-Vdc at idle to 14.5-Vdc as the engine speed is increased. To an oscillator in electronic equipment, that range might be intolerable.

Crystal oscillators are not immune to drift problems, even when they previously worked well. In addition to the problems with dc power supplies, it is also possible

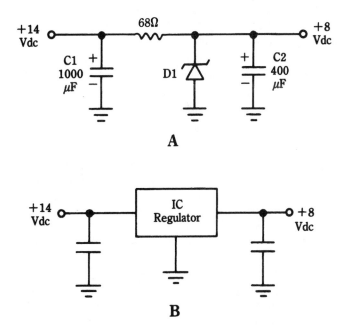

25-4 Voltage regulator circuits: (A) zener diode and (B) IC regulator.

for the crystal to become defective. I have seen several sets over the years in which a terrible drift problem resulted from defective crystal elements. Replacing the crystal solved the problem.

Problems with older equipment

Certain older amateur equipment exhibits a problem in frequency shift that is caused by the use of ferrite cores in the VFO and IF coils. Ferrite cores age with time and heat and as a result have a different permeability than when the equipment was first calibrated. I have seen that problem even on the legendary Collins permeability-tuned oscillator (PTO) used in their amateur and commercial communications equipment. In some cases, a simple realignment will suffice to bring the coil back to the correct resonant frequency. In other cases, replacement of the coil, or at least the ferrite core, is required.

Several years ago I had the opportunity to help a novice on the air with a 20-year-old Heath DX-60B that he bought for a song at the Gaithersburg Hamfest. He complained that keying was erratic: sometimes it keyed, other times it did not. It rapidly became apparent by looking at the grid and plate meter readings that the oscillator was not running all the time. It turned out that the DX-60B crystal oscillator circuit uses a tuned plate circuit. The ferrite core of the coil had dried out over the years and mistuned the oscillator. Readjusting the coil made the oscillator run properly.

Heat problems

Because the root of many drift problems is the temperature coefficient of capacitors and inductors, it seems obvious that temperature needs to be controlled in radio equipment. In the past, several otherwise well-regarded pieces of equipment suffered drift because of the tremendous heat inside the cabinet. Ventilation and a blower might help in some cases. In other cases, using a little insulation in and around the offending oscillator is also helpful. In certain crystal oscillator circuits, you can make progress by designing in a crystal oven to keep the crystal temperature constant.

Equipment modifications

Amateurs have a long tradition of modifying commercial equipment. Although many mods are ill-advised, some are certainly worthwhile and well-engineered. The process is a lot less dangerous, incidentally, if you make good notes so that the rig can be restored to original condition if the mod doesn't work out as expected. For most equipment, the first place to start is ensuring the power-supply voltage to the oscillator is stable. Furthermore, make certain that the printed circuit, its mounting, and the individual components are solidly anchored. Finally, be sure the circuit is not overheating.

If those methods fail to solve the problem, then and only then should you dive into the circuit to attempt temperature compensation. If you have any insight on procedures, techniques, and so forth then please communicate them to me so that they can be shared with others.

Frequency shift and drift problems are not easily found in many cases. Understanding the causes and potential solutions of these problems goes a long way toward finding the fault in any particular case.

26
CHAPTER

The Smith chart

The mathematics of transmission lines, and certain other devices, becomes cumbersome at times, especially when dealing with complex impedances and "nonstandard" situations. In 1939, Philip H. Smith published a graphical device for solving these problems, followed in 1945 by an improved version of the chart. That graphic aid, somewhat modified over time, is still in constant use in microwave electronics and other fields where complex impedances and transmission line problems are found. The Smith chart is indeed a powerful tool for the RF designer.

Smith chart components

The modern Smith chart is shown in Fig. 26-1 and consists of a series of overlapping orthogonal circles (i.e., circles that intersect each other at right angles). This chapter will dissect the Smith chart so that the origin and use of these circles is apparent. The set of orthogonal circles makes up the basic structure of the Smith chart.

The normalized impedance line

A baseline is highlighted in Fig. 26-2 and it bisects the Smith chart outer circle. This line is called the *pure resistance line,* and it forms the reference for measurements made on the chart. Recall that a complex impedance contains both resistance and reactance and is expressed in the mathematical form:

$$Z = R \pm jX, \tag{26-1}$$

where
Z = the complex impedance
R = the resistive component of the impedance
X = the reactive component of the impedance

The pure resistance line represents the situation where $X = 0$ and the impedance is therefore equal to the resistive component only. In order to make the Smith chart universal, the impedances along the pure resistance line are normalized with reference to system impedance (e.g., Z_o in transmission lines); for most microwave RF systems the system impedance is standardized at 50 Ω. To normalize the actual

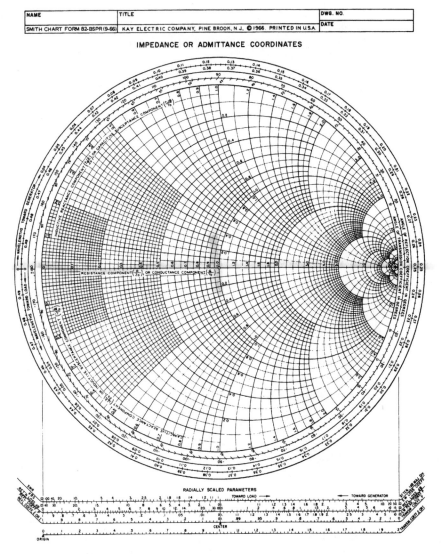

26-1 The Smith chart.

impedance, divide it by the system impedance. For example, if the load impedance of a transmission line is Z_L and the characteristic impedance of the line is Z_o then $Z = Z_L/Z_o$. In other words:

$$Z = \frac{R \pm jX}{Z_o}. \tag{26-2}$$

The pure resistance line is structured such that the system standard impedance is in the center of the chart and has a normalized value of 1.0 (see point "A" in Fig. 26-2). This value derives from $Z_o/Z_o = 1.0$.

To the left of the 1.0 point are decimal fraction values used to denote impedances less than the system impedance. For example, in a 50-Ω transmission-line system

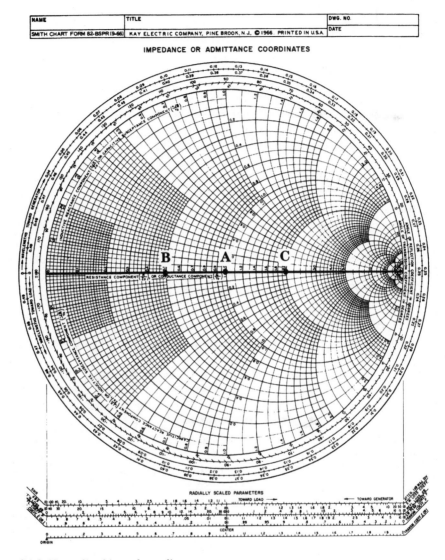

NAME	TITLE	DWG. NO.	
SMITH CHART FORM 82-BSPR (9-66)	KAY ELECTRIC COMPANY, PINE BROOK, N.J. © 1966. PRINTED IN U.S.A.	DATE	

IMPEDANCE OR ADMITTANCE COORDINATES

26-2 Normalized impedance line.

with a 25-Ω load impedance, the normalized value of impedance is 25 Ω/50 Ω or 0.50 ("B" in Fig. 26-2). Similarly, points to the right of 1.0 are greater than 1 and denote impedances that are higher than the system impedance. For example, in a 50-Ω system connected to a 100-Ω resistive load, the normalized impedance is 100 Ω/50 Ω, or 2.0: this value is shown as point "C" in Fig. 26-2. By using normalized impedances, you can use the Smith chart for almost any practical combination of system and load and/or source impedances, whether resistive, reactive, or complex.

Reconversion of the normalized impedance to actual impedance values is done by multiplying the normalized impedance by the system impedance. For example, if the resistive component of a normalized impedance is 0.45 then the actual impedance is:

$$Z = (Z_{\text{normal}})(Z_{\text{o}}) \tag{26-3}$$

$$Z = (0.45)(50\,\Omega) \tag{26-4}$$

$$Z = 22.5\,\Omega. \tag{26-5}$$

The constant resistance circles

The isoresistance circles, also called the *constant resistance circles,* represent points of equal resistance. Several of these circles are shown highlighted in Fig. 26-3. These circles are all tangent to the point at the righthand extreme of the pure resistance line and are bisected by that line. When you construct complex impedances (for which X = nonzero) on the Smith chart, the points on these circles will all have the same resistive component. Circle "A," for example, passes through the center of the chart, so it has a normalized constant resistance of 1.0. Notice that impedances that are pure resistances (i.e., $Z = R + j0$) will fall at the intersection of a constant resistance circle and the pure resistance line and complex impedances (i.e., X not equal to zero) will appear at any other points on the circle. In Fig. 26-2, circle "A" passes through the center of the chart so it represents all points on the chart with a normalized resistance of 1.0. This particular circle is sometimes called the *unity resistance circle.*

The constant reactance circles

Constant reactance circles are highlighted in Fig. 26-4. The circles (or circle segments) above the pure resistance line (Fig. 26-4A) represent the inductive reactance $(+X)$ and the circles (or segments) below the pure resistance line (Fig. 26-4B) represent capacitive reactance $(-X)$. In both cases, circle "A" represents a normalized reactance of 0.80. One of the outer circles (i.e., circle "A" in Fig. 26-4C) is called the *pure reactance circle.*

Points along circle "A" represent reactance only; in other words, an impedance of $Z = 0 \pm jX$ $(R = 0)$. Figure 26-4D shows how to plot impedance and admittance on the Smith chart. Consider an example in which system impedance Z_{o} is 50 Ω and the load impedance is $Z_{\text{L}} = 95 + j55\,\Omega$. This load impedance is normalized to:

$$Z = \frac{Z_{\text{L}}}{Z_{\text{o}}} \tag{26-6}$$

$$Z = \frac{95 + j55\,\Omega}{50\,\Omega} \tag{26-7}$$

$$Z = 1.9 + j1.1. \tag{26-8}$$

An impedance radius is constructed by drawing a line from the point represented by the normalized load impedance. $1.9 + j1.1$, to the point represented by the normalized system impedance (1.0) in the center of the chart. A circle is constructed from this radius and is called the VSWR circle.

Admittance is the reciprocal of impedance, so it is found from:

$$Y = \frac{1}{Z}. \tag{26-9}$$

Because impedances in transmission lines are rarely pure resistive, but rather contain a reactive component also, impedances are expressed using complex notation:

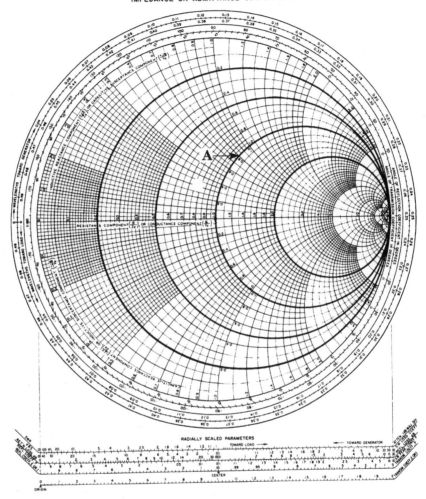

IMPEDANCE OR ADMITTANCE COORDINATES

26-3 Constant resistance circles.

$$Z = R \pm jX, \tag{26-10}$$

where

Z = the complex impedance
R = the resistive component
X = the reactive component.

To find the complex admittance, take the reciprocal of the complex impedance by multiplying the simple reciprocal by the complex conjugate of the impedance. For example, when the normalized impedance is $1.9 + j1.1$, the normalized admittance will be:

$$Y = \frac{1}{Z} \tag{26-11}$$

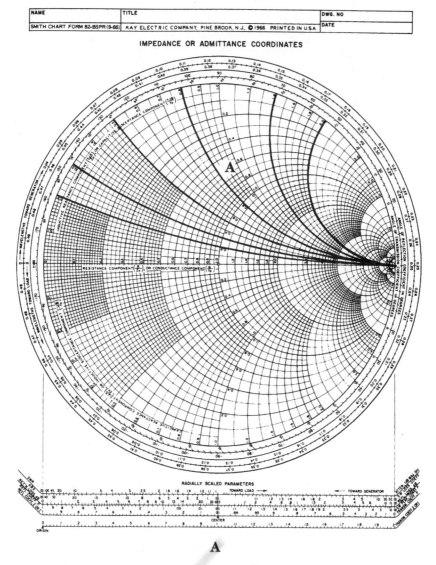

IMPEDANCE OR ADMITTANCE COORDINATES

RADIALLY SCALED PARAMETERS

A

26-4 (A) Constant inductive reactance lines, (B) constant capacitive reactance lines, (C) angle of transmission coefficient circle, and (D) VSWR circles.

$$Y = \frac{1}{1.9 + j1.1} \times \frac{1.9 - j1.1}{1.9 - j1.1} \tag{26-12}$$

$$Y = \frac{1.9 - j1.1}{3.6 + 1.2} \tag{26-13}$$

$$Y = \frac{1.9 - j1.1}{4.8} = 0.39 - j0.23. \tag{26-14}$$

One of the delights of the Smith chart is that this calculation is reduced to a quick graphical interpretation! Simply extend the impedance radius through the 1.0

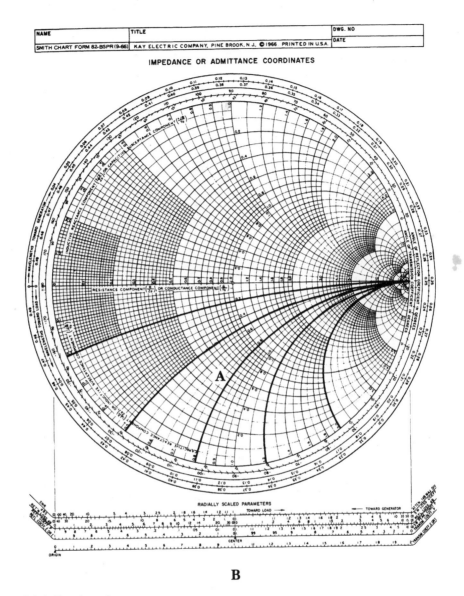

NAME	TITLE	DWG. NO
SMITH CHART FORM 82-BSPR(9-66)	KAY ELECTRIC COMPANY, PINE BROOK, N.J. © 1966 PRINTED IN U.S.A.	DATE

IMPEDANCE OR ADMITTANCE COORDINATES

RADIALLY SCALED PARAMETERS

B

26-4 *Continued.*

center point until it intersects the VSWR circle again. This point of intersection represents the normalized admittance of the load.

Outer circle parameters

The standard Smith chart shown in Fig. 26-4C contains three concentric calibrated circles on the outer perimeter of the chart. Circle "A" has already been covered and it is the pure reactance circle. The other two circles define the wavelength distance ("B") relative to either the load or generator end of the transmission line and either the transmission or reflection coefficient angle in degrees ("C").

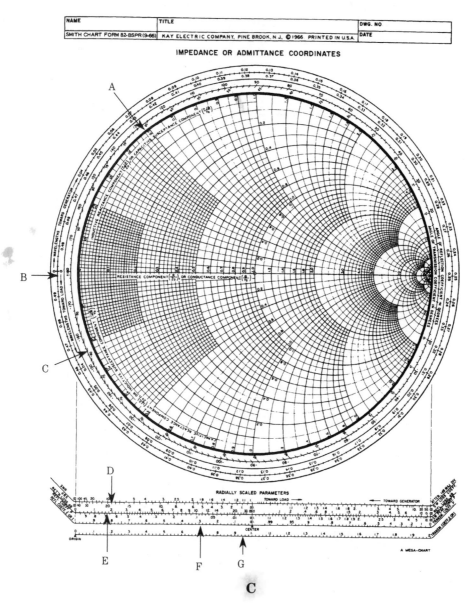

26-4 *Continued.*

There are two scales on the wavelengths circle ("B" in Fig. 26-4C) and both have their zero origin on the left-hand extreme of the pure resistance line. Both scales represent one-half wavelength for one entire revolution and are calibrated from 0 through 0.50 such that these two points are identical with each other on the circle. In other words, starting at the zero point and traveling 360 degrees around the circle brings one back to zero, which represents one-half wavelength, or 0.5 λ.

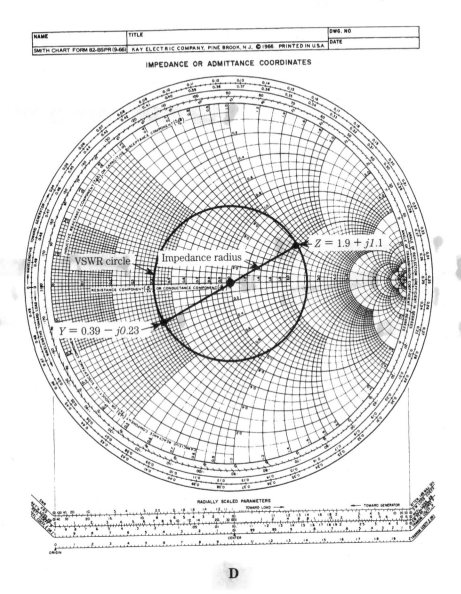

| NAME | | TITLE | | | DWG. NO | |
| SMITH CHART FORM 82-BSPR (9-66) | KAY ELECTRIC COMPANY, PINE BROOK, N.J. © 1966. PRINTED IN U.S.A. | | | | DATE | |

IMPEDANCE OR ADMITTANCE COORDINATES

$Z = 1.9 + j1.1$

VSWR circle   Impedance radius

$Y = 0.39 - j0.23$

RADIALLY SCALED PARAMETERS

TOWARD LOAD →   ← TOWARD GENERATOR

CENTER

ORIGIN

D

26-4 *Continued.*

Although both wavelength scales are of the same magnitude (0–0.50), they are opposite in direction. The outer scale is calibrated clockwise and it represents wavelengths toward the generator; the inner scale is calibrated counterclockwise and it represents wavelengths toward the load. These two scales are complementary at all points. Thus, 0.12 on the outer scale corresponds to (0.50–0.12) or 0.38 on the inner scale.

The angle of transmission coefficient and angle of reflection coefficient scales are shown in circle "C" in Fig. 26-4C. These scales are the relative phase angle between reflected and incident waves. Recall from transmission line theory that a short

(0 Ω = short
reflects back at 180°)

circuit at the load end of the line reflects the signal back toward the generator 180° out of phase with the incident signal; an open line (i.e., infinite impedance) reflects the signal back to the generator in phase (i.e., 0°) with the incident signal. This is shown on the Smith chart because both scales start at 0° on the right-hand end of the pure resistance line, which corresponds to an infinite resistance, and it goes half-way around the circle to 180° at the 0-end of the pure resistance line. Notice that the upper half-circle is calibrated 0 to +180° and the bottom half-circle is calibrated 0 to +180°, reflecting inductive or capacitive reactance situations, respectively.

Radially scaled parameters

There are six scales laid out on five lines ("D" through "G" in Fig. 26-4C and in expanded form in Fig. 26-5) at the bottom of the Smith chart. These scales are called the *radially scaled parameters* and they are both very important and often overlooked. With these scales, you can determine such factors as VSWR (both as a ratio and in decibels), return loss in decibels, voltage or current reflection coefficient, and the power reflection coefficient.

The reflection coefficient (Γ) is defined as the ratio of the reflected signal to the incident signal. For voltage or current:

$$\Gamma = \frac{E_{ref}}{E_{inc}} \qquad (26\text{-}15)$$

and

$$\Gamma = \frac{I_{ref}}{I_{inc}}. \qquad (26\text{-}16)$$

Power is proportional to the square of voltage or current, so:

$$P_{pwr} = \Gamma^2 \qquad (26\text{-}17)$$

or

$$\Gamma_{pwr} = \frac{P_{ref}}{P_{inc}}. \qquad (26\text{-}18)$$

Example: Ten watts of microwave RF power is applied to a lossless transmission line, of which 2.8 W is reflected from the mismatched load. Calculate the reflection coefficient:

$$\Gamma_{pwr} = \frac{P_{ref}}{P_{inc}} \qquad (26\text{-}19)$$

$$\Gamma_{pwr} = \frac{2.8\,W}{10\,W} \qquad (26\text{-}20)$$

$$\Gamma_{pwr} = 0.28. \qquad (26\text{-}21)$$

The voltage reflection coefficient (Γ) is found by taking the square root of the power reflection coefficient, so in this example it is equal to 0.529. These points are plotted at "A" and "B" in Fig. 26-5.

Standing wave ratio (SWR) can be defined in terms of reflection coefficient:

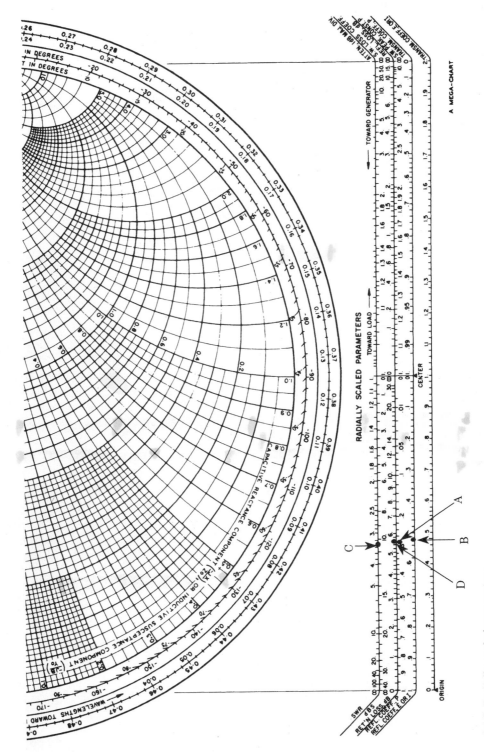

26-5 Radially scaled parameters.

$$\text{VSWR} = \frac{1 + \Gamma}{1 - \Gamma} \qquad (26\text{-}22)$$

or

$$\text{VSWR} = \frac{1 + \sqrt{\Gamma_{\text{pwr}}}}{1 - \sqrt{\Gamma_{\text{pwr}}}}, \qquad (26\text{-}23)$$

or in our example:

$$\text{VSWR} = \frac{1 + \sqrt{0.28}}{1 - \sqrt{0.28}} \qquad (26\text{-}24)$$

$$\text{VSWR} = \frac{1 + 0.529}{1 - 0.529} \qquad (26\text{-}25)$$

$$\text{VSWR} = \frac{1.529}{0.471} = 3.25{:}1, \qquad (26\text{-}26)$$

or in decibel form:

$$\text{VSWR}_{\text{dB}} = 20 \log (\text{VSWR}) \qquad (26\text{-}27)$$

$$\text{VSWR}_{\text{dB}} = 20 \log (20) \qquad (26\text{-}28)$$

$$\text{VSWR}_{\text{dB}} = (20)(0.510) = 10.2 \, \text{dB}. \qquad (26\text{-}29)$$

These points are plotted at "C" in Fig. 26-5. Shortly, you will work an example to show how these factors are calculated in a transmission-line problem from a known complex load impedance.

Transmission loss is a measure of the one-way loss of power in a transmission line because of reflection from the load.

Return loss represents the two-way loss so it is exactly twice the transmission loss. Return loss is found from:

$$\text{Loss}_{\text{ret}} = 10 \log (\Gamma_{\text{pwr}}) \qquad (26\text{-}30)$$

and for our example, in which $\Gamma_{\text{pwr}} = 0.28$:

$$\text{Loss}_{\text{ret}} = 10 \log (0.28) \qquad (26\text{-}31)$$

$$\text{Loss}_{\text{ret}} = (10)(-0.553) = -5.53 \, \text{dB}. \qquad (26\text{-}32)$$

This point is shown as "D" in Fig. 26-5. The transmission loss coefficient can be calculated from:

$$\text{TLC} = \frac{1 + \Gamma_{\text{pwr}}}{1 - \Gamma_{\text{pwr}}} \qquad (26\text{-}33)$$

or for our example:

$$\text{TLC} = \frac{1 + (0.28)}{1 - (0.28)} \qquad (26\text{-}34)$$

$$\text{TLC} = \frac{1.28}{0.72} = 1.78. \qquad (26\text{-}35)$$

The TLC is a correction factor that is used to calculate the attenuation caused by mismatched impedance in a lossy, as opposed to the ideal "lossless," line. The TLC is found from laying out the impedance radius on the Loss Coefficient scale on the radially scaled parameters at the bottom of the chart.

Smith chart applications

One of the best ways to demonstrate the usefulness of the Smith chart is by practical example. The following sections look at two general cases: transmission-line problems and stub-matching systems. $(\vdash 8^{-10})??$ $8\cdot 10^{-1}$

Transmission line problems

Figure 26-6 shows a 50-Ω transmission line connected to a complex load impedance, Z_L, of $36 + j40$ Ω. The transmission line has a velocity factor (v) of 0.80, which means that the wave propagates along the line at 8^{-10} the speed of light ($c = 300,000,000$ m/s). The length of the transmission line is 28 cm. The generator (V_{in}) is operated at a frequency of 4.5 GHz and produces a power output of 1.5 W. See what you can glean from the Smith chart (Fig. 26-7).

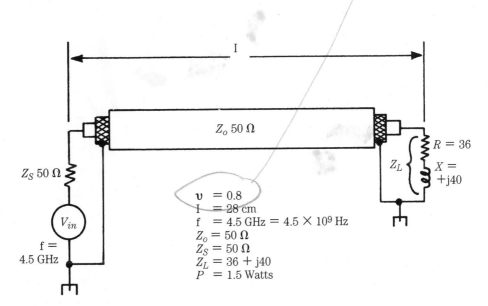

26-6 Transmission line and load circuit.

First, normalize the load impedance. This is done by dividing the load impedance by the systems impedance (in this case $Z_o = 50$ Ω):

$$Z = \frac{36 + j40 \ \Omega}{50 \ \Omega} \tag{26-36}$$

$$Z = 0.72 + j0.8. \tag{26-37}$$

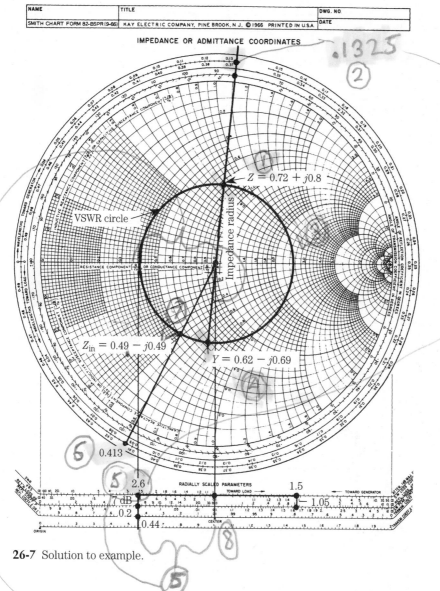

26-7 Solution to example.

The resistive component of impedance, Z, is located along the "0.72" pure resistance circle (see Fig. 26-7). Similarly, the reactive component of impedance Z is located by traversing the 0.72 constant resistance circle until the $+j0.8$ constant reactance circle is intersected. This point graphically represents the normalized load impedance $Z = 0.72 + j0.80$. A VSWR circle is constructed with an impedance radius equal to the line between "1.0" (in the center of the chart) and the "0.72 + $j0.8$" point. At a frequency of 4.5 GHz, the length of a wave propagating in the transmission line, assuming a velocity factor of 0.80, is:

$$\lambda_{\text{line}} = \frac{c\,v}{F_{\text{HZ}}} \qquad (26\text{-}38)$$

$$\lambda_{\text{line}} = \frac{(3 \times 10^8 \,\text{m/s})(0.80)}{4.5 \times 10^9 \,\text{Hz}} \tag{26-39}$$

$$\lambda_{\text{line}} = \frac{2.4 \times 10^8 \,\text{m/s}}{4.5 \times 10^9 \,\text{Hz}} \tag{26-40}$$

wavelengths toward generator

$$\lambda_{\text{line}} = 0.053 \,\text{m} \times \frac{100 \,\text{cm}}{m} = 5.3 \,\text{cm}. \tag{26-41}$$

One wavelength is 5.3 cm, so a half-wavelength is 5.3 cm/2, or 2.65 cm. The 28-cm line is 28 cm/5.3 cm, or 5.28 wavelengths long. A line drawn from the center (1.0) to the load impedance is extended to the outer circle and it intersects the circle at 0.1325. Because one complete revolution around this circle represents one-half wavelength, 5.28 wavelengths from this point represents 10 revolutions plus 0.28 more. The residual 0.28 wavelengths is added to 0.1325 to form a value of (0.1325 + 0.28) = 0.413. The point "0.413" is located on the circle and is marked. A line is then drawn from 0.413 to the center of the circle and it intersects the VSWR circle at $0.49 - j0.49$, which represents the input impedance (Z_{in}) looking into the line. To find the actual impedance represented by the normalized input impedance, you have to "denormalize" the Smith chart impedance by multiplying the result by Z_0:

$$Z_{\text{in}} = (0.49 - j0.49)(50 \,\Omega) \tag{26-42}$$

$$Z_{\text{in}} = 24.5 - j24.5 \,\Omega. \tag{26-43}$$

This impedance must be matched at the generator by a conjugate matching network. The admittance represented by the load impedance is the reciprocal of the load impedance and is found by extending the impedance radius through the center of the VSWR circle until it intersects the circle again. This point is found and represents the admittance $Y = 0.62 - j0.69$. Confirming the solution mathematically:

$$Y = \frac{1}{Z} \tag{26-44}$$

$$Y = \frac{1}{0.72 + j0.80} \times \frac{0.72 - j0.80}{0.72 - j0.80} \tag{26-45}$$

$$Y = \frac{0.72 - j0.80}{1.16} = 0.62 - j0.69. \tag{26-46}$$

The VSWR is found by transferring the "impedance radius" of the VSWR circle to the radial scales. The radius $(0.72 - 0.80)$ is laid out on the VSWR scale (topmost of the radially scaled parameters) with a pair of dividers from the center mark, and you find that the VSWR is approximately 2.6:1. The decibel form of VSWR is 8.3 dB (next scale down from VSWR) and this is confirmed by:

$$\text{VSWR}_{\text{dB}} = 20 \log (\text{VSWR}) \tag{26-47}$$

$$\text{VSWR}_{\text{dB}} = (20) \log (2.6) \tag{26-48}$$

$$\text{VSWR}_{\text{dB}} = (20)(0.431) = 8.3 \,\text{dB}. \tag{26-49}$$

The transmission loss coefficient is found in a manner similar to the VSWR, using the radially scaled parameter scales. In practice, once you have found the VSWR, you need only drop a perpendicular line from the 2.6:1 VSWR line across the other scales. In this case, the line intersects the voltage reflection coefficient at 0.44. The power reflection coefficient (G_{pwr}) is found from the scale and is equal to G_2. The perpendicular line intersects the power reflection coefficient line at 0.20. The angle of reflection coefficient is found from the outer circles of the Smith chart. The line from the center to the load impedance ($Z = 0.72 + j0.80$) is extended to the Angle of Reflection Coefficient in Degrees circle and intersects it at approximately 84°. The reflection coefficient is therefore 0.44/84°. The transmission loss coefficient (TLC) is found from the radially scaled parameter scales also. In this case, the impedance radius is laid out on the Loss Coefficient scale, where it is found to be 1.5. This value is confirmed from:

$$\text{TLC} = \frac{1 + \Gamma_{pwr}}{1 - \Gamma_{pwr}} \tag{26-50}$$

$$\text{TLC} = \frac{1 + (0.20)}{1 - (0.21)} \tag{26-51}$$

$$\text{TLC} = \frac{1.20}{0.79} = 1.5. \tag{26-52}$$

The Return Loss is also found by dropping the perpendicular from the VSWR point to the RET'N LOSS, dB line, and the value is found to be approximately 7 dB, which is confirmed by:

$$\text{Loss}_{ret} = 10 \log (\Gamma_{pwr}) \text{dB} \tag{26-53}$$

$$\text{Loss}_{ret} = 10 \log (0.21) \text{dB} \tag{26-54}$$

$$\text{Loss}_{ret} = (10) (-0.677) \text{dB} \tag{26-55}$$

$$\text{Loss}_{ret} = 6.77 \text{dB} = -6.9897 \text{dB}. \tag{26-56}$$

The reflection loss is the amount of RF power reflected back down the transmission line from the load. The difference between incident power supplied by the generator (1.5 W, in this example), $P_{inc} - P_{ref} = P_{abs}$, and the reflected power is the absorbed power (P_a) or, in the case of an antenna, the radiated power. The reflection loss is found graphically by dropping a perpendicular from the TLC point (or by laying out the impedance radius on the R_{EFL}. Loss, dB scale) and in this example (Fig. 26-7) is −1.05 dB. You can check the calculations: The return loss was −7 dB, so:

$$-7 \text{dB} = 10 \log \left(\frac{P_{ref}}{P_{inc}}\right) \tag{26-57}$$

$$-7 = 10 \log \left(\frac{P_{ref}}{1.5 \text{W}}\right) \tag{26-58}$$

$$\frac{-7}{10} = \log \left(\frac{P_{ref}}{1.5 \text{W}}\right) \tag{26-59}$$

$$10^{\left(\frac{-7}{10}\right)} = \frac{P_{ref}}{1.5 \text{ W}} \tag{26-60}$$

$$0.2 = \frac{P_{ref}}{1.5 \text{W}} \tag{26-61}$$

$$(0.2) (1.5 \text{W}) = P_{ref} \tag{26-62}$$

$$0.3 \text{W} = P_{ref}. \tag{26-63}$$

The power absorbed by the load (P_a) is the difference between incident power (P_{inc}) and reflected power (P_{ref}). If 0.3 W is reflected, the absorbed power is $(1.5 - 0.3)$, or 1.2 W. The reflection loss is -1.05 dB and can be checked from:

$$-1.05\,\text{dB} = 10\log\left(\frac{P_a}{P_{inc}}\right) \qquad (26\text{-}64)$$

$$\frac{-1.05}{10} = \log\left(\frac{P_a}{1.5\,\text{W}}\right) \qquad (26\text{-}65)$$

$$10^{\left(\frac{-1.05}{10}\right)} = \frac{P_a}{1.5\,\text{W}} \qquad (26\text{-}66)$$

$$0.785 = \frac{P_a}{1.5\,\text{W}} \qquad (26\text{-}67)$$

$$(1.5\,\text{W}) \times (0.785) = P_a \qquad (26\text{-}68)$$

$$1.2\,\text{W} = P_a. \qquad (26\text{-}69)$$

Now check what you have learned from the Smith chart. Recall that 1.5 W of 4.5-GHz microwave RF signal were input to a 50-Ω transmission line that was 28 cm long. The load connected to the transmission line has an impedance of $36 + j40$. From the Smith chart:

Admittance (load):	$0.62 - j0.69$
VSWR:	2.6:1 VSWR (dB): 8.3 dB
Refl. coef. (E):	0.44
Refl. coef. (P):	0.2
Refl. coef. angle:	84°
Return loss:	-7 dB
Refl. loss:	-1.05 dB
Trans. loss. coef.:	1.5

Notice that in all cases, the mathematical interpretation corresponds to the graphical interpretation of the problem, within the limits of accuracy of the graphical method.

Stub matching systems

A properly designed matching system will provide a conjugate match to a complex impedance. Some sort of matching system or network is needed any time the load impedance (Z_L) is not equal to the characteristic impedance (Z_o) of the transmission line. In a transmission-line system, it is possible to use a shorted stub connected in parallel with the line, at a critical distance back from the mismatched load, to affect a match. The stub is merely a section of transmission line that is shorted at the end not connected to the main transmission line. The reactance (hence also susceptance) of a shorted line can vary from $-\lambda$ to $+\lambda$, depending on length, so you can use a line of critical length L_2 to cancel the reactive component of the load impedance. Because the stub is connected in parallel with the line, it is a bit easier to work with admittance parameters rather than impedance.

Consider the example of Fig. 26-8, in which the load impedance is $Z = 100 + j60$, which is normalized to $2.0 + j1.2$. This impedance is plotted on the Smith chart in Fig. 26-9 and a VSWR circle is constructed. The admittance is found on the chart at point $Y = 0.37 - j0.22$.

To provide a properly designed matching stub, you need to find two lengths. L_1 is the length (relative to wavelength) from the load toward the generator (see L_1 in Fig. 26-8); L_2 is the length of the stub itself.

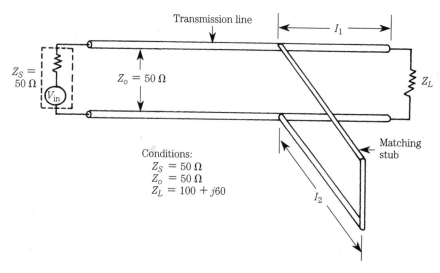

Transmission line

26-8 Matching stub length and position.

The first step in finding a solution to the problem is to find the points where the unit conductance line (1.0 at the chart center) intersects the VSWR circle; there are two such points shown in Fig. 26-9: $1.0 + j1.1$ and $1.0 - j1.1$. Select one of these (choose $1.0 + j1.1$) and extend a line from the center 1.0 point through the $1.0 + j1.1$ point to the outer circle (WAVELENGTHS TOWARD GENERATOR). Similarly, a line is drawn from the center through the admittance point $0.37 - 0.22$ to the outer circle. These two lines intersect the outer circle at the points 0.165 and 0.461. The distance of the stub back toward the generator is found from:

$$L_1 = 0.165 + (0.500 + 0.461)\lambda \qquad (26\text{-}70)$$
$$L_1 = 0.165 + 0.039\lambda \qquad (26\text{-}71)$$
$$L_1 = 0.204\lambda. \qquad (26\text{-}72)$$

The next step is to find the length of the stub required. This is done by finding two points on the Smith chart. First, locate the point where admittance is infinite (far right side of the pure conductance line); second, locate the point where the admittance is $0 - j1.1$ (notice that the susceptance portion is the same as that found where the unit conductance circle crossed the VSWR circle). Because the conductance component of this new point is 0, the point will lie on the $-j1.1$ circle at the intersection with the outer circle. Now draw lines from the center of the chart through each of these points to the outer circle. These lines intersect the outer circle at 0.368 and 0.250. The length of the stub is found from:

$$L_1 = (0.368 - 0.250)\lambda \qquad (26\text{-}73)$$
$$L_1 = 0.118\lambda. \qquad (26\text{-}74)$$

From this analysis, you can see that the impedance, $Z = 100 + j60$, can be matched by placing a stub of a length 0.118λ at a distance 0.204λ back from the load.

The Smith chart in lossy circuits

Thus far, you have dealt with situations in which loss is either zero (i.e., ideal transmission lines) or so small as to be negligible. In situations where there is appreciable loss in the circuit or line, however, you see a slightly modified situation. The VSWR circle, in that case, is actually a spiral, rather than a circle.

Figure 26-10 shows a typical situation. Assume that the transmission line is 0.60λ long and is connected to a normalized load impedance of $Z = 1.2 + j1.2$. An

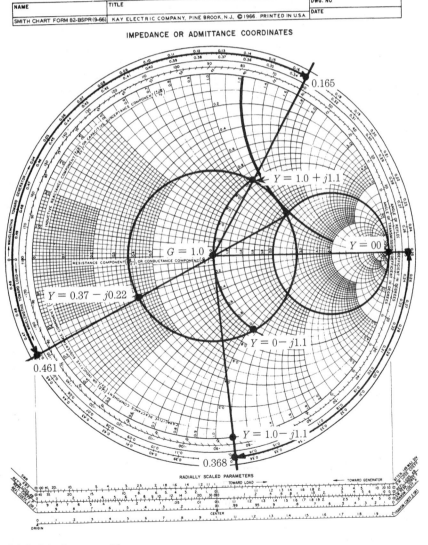

26-9 Solution to problem.

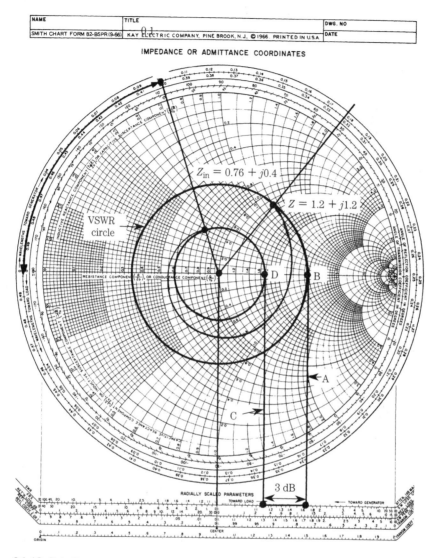

IMPEDANCE OR ADMITTANCE COORDINATES

$Z_{in} = 0.76 + j0.4$

$Z = 1.2 + j1.2$

VSWR circle

3 dB

26-10 Solution.

"ideal" VSWR circle is constructed on the impedance radius represented by $1.2 + j1.2$. A line ("A") is drawn, from the point where this circle intersects the pure resistance baseline ("B"), perpendicularly to the ATTEN 1 dB/MAJ. DIV. line on the radially scaled parameters. A distance representing the loss (3 dB) is stepped off on this scale. A second perpendicular line is drawn from the -3-dB point back to the pure resistance line ("C"). The point where "C" intersects the pure resistance line becomes the radius for a new circle that contains the actual input impedance of the line. The length of the line is 0.60λ, so you must step back $(0.60 - 0.50)\lambda$ or 0.1λ. This point is located on the WAVELENGTHS TOWARD GENERATOR outer circle. A line is drawn from this point to the 1.0 center point. The point where this new line intersects the new circle is the actual input impedance (Z_{in}). The intersection occurs at 0.76 + j0.4, which (when denormalized) represents an input impedance of 38 + j20 Ω.

Frequency on the Smith chart

A complex network may contain resistive, inductive reactance, and capacitive reactance components. Because the reactance component of such impedances is a function of frequency, the network or component tends to also be frequency-sensitive. You can use the Smith chart to plot the performance of such a network with respect to various frequencies. Consider the load impedance connected to a 50-Ω transmission line in Fig. 26-11. In this case, the resistance is in series with a 2.2-pF capacitor, which will exhibit a different reactance at each frequency. The impedance of this network is:

$$Z = R - j\left(\frac{1}{\omega C}\right) \tag{26-75}$$

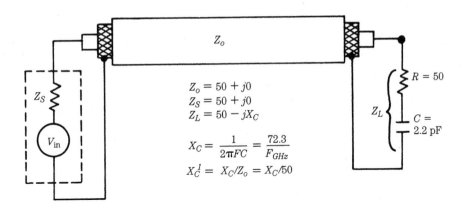

Freq. (GHz)	X_C	
1	$-j72.3$	$-j1.45$
2	$-j36.2$	$-j0.72$
3	$-j24.1$	$-j0.48$
4	$-j18$	$-j0.36$
5	$-j14.5$	$-j0.29$
6	$-j12$	$-j0.24$

26-11 Load and source-impedance transmission-line circuit.

or

$$Z = 50 - j\left(\frac{1}{(2\pi FC)}\right), \qquad (26\text{-}76)$$

and, in normalized form

$$Z' = 1.0 - \left(\frac{j}{(2\pi FC) \times 50}\right) \cdot \qquad (26\text{-}77)$$

$$Z' = 1.0 - \frac{j}{(6.9 \times 10^{-10}\,\text{F})} \qquad (26\text{-}78)$$

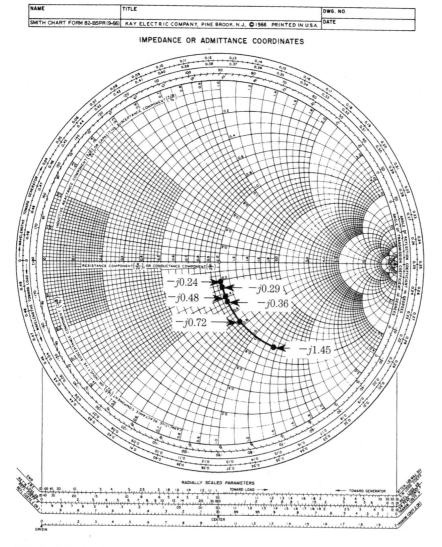

26-12 Plotted points.

$$Z' = 1.0 - \left(\frac{j \times 7.23 \times 10^{10}}{F} \right), \tag{26-79}$$

or, converted to GHz:

$$Z' = 1.0 - \frac{j72.3}{F_{\text{GHz}}}. \tag{26-80}$$

The normalized impedances for the sweep of frequencies from 1 to 6 GHz are therefore:

$$Z = 1.0 - j1.45 \tag{26-81}$$
$$Z = 1.0 - j0.72 \tag{26-82}$$
$$Z = 1.0 - j0.48 \tag{26-83}$$
$$Z = 1.0 - j0.36 \tag{26-84}$$
$$Z = 1.0 - j0.29 \tag{26-85}$$
$$Z = 1.0 - j0.24 \tag{26-86}$$

These points are plotted on the Smith chart in Fig. 26-12. For complex networks, in which both inductive and capacitive reactance exist, take the difference between the two reactances (i.e., $X = X_L - X_C$).

27
CHAPTER

Detector and demodulator circuits

The purpose of the detector or demodulator circuits is to recover the intelligence impressed on the radio carrier wave at the transmitter. This process is called *demodulation* and the circuits used to accomplish this are called *demodulators*. They are also called *second detectors* in superheterodyne receivers.

In a superheterodyne receiver the detector or demodulator circuit is placed between the IF amplifier and the audio amplifier (Fig. 27-1). This position is the same in AM, FM, pulse modulation, and digital receivers (although in digital receivers the demodulator might be in a circuit called a MODEM).

AM envelope detectors

An amplitude modulation (AM) signal consists of a slow audio signal which revolves around an average radio frequency (RF) carrier signal. It is essentially a multiplication or mixing process in which the RF carrier and AF signals are both output, along with the (RF − AF) and (RF + AF) signals. Because of the selectivity of the transmitter circuits, only the RF carrier and the sum and difference signals appear in the output. The AF signal is suppressed. The sum signal (RF + AF) is known as the *upper sideband* (USB) while the difference signal (RF − AF) is known as the *lower sideband* (LSB). Because of this action, the bandwidth of the AM signal is determined by the highest audio frequency transmitted and is equal to twice that frequency. A total of 66.67% of the RF power in an AM signal is in the carrier so only 33.33% is split between two sidebands.

Figure 27-2 shows a simple AM *envelope detector* circuit while Fig. 27-3 shows the waveforms associated with this circuit. The circuit consists of a signal diode rectifier connected to the output of an IF amplifier. There is a capacitor (C_1) and resistive load connected to the rectifier. When the input signal is received (Fig. 27-3A) it is rectified (Fig. 27-3B), producing an average current output that translates to the voltage waveform of Fig. 27-3C.

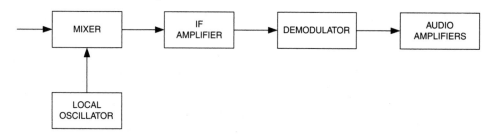

27-1 Location of the demodulator circuit.

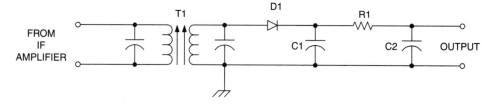

27-2 Simple envelope detector.

The capacitor is charged to a value equal to 0.637 times the peak voltage (which is the average value) and then scales upward with modulation to the full peak value. Low-pass filter R_1C_2 takes out the residual RF/IF signal.

The important attribute of a demodulator is for it to have a nonlinear response, preferably with a sharp cutoff. Vacuum tubes, bipolar transistors, and field-effect transistors possess these characteristics, but so does the simple diode rectifier. Figure 27-4 shows the input and output characteristics compared with the diode's *I-vs-V* curve. At low signal levels the diode acts like a *square law detector*, but at higher signal levels the operation is somewhat linear.

Consider a diode with an I-vs-V characteristic of:

$$i_d = a_0 + a_1 V_d + a_2 V_d^2. \tag{27-1}$$

Let:

$$V_d = A\left[1 + ms(t)\right]\cos(2\pi f_c t). \tag{27-2}$$

We can then write:

$$i_d = a_0 + a_1 A\left[1 + ms(t)\right]\cos(2\pi f_c t)$$
$$+ a_2 A^2\left[1 + ms(t)\right]^2\cos^2(2\pi f_c t) \tag{27-3}$$

$$i_d = a_0 + a_1 A\left[1 + ms(t)\right]\cos(2\pi f_c t) + \frac{a_2 A^2}{2}$$

$$+ \cdots + \left(a_2 A^2\, ms(t) + \frac{a_2 A^2}{2} m_2 S^2(t)\right)$$

$$+ \left(\frac{a_2 A^2}{2}\left[1 + ms(t)\right]^2\cos(4\pi f_c t)\right) \tag{27-4}$$

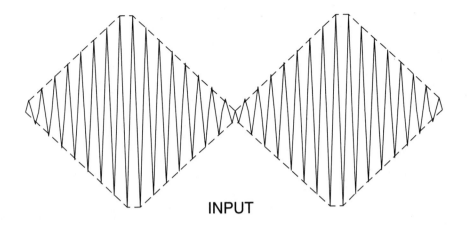

INPUT

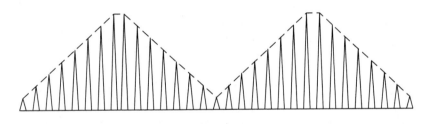

RECTIFIED OUTPUT

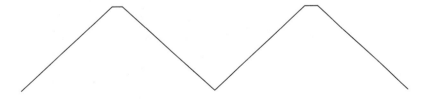

FILTERED OUTPUT

27-3 AM detector outputs.

where:
    m is the modulation index
    f_c is the carrier frequency
    A is the peak amplitude
    a_0, a_1, and a_2 are constants.

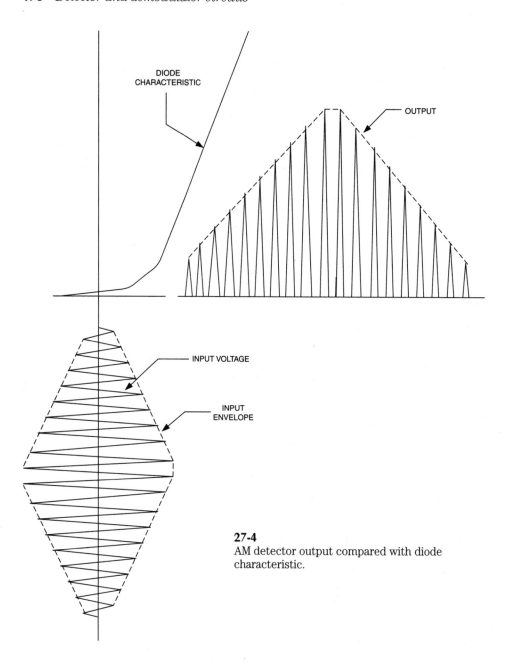

27-4
AM detector output compared with diode
characteristic.

The terms in the brackets are the modulation and distortion products. The second-order terms are modulation and distortion, where the higher-order terms are distortion only. What falls out of the equation is that to keep distortion low, modulation index (m) must be low as well.

Figure 27-5 shows what happens at the capacitor. The dotted lines represent the output of the diode, which is a half-wave rectified RF/IF signal; the heavy line repre-

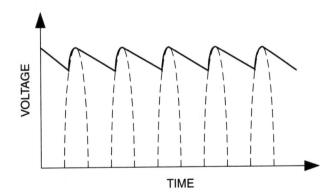

27-5 Unfiltered AM output.

sents the capacitor voltage (V_c). The capacitor charges to the peak value and then the diode cuts off. The voltage across the capacitor drops slightly to the point where its voltage is equal to the input voltage, when it turns on again. The diode may be modeled as a switch with resistance. During the nonconduction period, the switch is open and the capacitor discharges. During the conduction period, the "switch" conducts and charges the capacitor. The waveform across the capacitor when there is no modulation is close to a sawtooth at the carrier frequency (f_c) and it represents the residual RF. These are eliminated in the RC filter to follow the envelope detector (R_1C_2) and the response of the amplifiers to follow the detector. These components are typically about 30 dB below the carrier level.

The maximum time constant of the filtering action of C_1, plus ($R_1 + R$), where R is the resistance to ground (typically a volume control), that can be accommodated depends on the maximum audio frequency that must be processed. A frequency of 3000 Hz and a time constant of 10 μS yields:

$$2\pi f_m T_c = 2\pi\ 3000 \times 10^{-5}\ \text{s} = 0.1884 \qquad (27\text{-}5)$$

which produces an output reduction of

$$\sqrt{1 + (2\pi f_m T_c)^2} = 1.018, \qquad (27\text{-}6)$$

which is 0.16 dB. At 10 KHz these values are 0.628 and 1.18, or 1.44 dB.

Figure 27-6 shows a version of the envelope detector that uses a high-pass filter at the output. The time constant referred to above is the time constant R_1C_1 and eliminates the residual RF/IF signal. The time constant R_2C_2 is set to eliminate low-frequency hum and noise. The speech requirement in communications receivers is 300 to 3000 Hz so a 60-Hz hum is easily accommodated. For broadcast receivers, the low-frequency audio is on the order of 100 Hz so this method is less workable.

In the linear mode of operation of the diode, the modulation index for which distortion begins is:

$$m = \frac{(R_1 + R_2)R_d + R_1R_2}{(R_1 + R_2)(R_1 + R_d)}. \qquad (27\text{-}7)$$

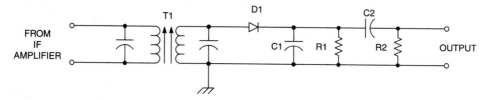

27-6 Simple envelope detector.

The distortion will be small if

$$\frac{|Z_\mathrm{m}|}{R_1} < m, \qquad (27\text{-}8)$$

where

Z_m is the impedance of the circuit at the
modulating frequency

Equation (27-7) reduces to this form when $R_\mathrm{d} << R_1$ and $Z_\mathrm{m} = R_2$.

Figure 27-7 shows three situations of input waveform and output waveform. Figure 27-7A is the situation in which there is no distortion of the output waveform. Although this waveform never occurs in real circuits, it is included for the sake of making the comparison. In Fig. 27-7B, the circuit clips the negative peaks of the modulating signal. In Fig. 27-7C we see an example of *diagonal clipping*. This form of clipping occurs when Z_m is not resistive.

Another form of distortion occurs when the RF/IF waveform is distorted. Both

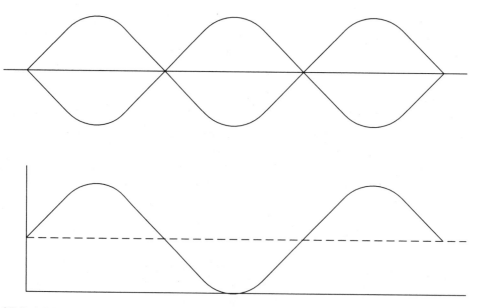

27-7 (A) Proper AM detection; (B) Clipping evident; (C) odd form of clipping present.

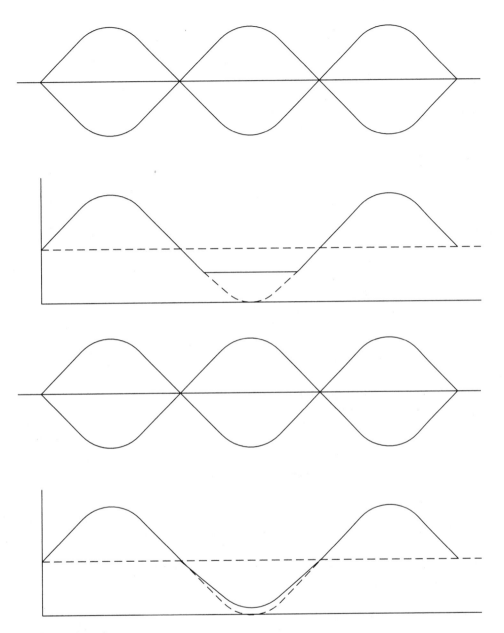

27-7 *Continued*

in-phase and quadrature distortion can occur in the RF/IF waveform, especially if the bandpass filters used in the RF/IF circuit have complex poles and zeros distributed asymmetrically about the filter center frequency or when the carrier is not tuned to the exact center of the filter passband.

Noise

All radio reception is basically a game of signal-to-noise ratio (SNR). If the SNR is not advantageous, then the receiver ultimately fails. At low signal levels (Fig. 27-8) the SNR is poor and the noise dominates the output. As the signal level increases, however, the noise level increases but at a slower rate than the signal output. The noise level comes to rest (see dotted line in Fig. 27-8) about 3.7 dB over the no-carrier state but the signal level continues upward.

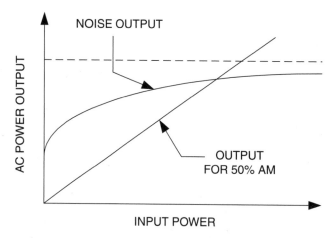

27-8 Noise vs. output for 50% AM modulation.

Balanced demodulators

Figure 27-9 shows two examples of balanced or full-wave demodulator circuits. These circuits work generally better than the half-wave version shown earlier. Figure 27-9A shows the circuit for a conventional full-wave circuit. It depends on two diodes and a center-tapped transformer (T_1). It works like the power-supply circuit of the same type. The center tap on the transformer establishes the zero point, or common, so the polarities of the voltages at the ends of the secondary of T_1 are equal but opposite. When the top of the secondary is positive with respect to the bottom, diode D_1 conducts and charges capacitor C_1. On the opposite half-cycle, the opposite occurs. In that instance, the secondary is reverse polarity, so the bottom is more positive than the top. This turns off D_1 and turns on D_2, causing it to conduct and charge C_1.

Figure 27-9B shows a bridge rectified version of the AM envelope detector. When the top of the T_1 secondary is positive with respect to the bottom, diodes D_2 and D_3 conduct, charging C_1. When the bottom of the T_1 secondary is positive with respect to the top, diodes D_1 and D_4 conduct, charging the capacitor. Hence, since these conditions occur on alternate half-cycles of the input waveform, full-wave rectification occurs.

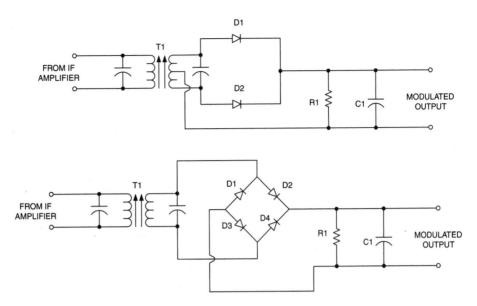

27-9 (A) Full wave AM envelope detector; (B) full wave bridge type AM envelope detector.

Synchronous AM demodulation

One of the factors that controls the comfort level of listening to demodulated AM transmissions is that the carrier and two sidebands fade out of phase with each other. This can be overcome with quasisynchronous demodulation or synchronous demodulation. Both require that the incoming carrier be eliminated. The difference is that in quasisynchronous demodulation the reinserted carrier is not in phase with the original, whereas in synchronous demodulation it is.

Quasisynchronous demodulation is much like the demodulators discussed below under single-sideband demodulators. They have a *beat frequency oscillator* to replace the carrier with an out-of-phase version. As long as the signal does not drop to zero, the process works fine. But when the signal drops to zero, as it does under deep fading, then the process falls down and synchronous demodulation wins. In synchronous demodulation the reinserted carrier is in phase with the original carrier signal. The circuit must phase lock to the original carrier.

Double-sideband (DSBSC) and single-sideband (SSBSC) suppressed carrier demodulators

Double- and single-sideband suppressed carriers are a lot more efficient than straight AM. In straight AM, the carrier contains two-thirds of the RF power, with one-third split between the two sidebands. Interestingly enough, the entire intelligent content of the speech waveform is fully contained within one sideband, so the other sideband (and carrier) are superfluous.

Figure 27-10 shows a single-sideband suppressed carrier (SSBSC, usually shortened to "SSB") transmitter. The heart of the circuit is a *balanced modulator* circuit. This circuit is balanced to produce an RF output from the crystal oscillator only when the audio frequency (AF) signal is present. As a result, the carrier is suppressed in the double-sideband output.

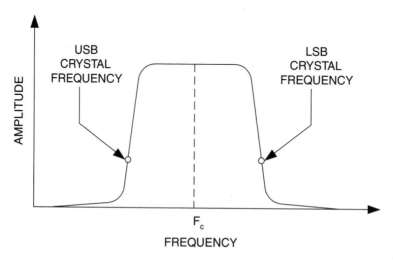

27-10 Location of SSB crystal frequencies.

The next stage is a symmetrical bandpass filter circuit that removes the unwanted sideband, leaving only the desired sideband. The crystal oscillator determines which sideband is generated. By positioning the frequency of the oscillator on the lower skirt of the filter, an *upper sideband* (USB) signal is generated. By positioning the frequency of the oscillator on the upper skirt of the filter, the *lower sideband* (LSB) is generated. This is shown in Fig. 27-11. In this figure, F_c is the center frequency of the filter while the LSB and USB frequencies are shown.

Following the filter circuit is any amplification or frequency mixing needed to accomplish the purposes of the transmitter. Contrary to AM transmitters, all stages following the balanced modulator are expected to be linear amplifiers. This is because nonlinear stages will distort the envelope of the SSB signal. This means that heterodyning must be used rather than multipliers or other nonlinear means in order to translate the frequencies. It is generally the way of SSB transmitter designers to generate the SSB signal at a fixed pair of frequencies and then heterodyne them to the desired operating frequency. There is also a phasing method of generating an SSB signal. Double-sideband transmitters are the same as Fig. 27-10, except that the filter is not present.

The basis for SSB and DSB demodulation is the *product detector* circuit. Figure 27-12 shows basis for product detection. In Fig. 27-12A we see the SSB or DSB signal, with the carrier suppressed (it's actually a DSB signal, but an SSB signal

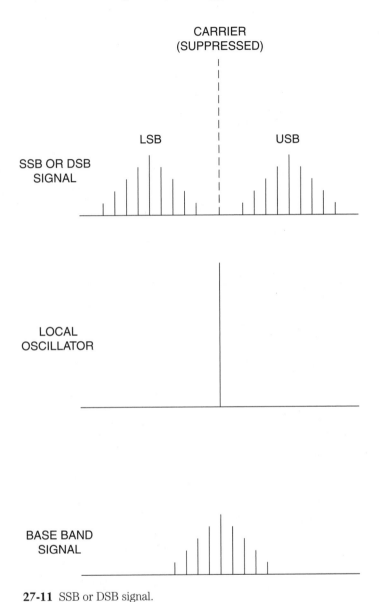

27-11 SSB or DSB signal.

would lack the other sideband). This signal is combined with a strong local oscillator signal (called a *beat frequency oscillator* or *BFO;* Fig. 27-12B) to produce the base band signal (Fig. 27-12C).

Figure 27-13 shows a simple form of product detector circuit. It is like the envelope detector except for the extra diode (D_2) and a carrier regeneration oscillator (also called a BFO). The circuit works by switching the diodes into and out of conduction with the oscillator signal. When the output of the oscillator is negative, the diodes will conduct, passing signal to the output. But on positive excursions of the

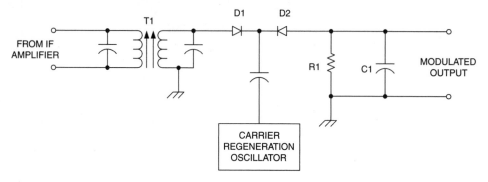

27-12 Simple product detector.

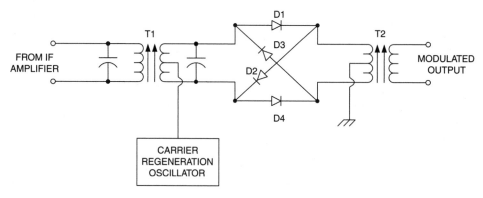

27-13 Four-diode product detector.

local oscillator signal the diodes are blocked from conducting by the bias produced by the oscillator signal. The residual RF/IF signal is filtered out by capacitor C_1.

A superior circuit is shown in Fig. 27-14. This circuit is a balanced product detector. It consists of a balanced ring demodulator coupled through a pair of center tapped transformers (T_1 and T_2). Transformer T_1 is the last IF transformer. It has a center-tapped secondary to receive the strong local oscillator signal used for carrier regeneration. The output transformer (T_2) is also center-tapped but in that case the center-tap is grounded.

The circuit works by switching in and out of the circuit pairs of diodes on alternate half-cycles of the SSB waveform. Consider first circuit action when no signal is present. That leaves only the local oscillator signal, which is applied to the common point on the transformer T_1 secondary. That means that the ends of the transformer will be positive and negative at the same time. Under this circumstance, the diodes D_1–D_4 conduct and then diodes D_2–D_3 conduct, but both signals are nulled in the primary winding of T_2. This occurs because both signals are the same.

When the top of the T_1 secondary is positive with respect to the bottom, and the local oscillator signal is positive, then diodes D_1 and D_2 conduct, creating an unbalanced situation, which results in an output. Similarly, when the situation is reversed

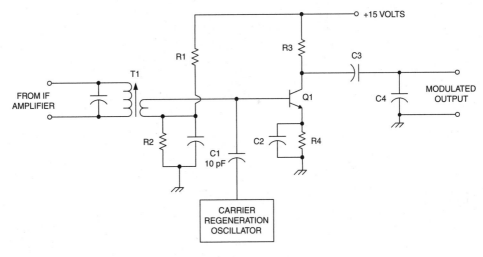

27-14 Transistor product detector.

only the D_2–D_3 diodes conduct and the others are cut off. Thus, the local oscillator controls the output of the product detector.

Figure 27-15 shows a circuit in which a bipolar transistor is used as a product detector. The circuit would act like any amplifier, except that the base of the transistor is controlled by the local oscillator. The SSB signal is applied through the low-impedance secondary of T_1 and would ordinarily be amplified by Q_1. But the quenching action of the local oscillator prohibits this action. The transistor is alternately cut off and cut on by the local oscillator circuit and that creates the nonlinearity needed to demodulate the waveform.

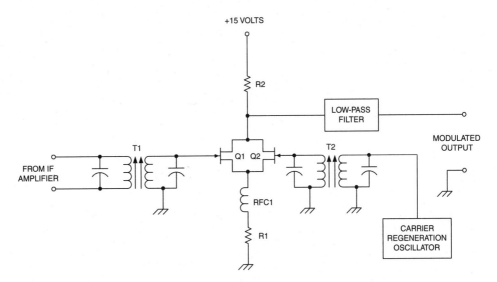

27-15 Dual JFET product detector.

A differential pair of junction field effect transistors (JFETs) is used in Fig. 27-16 to produce the product detection. The SSB signal is applied to the gate of Q_1, while the local oscillator signal is used to disrupt the operation of the circuit from the Q_2 side. Keep in mind that the local oscillator signal is very much greater in amplitude than the SSB signal. A low-pass filter tuned to the spectrum that is to be recovered (typically audio) is connected to the common drain circuits and thence to the output.

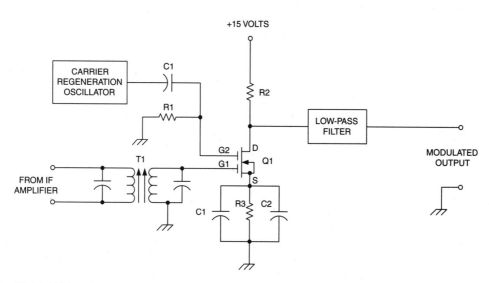

27-16 MOSFET product detector.

A dual-gate MOSFET transistor is the subject of Fig. 27-17. The normal signal input of the MOSFET, gate 1, is used to receive the SSB signal from the IF amplifier. The MOSFET is turned on and off by the local oscillator signal applied to gate 2. Again, an audio low-pass filter is present at the output (drain) circuit to limit the residual IF signal that gets through to the modulated output.

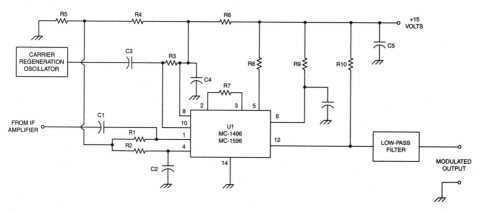

27-17 MC-1496 product detector.

Notice in this circuit that there are two capacitors in the source circuit of the MOSFET transistor (C_1 and C_2). Typically, one of these will be for RF and the other for AF, although with modern capacitors it might not be strictly necessary.

An integrated circuit SSB product detector is shown in Fig. 27-18. This circuit is based on the MC-1496 analog multiplier chip. It contains a transconductance cell demodulator that is switched on and off by the action of the local oscillator. The SSB IF signal is input through pin no. 1 and the local oscillator through pin no. 10. The alternate pins in each case are biased and not otherwise used.

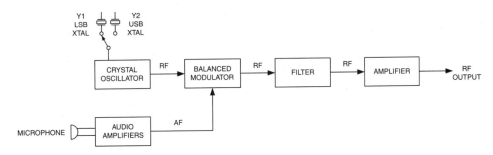

27-18 Filter-type SSB transmitter.

Phasing method

The methods for demodulating an SSB or DSB signal thus far presented have been product detectors. There is also a phasing method used in some cases. Figure 27-19 shows this method in block diagram form. This method splits the SSB IF signal into two paths, I and Q. They are mixed with a pair of signals that are the same except for phasing. The I signal is mixed with the $\cos (2\pi nF_c/F)$ version of the local oscillator signal while the Q is mixed with the $-\sin (2\pi nF_c/F)$ version of the same signal. The Q channel is passed through a Hilbert transformer, which has the effect of further phase shifting it by 90°. The I channel signal is delayed an amount equal to the delay of the Q channel signal or $(N - 1)/2$ samples. The two signals are then summed in a linear mixer circuit to produce the output. This method will yield the lower sideband part of the signal. If we subtract the two signals instead of adding them we will yield the upper sideband signal.

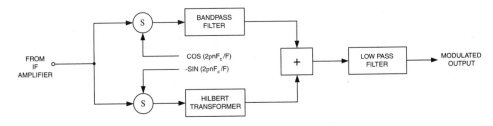

27-19 Phasing type SSB transmitter.

FM and PM demodulator circuits

Frequency modulation and *phase modulation* are examples of *angle modulation*. Figure 27-20 shows this action graphically. The audio signal causes the frequency (or phase) to shift plus and minus from the quiescent value, which exists when there is no modulation present. Frequency and phase modulation are different but similar enough to make the demodulation schemes the same. The difference between FM and PM is that the phase modulation needs no preemphasis curve to the audio waveform (it does it naturally) while the frequency modulation transmitter is preemphasized for noise abatement.

Another difference between FM and PM transmitters is the location of the *reactance modulator* used to generate the modulated signal. In the FM transmitter the

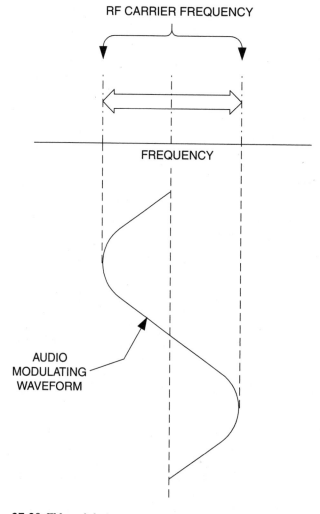

27-20 FM modulation.

reactance modulator is part of the frequency determining circuitry, whereas in the PM transmitter it follows that circuitry.

Discriminator circuits

One of the classic FM discriminator circuits is shown in Fig. 27-21. This circuit uses a special transformer that has two secondary windings. One secondary winding is tuned slightly above the IF frequency while the other is tuned the same amount below the IF frequency. The two frequencies are spaced slightly more than the expected transmitter swing. Their outputs are combined in a differential pair of diodes (D_1 and D_2). The outputs of the diodes are connected to load resistors R_1 and R_2. Normally, when the signal is unmodulated, the algebraic sum of the two diodes outputs is zero, resulting in zero output. When the frequency or phase swings above the quiescent value, one diode will conduct higher than the other, resulting in an imbalance across R_1 and R_2. That produces an output. Similarly, when the frequency or phase drops below the quiescent value, then the opposite situation exists. The other diode will conduct harder and produce an output of the opposite polarity.

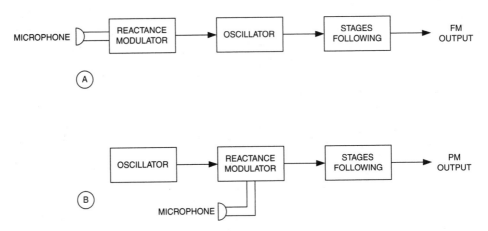

27-21 Two forms of angular modulation (A) FM, (B) PM.

Foster–Seeley discriminator circuits

The Foster–Seeley discriminator circuit is shown in Fig. 27-22, while the waveforms are shown in Fig. 27-23. This circuit requires only two tuned circuits rather than the three required by the previous circuit. The output voltage is the algebraic sum of the voltages developed across the R_2 and R_3 load resistances. Figure 27-23A shows the relationship of the output voltage and the frequency.

The primary tuned circuit is in series with both halves of the secondary winding. When the signal is unmodulated, the IF voltage across the secondary is 90° out of phase with the primary voltage. This makes the voltages applied to each diode equal but out of phase (Fig. 27-23B), resulting in zero output. But consider what happens when the frequency deviates (Fig. 27-23C). The voltages applied to the diodes are

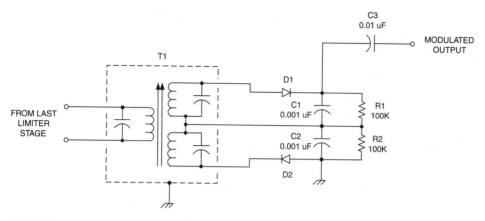

27-22 Dual-tuned discriminator.

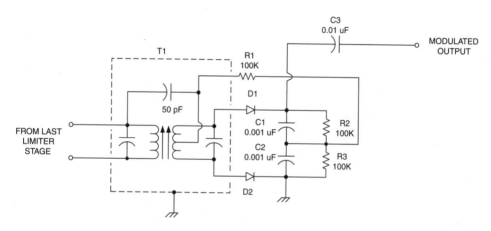

27-23 Single tuned discriminator.

no longer equal but opposite, and that creates an output from the detector that is frequency- or phase-sensitive. As the input frequency deviates back and forth across the frequency of the tuned circuit an audio signal is created equal to the modulated frequency.

In order for the FM/PM transmitter to be received "noise-free" it is necessary to precede the discriminator circuit with a *limiter circuit*. This circuit limits the positive and negative voltage excursions of the IF signal, thus clipping off AM noise.

Ratio detector circuits

The *ratio detector* circuit is shown in Fig. 27-24. This circuit uses a special transformer in which there is a small capacitor between the center tap on the primary winding and the center tap on the secondary winding. Note that the diodes are

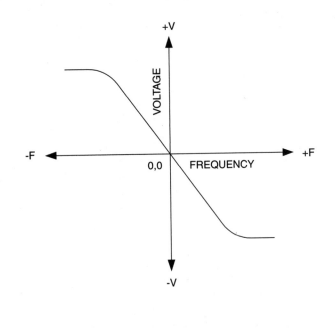

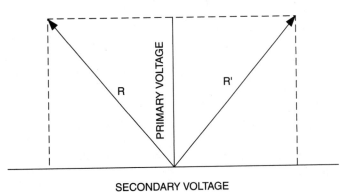

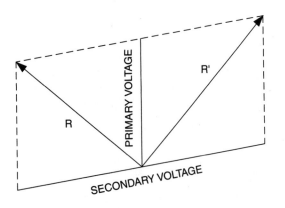

27-24 Waveforms for FM demodulation.

connected to aid each other rather than buck each other, as was the case in the Foster–Seeley discriminator circuit. When the signal is unmodulated, the voltage appearing across R_3 is one-half the AGC (automatic gain control) voltage appearing across R_2 because the contribution of each diode is the same. However, that situation changes as the input signal is modulated above or below the center frequency. In that case, the relative contribution of each diode changes. The total output voltage is equal to their ratio; hence the name ratio detector.

There are several advantages of a ratio detector over a Foster–Seeley discriminator. First, there is no need for a limiter amplifier ahead of the ratio detector, as there is with the Foster–Seeley discriminator. Furthermore, the circuit provides an AGC voltage, which can be used to control the gain of preceding RF or IF amplifier stages. However, the ratio detector is sensitive to AM variations of the incoming signal, so the AGC should be used on the stage preceding the ratio detector to limit those AM excursions. The capacitor, C_3, also helps eliminate the AM component of the signal, which is noise.

Pulse-counting detectors

The FM/PM detectors thus far considered have required special transformers to make the work. In this section we are going to look at a species of coil-less FM detector. The pulse-counting detector is shown in Fig. 27-25. This circuit uses two integrated circuits, a hex inverter and a dual J-K flip-flop. The hex inverter has six inverter stages. The first stage is used as an amplifier, while the next two are used to produce an output that is free of AM noise (most noise is AM). This is followed by a pair of divide-by-two (total divide-by-four) stages consisting of a pair of J-K flip-flops. An inverter at the output of the flip-flops is used to drive the input of a half-monostable multivibrator that has a period equal to about one-half the period of the unmodulated input signal. The output of the half-monostable is a time-varying pulse

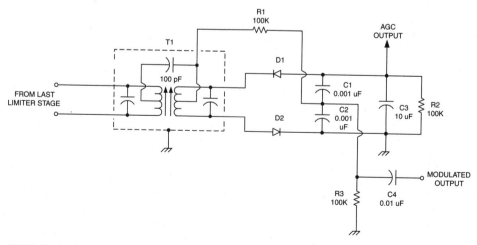

27-25 Ratio detector.

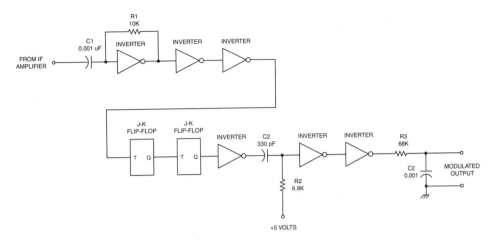

27-26 Coil-less FM detector.

train that varies with the audio modulation applied to the input signal. It is realized as audio in the low-pass filter consisting of R_3–C_2.

Another circuit, shown in block form, used a zero-crossing detector and a limiter amplifier to eliminate the AM excursions that are noise to an FM/PM signal. The output of the zero-crossing detector triggers a monostable multivibrator circuit. The output of the monostable is a pulse train that varies according to the modulating frequency applied. This is realized in a low-pass filter circuit.

Phase locked loop FM/PM detectors

The *phase locked loop* (PLL) circuit can be used as an FM demodulator if its control voltage is monitored. Figure 27-27A shows the basic PLL circuit. It consists of a voltage-controlled oscillator (VCO), phase detector, low-pass filter, and an amplifier. The FM signal from the IF amplifier is applied to one port of the phase detector, and the output of the VCO is connected to the other port. When the two frequencies are equal, there is no output from the circuit (or, the value is quiescent). When the FM IF signal deviates above or below the frequency of the VCO, there will be an error term generated. This error signal is processed in the low-pass filter and amplifier to control the VCO. Its purpose is to drive the VCO back on the right frequency. It's this error signal that becomes the modulated output of the PLL FM demodulator circuit. Figure 27-27B shows a PLL based on the NE-565 PLL integrated circuit. The resonant frequency is set by R_1 and C_1, which should be the center frequency of the FM signal. As the signal deviates up and down, the error voltage is monitored and becomes the modulated output signal.

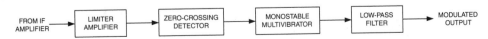

27-27 Block diagram of coil-less FM detector.

Quadrature detector

Figure 27-28 shows the *quadrature detector* circuit. This circuit is implemented in integrated circuit form (e.g., MC-1357P and CA-3089) and uses a single

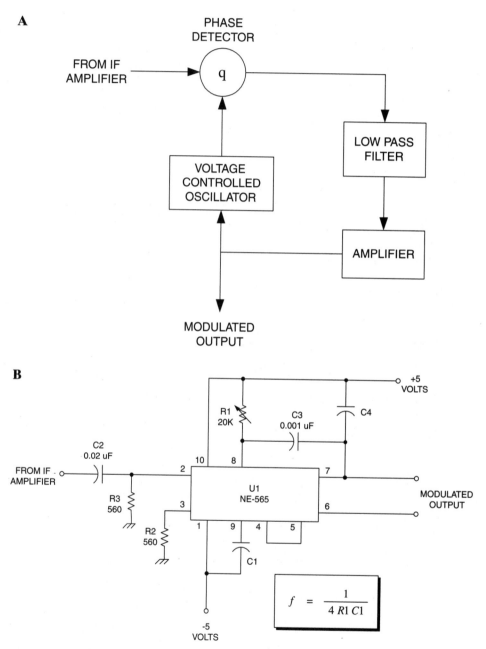

27-28 (A) Phase locked loop circuit; (B) NE-565 FM detector.

phase-shifting external coil to accomplish its goals. This is probably the most widely used form of FM demodulator in use today.

The typical quadrature detector IC uses a series of wideband amplifiers to boost the signal and limit it, eliminating the AM noise modulation that often rides on the signal. This signal is applied to the signal splitter. The two outputs of the signal splitter are applied to a gated synchronous detector, but one is phase-shifted 90°. The output of the gated synchronous detector is the modulated audio.

Index

Note: Boldface numbers indicate illustrations; *t* indicates a table.

About the author

Joseph J. Carr is a senior electronics engineer with 16 years of practical bench experience as an electronics technician. He is the author of *Mastering Radio Frequency Circuits Through Projects & Experiments, Mastering Oscillator Circuits Through Projects and Experiments,* and *Practical Antenna Handbook, Third Edition,* all published by McGraw-Hill. Mr. Carr has written over 800 articles for technical journals and writes monthly columns for *Popular Communications, Popular Electronics,* and other magazines. The author holds Certified Electronics Technician (CET) certificates in both consumer electronics and communications and helped write the examination for the CET in medical electronics. He lives in Annandale, Virginia.